SUSTAINING ROCKY MOUNTAIN LANDSCAPES

Science, Policy, and Management for the Crown of the Continent Ecosystem

Edited By
Tony Prato
and Dan Fagre

Resources for the Future
Washington, DC, USA

QH
77
.R57
S87
2007

Copyright © 2007 by Resources for the Future. All rights reserved.

Printed in the United States of America

No part of this publication may be reproduced by any means, whether electronic or mechanical, without written permission. Requests to photocopy items for classroom or other educational use should be sent to the Copyright Clearance Center, Inc., Suite 910, 222 Rosewood Drive, Danvers, MA 01923, USA (fax +1 978 646 8600; www.copyright.com). All other permissions requests should be sent directly to the publisher at the address below.

An RFF Press book
Published by Resources for the Future
1616 P Street, NW
Washington, DC 20036–1400
USA
www.rffpress.org

Library of Congress Cataloging-in-Publication Data

Sustaining Rocky Mountain landscapes : science, policy, and management for the crown of the continent ecosystem / Tony Prato and Dan Fagre, editors. — 1st ed.
   p. cm.
  ISBN 978-1-933115-45-0 (hardcover : alk. paper) — ISBN 978-1-933115-46-7 (pbk. : alk. paper) 1. Ecosystem management—Rocky Mountains. I. Prato, Tony. II. Fagre, Dan.
 QH77.R57S87 2007
 333.95'0978—dc22   2006036693

The paper in this book meets the guidelines for permanence and durability of the Committee on Production Guidelines for Book Longevity of the Council on Library Resources. This book was typeset by Peter Lindeman. It was copyedited by René Howard. The cover was designed by Joseph Petraglia. Cover photo taken by Catherine Cunningham, Nature's Reflection.

> The findings, interpretations, and conclusions offered in this publication are those of the contributors. They do not necessarily represent the views of Resources for the Future, its directors, or its officers.

ISBN 978-1-933115-45-0 (cloth)   ISBN 978-1-933115-46-7 (paper)

About Resources for the Future *and* RFF Press

RESOURCES FOR THE FUTURE (**RFF**) improves environmental and natural resource policymaking worldwide through independent social science research of the highest caliber. Founded in 1952, RFF pioneered the application of economics as a tool for developing more effective policy about the use and conservation of natural resources. Its scholars continue to employ social science methods to analyze critical issues concerning pollution control, energy policy, land and water use, hazardous waste, climate change, biodiversity, and the environmental challenges of developing countries.

RFF PRESS supports the mission of RFF by publishing book-length works that present a broad range of approaches to the study of natural resources and the environment. Its authors and editors include RFF staff, researchers from the larger academic and policy communities, and journalists. Audiences for publications by RFF Press include all of the participants in the policymaking process—scholars, the media, advocacy groups, NGOs, professionals in business and government, and the public.

Resources for the Future

Directors

Catherine G. Abbott
Vicky A. Bailey
Michael J. Bean
Preston Chiaro
Norman L. Christensen, Jr.
Maureen L. Cropper
W. Bowman Cutter
John M. Deutch

E. Linn Draper, Jr.
Mohamed T. El-Ashry
J. Andres Espinosa
Daniel C. Esty
Linda J. Fisher
Dod A. Fraser
Kathryn S. Fuller
Mary A. Gade

James C. Greenwood
David G. Hawkins
R. Glenn Hubbard
Charles F. Kalmbach
Michael A. Mantell
Steven W. Percy
Matthew R. Simmons
Robert N. Stavins

Officers

Lawrence H. Linden, *Chair*
Frank E. Loy, *Vice Chair*
Philip R. Sharp, *President*
Edward F. Hand, *Vice President–Finance and Administration*
Lesli A. Creedon, *Vice President–External Affairs*

Editorial Advisers for RFF Press

Walter A. Rosenbaum, *University of Florida*
Jeffrey K. Stine, *Smithsonian Institution*

Contents

Figures and Tables vii

Foreword ix
    Ralph Waldt

Preface xiii

Acknowledgments xvii

Contributors xix

PART I. Introduction

1. The Crown of the Continent: Striving for Ecosystem Sustainability 3
   Tony Prato and Dan Fagre

2. The Crown of the Continent Ecosystem: 17
   Profile of a Treasured Landscape
   Ben Long

PART II. Social Dimensions

3. Native Peoples and Archaeology of Waterton Glacier International 39
   Peace Park
   Brian O.K. Reeves

4. Economic Growth and Landscape Change 55
   Tony Prato, Dan Fagre, and Ramanathan Sugumaran

5. Sustaining Wildland Recreation in the CCE: Issues, Challenges, ... 67
   and Opportunities
   Stephen F. McCool and John C. Adams

PART III. Biophysical Dimensions

6. Alpine Ecosystem Dynamics and Change: A View from the Heights 85
   George P. Malanson, David R. Butler, and Dan Fagre

7. Conserving Biodiversity 102
   Michael Quinn and Len Broberg

8. Aquatic Ecosystem Health 117
   F. Richard Hauer, Jack A. Stanford, Mark S. Lorang,
   Bonnie K. Ellis, and James A. Craft

9. Conserving Water Resources in the CCE 135
   James M. Byrne and Stefan Kienzle

PART IV. Ecosystem Dynamics

10. Paleoperspectives on Climate and Ecosystem Change 151
    Greg Pederson, Cathy Whitlock, Emma Watson,
    Brian Luckman, and Lisa Graumlich

11. Modeling and Monitoring Biophysical Dynamics and 171
    Change in the CCE
    Dan Fagre

12. Ecosystem Responses to Global Climate Change 187
    Dan Fagre

13. CCE Fire Regimes and Their Management 201
    Robert E. Keane and Carl Key

PART V. Management Issues and Challenges

14. Cumulative Effects Analysis and the Crown Managers Partnership 215
    Michael Quinn, Danah Duke, and Guy Greenaway

15. Transboundary Conservation and the Yellowstone to Yukon 229
    Conservation Initiative
    Marguerite H. Mahr

16. Adaptive Ecosystem Management 249
    Tony Prato

17. Challenges of Managing Glacier National Park in a Regional Context ...260
    Tara Carolin, Steve Gniadek, Sallie Hejl, Joyce Lapp, Dawn LaFleur,
    Leo Marnell, Richard Menicke, and Jack Potter

18. Resolving Transboundary Conflicts: The Role of 285
    Community-Based Advocacy
    Steve Thompson and David Thomas

19. Achieving Ecosystem Sustainability 302
    Tony Prato and Dan Fagre

Index 313

Figures and Tables

Front map   The Crown of the Continent Ecosystem xxviii

Figure 3-1   Native Culture History of Waterton-Glacier 44
Figure 3-2   Waterton-Glacier Valley Floor and Alpine Culturally
Assignable Sites 47
Figure 4-1   Schematic of Flathead Landscape Ecosystem Model 58
Figure 4-2   Hypothetical Wildlife and Economic Benefits
for Nine Alternative Futures 62
Table 5-1   Areas of CCE in Protected Status, 2005 72
Figure 6-1   Four Topographic Types of Alpine Tundra 89
Figure 6-2   Alpine Glades in Preston Park, Glacier National Park 97
Figure 8-1   Annual Primary Productivity in Flathead Lake, 1978–2003 123
Figure 8-2   Total Phosphorous Concentration Entering and
Departing Lake McDonald 128
Figure 8-3   Total Nitrogen Concentration Entering and
Departing Lake McDonald 129
Figure 8-4   Effects of Logging on Annual Algal Biomass in
Watersheds of the CCE 130
Figure 10-1   Reconstructing Environmental History 154
Figure 10-2   Pollen-percentage Diagram from Johns Lake, Glacier National Park 158
Figure 10-3   Select Tree-ring-based Temperature and Moisture Reconstructions
Available in the CCE 162
Figure 10-4   Tree-ring-based Reconstructions of Seasonal and Net Mass Balance for
Peyto Glacier and the Glaciers of Glacier National Park 166
Figure 11-1   Pacific Decadal Oscillation and Snow Water Equivalent in Glacier
National Park 177
Figure 12-1   Shepard Glacier in Glacier National Park, Montana, 1913 and 2005 .190

Figure 12-2	Summer Average Minimum Temperature, Kalispell, Montana, 1900–2005	192
Figure 13-1	Ponderosa Pine Savanna Ecosystem of Big Prairie, Bob Marshall Wilderness Area, Montana	202
Figure 13-2	Characteristics of Fire Regimes along Elevation and Aspect Gradients	205
Figure 15-1	Crown of the Continent Ecosystem on the Yellowstone to Yukon Corridor	231
Table 15-1	Transboundary CCE Species at Risk in Canada and the United States	234
Figure 15-2	Grizzly Bear Populations from Northern Wyoming to Northern British Columbia	236
Figure 15-3	Partners in Flight Priority Bird Species Richness in Western Montana	240
Figure 15-4	Historical and Current Fish Faunal Groups in the Y2Y Corridor	243
Table 16-1	Hypothetical Most Likely Bear Habitat Condition and Optimal Bear Management Decision	255
Table 16-2	Hypothetical Expected Gains and Losses with Two Road Management Actions and Two Habitat Conditions	256
Table 17-1	Native Fish Species in Glacier National Park	264
Table 17-2	Non-Native Fish Species in Glacier National Park	265

Foreword

Tucked away within the continental folds of the great Rocky Mountain chain in southern Canada and western America, one last, truly wild place remains. Sequestered in southern British Columbia, Alberta, and northern Montana is a landscape that has escaped the heavy hand of modern humans, a place so vast and undeveloped that much of it is still roadless wilderness. For more than 170 miles, the Rockies rise from the land in a spectacular thrust of sharp mountain peaks covered with great swaths of forest. Whole river systems still function here without the impediment of dams, flowing seaward with glass-clear, unpolluted waters. Cupped within the depths of the mountains are hundreds upon hundreds of shimmering lakes of remarkable color and transparency. This is a region so stunningly beautiful that it could have inspired a magnificent opus or a legendary painting. It is a place so wild that it is the stronghold for grizzly bears in the contiguous United States, so wild that it retains a more intact assemblage of native mammals than anywhere else from the Mexican border to west–central Canada. This landscape is a thriving ecosystem of remarkable richness and uncommon biological diversity. This is truly a place like no other, a piece of Eden rising above an industrialized world. It is a place known fittingly as the Crown of the Continent Ecosystem (CCE).

The CCE is a vast area. Within the United States' portion are five federally protected wilderness areas and a national park. Within the Canadian portion are another national park and an adjoining provincial park. The presence of these core protected areas attests powerfully to the uniqueness and value of the ecosystem. Among the CCE's many additional attributes is a rich plant community, a rare blending of four major floral regions. The continental position of the region, roughly midway along the Rocky Mountain chain and on the flank of the western edge of the Great Plains, gives it exceptional botanical diversity. Plant species native to the Pacific Northwest reach the eastern edge of their ranges in the CCE; plants from the Great Plains reach their western limits. Boreal plants from the northern Rockies reach the southern limit of their ranges within the CCE; other plants from the middle Rockies reach their northern limits. With respect to wildlife, the CCE literally stands alone as a refuge for vanishing species. The last population of grizzly bears that still relies on prairie habitat makes its home in the area, as does the densest population of grizzlies found anywhere south of Canada. Viable populations of other rare mammals—including wolves, lynx, fishers, and wolverines—find sanctuary in the CCE.

The presence of these species brings to light another highly important attribute of the ecosystem. The CCE is *connected* with Canadian wildlands farther north, allowing travel corridors and genetic interchange for populations of wide-ranging mammals. In most other ecosystems within the American and Canadian west, such connectivity has been lost.

Like all wild places in the twenty-first century, the CCE faces a battery of threats to its existence. In western Montana, a burgeoning population puts pressure on the land in terms of rural subdivisions and developments, increasing recreational use, industrial development, timber harvesting, and a variety of other factors. Along Montana's Rocky Mountain Front, the potential development of oil and natural gas reserves poses a substantial threat to a vast landscape that has known relatively little extraction of these resources. In the Canadian portions of the CCE, the possibility of coal-bed methane extraction looms as a force that could have devastating effects on the ecological integrity of one of the ecosystem's most biologically important areas.

If we are to sustain this grand ecosystem in a manner that protects the scenic and wildlife values inherent to it—while accommodating human needs—we must have the right set of tools to do the job effectively. Enter this book. *Sustaining Rocky Mountain Landscapes* is a valuable collection of expertise and methodology, a synergistic compilation of information written by a distinguished and respected cadre of professors, agency personnel, land managers, naturalists, scientists, and more. This book is the first of its kind. Never before have experts applied such a needed synthesis of social, biological, and managerial dynamics to the CCE. And never before has the need for such a book been so pressing, with the CCE facing an uncertain future brought about by steadily increasing pressures from humanity.

Most of the wilderness that once so characterized the North American continent is now gone. Much of the loss can be attributed to humankind's need to sustain an economy. We cannot argue with the fact that no matter where people live in modern times, they must realize an economic return from the landscapes they inhabit. For decades, extractive industries formed one of the mainstays for the economy in the CCE. Logging and mining, for example, generated income for residents, but were rough on the land and its wildlife. Now people realize that other industries, like tourism and backcountry outfitting, are capable of driving the local economy. The difference between these two economies is that one is, by its very nature, short-lived. The other can go on indefinitely as long as the landscape that enables it endures. This book shows us viable ways to encourage a lasting economy from large tracts of wildlands without destroying them.

Sustaining this rare ecosystem requires a combination of educating the public, protecting existing wild areas from development, and balancing sustainable land-use practices wisely in relation to an expanding human populace. Education is critical. If our progeny do not learn to value wildlands, our current efforts at protecting such places will prove almost pointless. The protection of existing wildlands is also critically important. Most of the rare animals that grace the CCE remain there today because of the large blocks of roadless wilderness, national park acreage, and private ranchlands within the area. In the long term, the most pivotal issue facing the CCE may be the economics of land use, and several chapters within this volume address this issue. *Sustaining Rocky Mountain Landscapes* should help our society forge a future that will embrace a combination of respect and value for the CCE. This combination translates into the ways that we choose to use the land for economic return.

During the forty-plus years that I have lived in the West, I have traveled widely, viewing many of its wild areas. I have hiked into stands of redwoods and sequoias, the rainforests of Washington, the slickrock canyonlands of Utah, the wild mountains of northern and central Idaho, Arizona, Wyoming, and more. With every one of these experiences came the realization that our nation still holds great beauty and wildness in many places. On my return home to Montana after each adventure, I kept realizing that the CCE is the wildest of all. As I grew older, I also realized that the CCE was more than a national treasure. It is a *world* treasure. A look around the globe in the twenty-first century will reveal very, very few places left in the temperate regions of our planet that are as large, wild, and biologically intact as the CCE.

As the future unfolds, I can only hope that the CCE remains as it is. A stellar product of billions of years of evolution, a wellspring of biodiversity, a landscape like no other. Keeping it this way demands a societal respect for the land and its life. Maintaining the ecosystem requires a wise combination of resource management, applied science, international cooperation, and sound economic practices. *Sustaining Rocky Mountain Landscapes* is a great place to start.

RALPH WALDT
The Nature Conservancy

Preface

Rocky Mountain landscapes in western America and Canada are endowed with public and private lands that afford residents and visitors environmental amenities and superb year-round outdoor recreational opportunities. These landscapes—which feature public wilderness areas and wildlife refuges, along with national forests, monuments, and parks—also sustain myriad ecosystem goods and services and enrich people's lives by offering a respite from the fast-paced, high-pressure, congested environments of metropolitan areas where most people reside. Rural gateway communities for public lands in the Rocky Mountain West generally have vibrant and growing economies because people find them to be desirable places to live, work, and play. A gap in income and standard of living is widening, however, between new and longtime residents.

Rocky Mountain landscapes exhibit dynamic interactions between people and the natural environment that have evolved over time and space. In the days of the Old West, extractive resource development primarily drove ecosystem change. During this exploitive era, regional landscapes were dramatically altered. Most old-growth forests that once existed outside protected areas were eliminated by timber harvesting. Mine tailings and logging polluted rivers and lakes and fire suppression caused woody biomass to accumulate, which later contributed to intense wildland fires. River systems were dramatically modified by hydroelectric power development and river floodplains were degraded by cattle grazing, timber harvesting, and human settlements. Wildlife habitat became fragmented by timber harvesting, oil and gas production, and rural development, and the natural balance of species and the number of invertebrates were altered by the introduction of exotic fish species. As environmental awareness increased, people demanded public policies to control the environmental impacts of growth and development and to protect natural resources. During the 1970s and 1980s, legislative bodies responded by passing numerous environmental laws.

In the 1990s, the Old West began giving way to the New West, in which high value is placed on preserving and experiencing the outstanding natural resources on public lands and the excellent quality of life in gateway communities. Ironically, the very environmental amenities and economic growth that birthed the New West have had negative environmental consequences, which take many forms. Conversion of agricultural lands to residential and commercial uses is reducing open spaces, eliminating and fragmenting fish and wildlife habitats, and making recovery of threatened and endangered species more difficult (although some successes have been achieved). Air and water quality are declining, rural residential development and drought have combined to increase wildland fire risk to human life and property, and ecosystem functions and processes are being impaired.

More recently, higher energy prices and political instability in key crude-oil-supplying nations have renewed political and economic interests in fossil fuel development, from tar sands in Alberta and coal and methane extraction in British Columbia to oil and gas development in Montana's Rocky Mountain Front (one of the few native prairies remaining in the northern Great Plains). Additional energy resource development and amenity-driven economic growth, although resulting in social and economic benefits, do pose further risks to natural resources and ecosystem services in the region's landscapes. A wild card has also entered the mix—the potentially negative human and ecological impacts of global climate change. These prospects require even greater diligence in protecting the human and ecological values sustained by Rocky Mountain landscapes.

Typically, most public policy and funding have been directed at ameliorating human disturbances in ecosystems outside the Rocky Mountain West. Prominent examples include the Great Lakes, Chesapeake Bay, the Everglades, Puget Sound, and the Columbia and Missouri rivers. Although ecological restoration is important and necessary, preserving relatively intact ecosystems is increasingly viewed as less costly and disruptive in human and ecological terms than restoring severely affected ecosystems. The Crown of the Continent Ecosystem (CCE), which straddles the border between Alberta and British Columbia in Canada and northwestern Montana in the United States, serves as a prime example of this type of preservation opportunity.

The CCE is named for the Triple Divide Peak area in Glacier National Park as well as the transboundary Waterton Glacier International Peace Park. Precipitation on Triple Divide Peak flows into three major river systems: the Columbia, Mississippi, and Saskatchewan. The Waterton Glacier International Peace Park (the world's first such park) is the ecological jewel of the CCE. The park's unique biological and cultural features have earned it the additional designations of World Heritage Site and biosphere reserve.

The CCE, however, is undergoing rapid economic growth and development on private and unprotected public lands, and human disturbances are having an adverse impact on the area's unique endowments of natural resources. Although these factors are bringing about major transformations, the CCE still contains one of the most intact assemblages of mammals of any region in southern Canada or the contiguous United States. It is home to 65 species of native mammals, 270 species of birds, 27 native fish, and 12 reptiles and amphibians. The North Fork of the Flathead River, which is the major river corridor in the northwestern sector of the CCE, has the highest density of inland grizzly bears and the most diverse association of ungulates in North America. In a 2005 workshop, 17 Canadian and 20 American scientists, managers, planners, and policymakers agreed that "… the region [CCE] is a globally unique nexus of converging ecosystems and biodiversity."

Participants in the same workshop posed two critical questions that are symptomatic of the issues facing the CCE: (1) "How do you provide access to a priceless resource without compromising the very thing, its pristine attributes, that makes it so valuable?" and (2) "What is needed to protect the natural environment, yet facilitate economic prosperity?" These questions are not easy to answer, but they require an improved understanding of how dynamic interactions among complex social, demographic, economic, and ecological processes influence the CCE. Furthermore, that improved understanding must be translated into management plans and policies that sustain the socioeconomic benefits of growth and development while minimizing negative impacts on ecological integrity. In short, the need to search for and discover ways to sustainably manage the CCE's human and natural resources is pressing.

The primary intended audience for this book spans planning staffs for local communities and counties/provinces; policymakers and public land managers; scientists and naturalists; park guides and interpreters; economic interest groups; educators and students; recreationists and tourists; and members of environmental, natural history, professional, and scientific associations. We hope that the information presented here will help readers to better understand the dynamic social, economic, and ecological forces that are shaping and transforming the CCE, allowing current and future generations to formulate effective plans and policies for sustaining the region's valuable environmental amenities and ecosystem services in perpetuity.

This book is organized in five parts. Part I (Introduction) introduces the book's subject and lays a foundation for the analysis it contains (Chapter 1), and gives a succinct, nontechnical overview of the CCE's human and natural resources (Chapter 2).

Part II (Social Dimensions) addresses the major cultural, historical, and economic forces that are shaping the CCE. It discusses native peoples and culture (Chapter 3); describes the impacts of economic growth and development on landscapes and wildlife habitat (Chapter 4); and evaluates the issues, challenges, and opportunities in wildland recreation and tourism (Chapter 5).

Part III (Biophysical Dimensions) assesses the biological and water resources of the CCE, covering alpine ecosystems (Chapter 6), biodiversity conservation (Chapter 7), aquatic ecosystem health (Chapter 8), and water conservation (Chapter 9).

Part IV (Ecosystem Dynamics) addresses the dynamic nature of ecosystem processes in the CCE in terms of paleoecological perspectives on climate and ecosystem change (Chapter 10), modeling and monitoring biophysical change (Chapter 11), ecosystem responses to global climate change (Chapter 12), and fire regimes and their management (Chapter 13).

Part V (Management Issues and Challenges) proposes ways to advance sustainable management of landscapes in the CCE. This part describes innovative and collaborative approaches to evaluating cumulative effects of human activities on natural resources (Chapter 14), an initiative to maintain and restore biological diversity and habitat connectivity in the Yellowstone to Yukon region (Chapter 15), an adaptive approach to ecosystem management (Chapter 16), contemporary challenges to managing Glacier National Park (Chapter 17), and the role of community-based advocacy groups in resolving conflicts in the transboundary CCE (Chapter 18). Finally, Chapter 19 proposes an approach for advancing sustainable landscape management in the CCE that draws on the concepts, knowledge, and insights presented in the book.

Tony Prato
Dan Fagre

Acknowledgments

The authors appreciate the useful comments and suggestions made on a draft of the book by three anonymous reviewers. We thank Don Reisman of RFF Press for his guidance, direction, and encouragement throughout the project. We particularly want to express our heartfelt gratitude to Lisa McKeon for going beyond the call of duty in helping with the myriad tasks of producing a book. Lisa was cheerful, organized, and resolute in her contributions. We extend a special thanks to René Howard for her outstanding copyediting.

Contributors

John C. Adams is a Ph.D candidate in the College of Forestry and Conservation at the University of Montana. His research focuses on stakeholders' competition to shape public land policy and management. He has written on public land issues for publications such as *The Denver Post*, *High Country News*, and *Montana Magazine*.

Len Broberg is a professor and program director for the environmental studies program at the University of Montana. He combines his nine years experience in the practice of law with his training in conservation biology to teach and research in the area of biodiversity conservation and environmental policy/law. His research has been published in *BioScience*, *Wildlife Society Bulletin*, and *Journal of Forestry*.

James M. Byrne is an associate professor of geography and director of the Water Resources Institute at the University of Lethbridge. He also is theme leader, Water Resources Management, and founding scientist and member, Research Management Committee, Canadian Water Network. He has earned several awards for his work as a producer of films and television series.

David R. Butler is a professor of geography and graduate program advisor at Texas State University-San Marcos. His research examines animals as geomorphic agents, and geomorphic and biogeographic processes in mountain environments, especially Glacier National Park. He received the Association of American Geographers Distinguished Career Award from the Mountain Geography Specialty Group, and the G.K. Gilbert Award for Excellence in Geomorphology from the Geomorphology Specialty Group for his book *Zoogeomorphology—Animals as Geomorphic Agents*.

Tara W. Carolin is an ecologist and research coordinator for the National Park Service in Glacier National Park. Her work for the National Park Service over the past 15 years has focused on inventory and monitoring of natural resources, whitebark and limber pine restoration, vegetation mapping, and environmental analysis and compliance.

James A. Craft is a research specialist in limnology for the Flathead Lake Biological Station at the University of Montana. His research has focused on the ecology of Flathead Lake, a large oligotrophic lake in Northwest Montana. Other recent research includes ecological effects of natural and anthropogenic disturbances on streams and lakes in Northwest Montana.

Danah Duke is the executive coordinator of the Miistakis Institute for the Rockies at the University of Calgary. As Executive Director, she oversees Miistakis' core research programs: facilitation of ecosystem-based research and management; transboundary geospatial analysis and research; and creation of tools to disseminate spatial and biological information. She has a strong background in the science and management of wildlife corridors in the Rocky Mountains.

Bonnie K. Ellis is a senior research scientist in limnology for the Flathead Lake Biological Station at the University of Montana. Her recent research has focused on a retrospective analysis of nutrient loading and food web change in Flathead Lake (1977-2004), with particular emphasis on the trophic cascade and the resulting alternate state. Recent publications have appeared in *Freshwater Biology, Verh. Internat. Verein. Limnol.*, and the *Journal of the North American Benthological Society*.

Daniel B. Fagre is a research ecologist for the U.S. Geological Survey Northern Rocky Mountain Science Center in Glacier National Park. He is a faculty affiliate at several universities and co-authored (with Tony Prato) the book *National Parks and Protected Areas: Approaches for Balancing Social, Economic, and Ecological Values*. He received the Director's Award for Natural Resource Research from the National Park Service and serves on the Montana governor's advisory board for climate change.

Lisa Graumlich is director of the School of Natural Resources, the University of Arizona, Tucson. As executive director of the Big Sky Institute at Montana State University, she facilitated the development of several research networks addressing mountain issues. She was elected as a fellow of the American Association for the Advancement of Science and has published numerous technical papers in journals such as *Climatic Change* and *Geophysical Research Letters*.

Steve Gniadek is a senior wildlife biologist for the National Park Service in Glacier National Park. His prior experience in wildlife conservation and management has been with the Peace Corps, Forest Service, Bureau of Land Management, Fish and Wildlife Service, and Yellowstone National Park. He has presented papers on wildlife management in national parks at conferences of The Wildlife Society, Society for Conservation Biology, and American Society of Mammalogists.

Guy Greenaway is a senior project manager for the Miistakis Institute for the Rockies at the University of Calgary. He has skills and experience in project management, conservation communication, ecosystem management, and private land conservation. He has worked as a consultant for numerous government agencies and non-profit organizations, and is a past executive director of the Southern Alberta Land Trust Society.

F. Richard Hauer is a professor of limnology for the Flathead Lake Biological Station at the University of Montana and holds the FLBS Endowed Chair in Limnology. His research has been coupling the vast array of ecological attributes of gravel-bed rivers and floodplains with remotely-sensed data from satellite and aircraft. He is co-editor/author of the widely acclaimed book *Methods in Stream Ecology* and has served as president of the North American Benthological Society.

Sallie Hejl is a resource education specialist for the National Park Service in Glacier National Park at the Crown of the Continent Research Learning Center. She educates park staff and the public about natural and cultural resources. Projects include citizen science on Common Loons, a fire ecology nature trail, and an invasive plant guide for the Crown of the Continent ecosystem. Past research focused on the effects of management actions and natural disturbances on Rocky Mountain birds.

Robert E. Keane is a research ecologist for the U.S. Department of Agriculture Forest Service, Rocky Mountain Research Station at the Missoula Fire Sciences Laboratory. His most recent research includes developing ecological computer models for exploring landscape, fire, and climate dynamics; the mapping of fuel characteristics; and investigating the ecology and restoration of whitebark pine. He is the project leader for the Fire Ecology Project.

Carl H. Key is a geographer for the U.S. Geological Survey Northern Rocky Mountain Science Center in Glacier National Park. Research includes development and national evaluation of remote sensing methodologies for mapping fire severity, and scientific guidance for Monitoring Trends in Burn Severity, a program supported by the Wildland Fire Leadership Council under the National Fire

Plan. Several science excellence and special achievement awards distinguish his 26-year career with the National Park Service and the U.S. Geological Survey.

Stefan W. Kienzle is an associate professor for Hydrology and GIS at the University of Lethbridge, Alberta. His research focuses on water resources impacts by climate and land use changes using complex hydrological simulation models. He has published in a variety of journals, including *Hydrological Processes, Water Resources Management*, and *Spatial Hydrology*.

Dawn LaFleur is a supervisory biologist for the National Park Service in Glacier National Park. She focuses on noxious weed management in wilderness areas and parks through monitoring and eradication. She collaborates with numerous local and state entities to provide a coordinated effort to prevent the spread of invasive plants.

Joyce Lapp is a supervisory horticulturist for the National Park Service in Glacier National Park. She has managed the Native Plant Restoration Program at the Park since 1987. She serves as consultant for the Federal Lands Highway Program out of Denver Service Center, developing restoration plans for other park units engaged in road reconstruction projects as well as other agencies needing restoration expertise.

Ben Long is a freelance writer and senior program director for Resource Media in Montana, a communications firm dedicated to "making the environment matter." He spent much of his career as a journalist, covering natural resource issues in the West. His book *Backtracking: By Foot, Canoe and Subaru on the Lewis and Clark Trail* received the Chinook Literary Prize and was named one of the top 10 outdoor books of the year by Amazon.com.

Mark S. Lorang is research assistant professor of geomorphology and physical ecology for the Flathead Lake Biological Station at the University of Montana. His research goal is to develop a better understanding of the links between fundamentally important geomorphic and ecosystem processes that can help guide conservation and restoration of river-lake ecosystems. This goal has lead him to investigate physical processes active in both regulated and unregulated rivers, as well as nearshore environments of lakes and ocean coasts.

Brian Luckman is a professor of geography at the University of Western Ontario. He has worked in the Canadian Rockies since 1968 and has published extensively on mass wasting processes, Holocene glacier fluctuations, and the environmental history of this region. His most recent work has focused on reconstructing the climate of the last millennium primarily using dendrochronology in both the Rockies and the Yukon.

Contributors

Marguerite (Marcy) H. Mahr is a conservation biologist consulting on wildlife conservation area design in the U.S. and Canadian Rockies. From 1999 to 2004 she was Conservation Science and Planning Coordinator for the Yellowstone to Yukon (Y2Y) Conservation Initiative leading multi-scale research on carnivores, birds, and aquatics. Her projects focus on identifying avian conservation areas within the Y2Y, and analyzing the role of agricultural land for wildlife habitat cores and corridors.

George P. Malanson is the Coleman-Miller professor in geography at the University of Iowa. His research on mountains and Glacier National Park have resulted in recent articles in *Geographical Analysis, Global Ecology & Biogeography, Ecological Modelling, Geomorphology,* and *Physical Geography.* He was elected a fellow of the American Association for the Advancement of Science and received the Parsons Distinguished Career Award, Association of American Geographers, Biogeography Specialty Group.

Leo Marnell is a retired supervisory biologist for the National Park Service in Glacier National Park. He conducted research on native fish, amphibians, and fish genetics over a career spanning several decades at the Park.

Stephen F. McCool is a professor of wildland recreation management at the College of Forestry and Conservation, the University of Montana. His research has emphasized issues associated with planning and management of recreation in wildland and protected area settings, including the design of planning and public engagement processes and the integration of different values. His books include *Tourism in National Parks and Protected Areas: Planning and Management* and *Tourism, Recreation and Sustainability: Linking Culture and the Environment.*

Richard Menicke is a geographer for the National Park Service in Glacier National Park. His work focuses on managing the Geographic Information System (GIS) program and applying GIS technology and other natural resource information to a broad range of park management and research questions.

Gregory T. Pederson is a research scientist for the U.S. Geological Survey Northern Rocky Mountain Science Center and the Big Sky Institute at Montana State University. His research has focused on climate variability as a driver of biological and physical components of mountainous ecosystems. Prior publications include a chapter in a book on glaciers entitled *Darkening Peaks,* and in scientific journals such as *Geophysical Research Letters, Earth Interactions,* and *Arctic, Antarctic, and Alpine Research.*

Jack Potter is a supervisory biologist and chief, Division of Science and Resources Management for the National Park Service in Glacier National Park. He oversees natural and cultural resource management and research permitting. He was awarded the National Park Service's Intermountain Region's Director's Award in 2003 for Excellence in Natural Resource Management and was awarded the NPS Honor Award for Superior Service in 2007. He has been very active in the six-park Rocky Mountain Inventory and Monitoring Network and wilderness management.

Tony Prato is professor of ecological economics and co-director of the Center for Agricultural, Resource and Environmental Systems at the University of Missouri-Columbia, and a faculty affiliate in the College of Forestry and Conservation at the University of Montana. He is coauthor (with Dan Fagre) of *National Parks and Protected Areas: Approaches for Balancing Social, Economic, and Ecological Values.*

Michael Quinn is an associate professor of environmental science and planning in the Faculty of Environmental Design and director of research and liaison for the Miistakis Institute for the Rockies at the University of Calgary, a collaborative ecosystem management research organization with a geographic focus on the Rocky Mountains and surrounding regions. His work has appeared in such journals as *Ecology and Society, Local Environment,* and *Western Humanities Review,* as well as contributions to several edited international books.

Brian O.K. Reeves is a professor emeritus at the University of Calgary and the president of Lifeways of Canada Ltd., Alberta's oldest cultural resource consulting and contracting company. He has over 30 years of research and professional experience and specializes in northern North American archaeology, ethnohistory, and cultural resource management.

Jack A. Stanford is the director and Jesse M. Bierman professor of ecology for the Flathead Lake Biological Station at the University of Montana. Jack has conducted research at FLBS since 1971 and became director in 1980. His research and education activities have taken him all over the world but his heart is in the Crown of the Continent Ecosystem where he has worked on everything from microbes to grizzly bears.

Ramanathan Sugumaran is director of the GeoTREE Center and is an associate professor of geography at the University of Northern Iowa. His research has focused on GeoInformatics application in water quality monitoring, tree species identification, and watershed modeling. He is a board member to the Iowa Geographical Information Council. His journal publications appeared in *Geographical Systems* and *Journal of Soil and Water Conservation.*

Contributors

David Thomas is communications director of Wildsight, a conservation group based in southeastern British Columbia. He assists program managers in the development of news releases, audio-visual presentations, and pages for www.wildsight.ca. He is also active in the forging of a transboundary coalition of conservation groups around the Crown of the Continent ecoregion.

Steven S. Thompson is the Glacier Program Manager for the National Parks Conservation Association. He is coordinating a community-based project with National Geographic Society to map the distinctive natural, cultural, and historical assets in the Crown of the Continent region. He received the 2004 National Achievement Award for Education from the Natural Resources Council of America for the report, *Gateway to Glacier: The Emerging Economy of Flathead County*.

Ralph Waldt was a resident naturalist at The Nature Conservancy's Pine Butte Preserve on the Rocky Mountain Front. He led more than 1,200 interpretive hikes and taught numerous courses. He is the author of *Crown of the Continent: The Last Great Wilderness of the Rocky Mountains*.

Emma Watson is a research scientist for the Meteorological Service of Canada, Environment Canada. Research for the chapter in this book was conducted during a visiting fellowship with the Climate Research Division of Environment Canada.

Cathy Whitlock is a professor of earth sciences at Montana State University. Her research has focused on the paleoenvironmental history of the northwestern U.S. and southern South America. Her studies concern the climate and vegetation changes that took place in the Rocky Mountain region during the last 11,000 years, and the role of fire as a catalyst of environmental change.

SUSTAINING ROCKY MOUNTAIN LANDSCAPES

Science, Policy, and Management for the Crown of the Continent Ecosystem

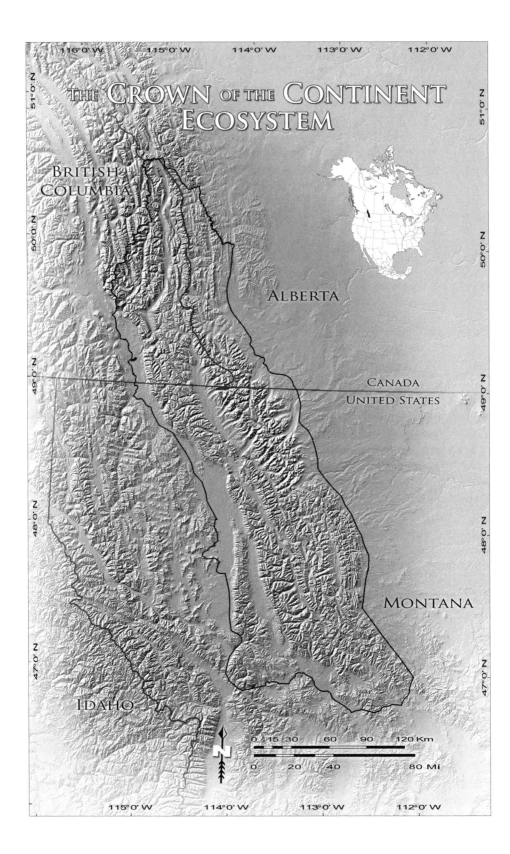

PART I
Introduction

1
The Crown of the Continent
Striving for Ecosystem Sustainability
Tony Prato and Dan Fagre

> *The Crown of the Continent is a place for two nations to protect, a place for all the world's people to cherish.*
>
> —Doug Cox

The Crown of the Continent Ecosystem (CCE) is a Rocky Mountain region that covers 43,700 km² in northwestern Montana, southwestern Alberta, and southeastern British Columbia (see map on p. xviii). Approximately 60% of the CCE is in the United States and 40% is in Canada (Waldt 2004). The region, which extends from the Highwood River south of Banff National Park (in Alberta) in the north to the Blackfoot River (in Montana) in the south, features many spectacular natural areas, including the Waterton Glacier International Peace Park in Alberta and Montana, the Castle Rock Wilderness and Elk River Valley in British Columbia, and Glacier National Park in Montana. The CCE is also home to the Bob Marshall, Great Bear, Scapegoat, Rattlesnake, and Mission Mountains wilderness areas. Mountains and plains converge on the eastern side of the CCE to form the Rocky Mountain Front, one of the few outstanding native prairies remaining in the northern Great Plains (TNC 2004).

Regional Growth and Change Escalate

Spectacular natural landscapes, diverse employment and outdoor recreation opportunities, and an excellent quality of life make the CCE an attractive place to live, work, and play. These amenities are driving demographic and economic growth and transitions in the region. For example, the Rocky Mountain West (which is home to the CCE) is one of the fastest growing areas in the United States and Canada. Between 1990 and 2000, the population of Colorado, Idaho, and Utah increased 23%, and the population of Alberta, British Columbia,

Montana, New Mexico, and Yukon Territory increased between 10 and 15% (Travis et al. 2002). During the 1990s, the population of the entire Rocky Mountain West grew more than 25% (Riebsame et al. 1997), and population growth in two-thirds of the region's counties exceeded the national average (Beyers and Nelson 2000). In the United States during the same period, 5 of the 10 fastest growing states and 9 of the 15 fastest growing counties were in the Rocky Mountain West (Fagre 2000).

Populations and economies in the Rocky Mountain West have grown vigorously, especially in counties near metropolitan areas, such as Calgary, Alberta, and Denver, Colorado. High-amenity communities located in protected areas (national parks and wilderness areas) and near resort towns, such as Aspen, Colorado; Banff, Alberta; and Kalispell, Montana have also seen brisk growth. For example, the population of the Greater Yellowstone ecosystem, which includes Yellowstone and Grand Teton national parks, portions of six national forests, and three national wildlife refuges, grew by 55% from 1970 to 1997 (Hansen et al. 1999). The recreational opportunities and environmental amenities afforded by public lands attract new people to the region and encourage long-term residents to stay.

Rocky Mountain landscapes are also being transformed by the shift from the traditional economic base of extractive resource industries of the Old West (i.e., mining, agriculture, and logging) to the service-oriented recreation and tourism industries of the New West (Riebsame et al. 1996; Power and Barrett 2001; Travis et al. 2002), which rely on environmental amenities.

Inflation-adjusted per capita income in the western United States has grown 60–75% faster in isolated rural counties that contain protected areas than in isolated rural counties that do not feature such areas (Rasker et al. 2004). The structure of regional economies has changed significantly over time, as evidenced by changes in the types of employment available and the sources of personal income. Service and professional jobs in technology development, communication, real estate, recreation, and tourism have become more numerous. Jobs in traditional resource-extraction industries (agriculture, logging, oil and gas development, and mining), however, have become more scarce. In nonmetropolitan (rural) areas of the American West from 1970 to 2000, (1) nonlabor income (money received from investments, retirement benefits, health care and disability payments, Medicare, Medicaid, and welfare) was the fastest growing source of personal income; (2) nonlabor income accounted for one-half of net growth in total personal income; (3) service-related industries were the second fastest growing source of personal income; and (4) income earned in resource-extraction industries decreased 20% (U.S. Department of Commerce 2001).

Flathead County, Montana, located in the northern U.S. portion of the CCE, illustrates the economic and demographic transformations taking place in the region. From 1990 to 2000, the total population of Flathead County increased

25.8%, compared to 12.9% for the state of Montana and 13.1% for the nation (U.S. Census Bureau 2006). Since 1980, Flathead County's population has doubled, making it the second fastest growing county in Montana. From 1987 to 1997, total personal income grew 45% and total employment rose 53% (Davis 2000). Employment growth in the county is heavily concentrated in the retail trade and services sectors. Included in the services sector are health care, business services, engineering and management, and social services. Currently, employment in the services sector accounts for nearly $295 million in labor earnings in the county, which represents about 30% of all labor earnings by county residents. The services and retail trade sectors accounted for more than 60% of all job growth in Flathead County during the 1990s. Rapid employment growth is also occurring in the construction, manufacturing, finance, insurance, and real estate sectors, as well as in sectors that benefit from travel and tourism.

In contrast, labor earnings for longstanding industries, such as lumber and wood products manufacturing, primary metals manufacturing, railroads, and agriculture, declined or grew slowly from 1980 to 2000. Specifically, labor earnings for these industries—as a percentage of all total labor earnings in the county—fell from 11.4% in the mid-1970s and 10.95% in the early 1980s to 6.8% in 2000. During the same period, these industries' share of total personal income fell from 9% to less than 5%. Such transformations are representative of the CCE's transition from the Old West to the New West (Flathead on the Move 2004).

Human Activities Have a Significant Impact

Because of rapid growth and development in many communities within the CCE, agricultural and forest lands have been converted to residential and commercial uses, with the following negative effects on the ecosystem: (1) loss or degradation in fish and wildlife habitats; (2) increased human–wildlife conflicts and associated increases in wildlife mortality; (3) proliferation of non-native invasive species; (4) loss of open spaces; (5) increased erosion and runoff; (6) higher temperatures in streams, lakes, and ponds; and (7) loss of environmental amenities. Many of these consequences can be seen in fertile river valleys and near forests where growth and development are concentrated.

Our discussion of sustainability is based on four premises related to human activity, described in the sections that follow.

1. Natural landscapes are worth preserving because they supply valuable consumer and ecosystem goods and services. Natural landscapes embody natural capital in the form of forests, soil, water, air, and minerals. To produce ecosystem goods, such as timber, fossil fuels (coal, crude oil, and natural gas), and natural fiber (cotton and wool), natural capital must be combined with human capital (labor) and manufactured capital (machinery, buildings,

and equipment). Ecosystem goods are used to produce intermediate goods (such as lumber, aluminum, and inorganic fertilizers) and consumer goods (such as homes, automobiles, and clothing). The values of ecosystem goods are derived from the market prices of the products made from those goods. For instance, the commercial value of a stand of timber is derived from the tree species in the stand and the prices of wood products manufactured from those species.

Natural landscapes supply ecosystem services including air and water purification, flood and drought mitigation, waste detoxification and decomposition, soil generation and renewal, biodiversity preservation, partial climate stabilization, nutrient cycling and services, pollination, and many others (Daily 1997).

Because market prices do not exist for system services, analysts estimate their value using nonmarket valuation methods. For example, the economic value of 17 ecosystem services yielded by 16 biomes has been estimated at $16 trillion to $54 trillion per year on a global basis, with an average value of $33 trillion per year (Costanza et al. 1997). The average value is three times that of global gross domestic product.

2. Population, economic growth, and environmental threats are expected to increase. Experts anticipate that population expansion and economic growth in the CCE will continue. Increases in the demand for recreation and tourism will be fueled by increases in per capita disposable income, higher rates of retirement, and higher numbers of urban refugees (people who move to rural areas to escape the crime, social problems, and higher cost of living in urban centers). As a consequence, threats to the CCE's ecological integrity are expected to increase, reducing its capacity to supply ecosystem goods and services. Growth rates may slow, particularly if the adverse environmental effects of growth and development discourage people from visiting or moving to the ecosystem, and cause others to leave. Another factor that may alter future growth in the CCE is its attractiveness relative to other ecosystems. For example, if real estate prices and environmental degradation grow faster (slower) in the Greater Yellowstone ecosystem (GYE) than in the CCE, the CCE could grow more rapidly (slowly) than the GYE.

3. Although past human activities have degraded natural landscapes, the ecosystem remains intact. Despite the negative environmental effects of growth and development, the CCE retains its standing as one of the most biologically intact ecosystems in the contiguous United States and western Canada. The U.S. portion of the CCE contains the only viable nontransplanted population of wolves, the largest native population of Rocky Mountain bighorn sheep, one of the largest elk herds, the largest mountain goat population, and the largest and densest population of grizzly bear in the American West (Waldt 2004). Other natural landscapes are less fortunate. The air pollution in the Great Smoky Mountains, Sequoia, Kings Canyon,

and Shenandoah national parks, for example, is substantially worse than in Waterton Glacier International Peace Park (Mansfield 2002). Negative impacts of snowmobile use on air quality and wildlife are a major concern in Yellowstone National Park, but snowmobiles are prohibited in Glacier National Park. Even though water quality has deteriorated in certain areas of the CCE, aquatic ecosystems are in good shape.
4. Sustainable landscape management is challenging because public lands dominate the region. As in other ecosystems in the Rocky Mountain West, the CCE is home to large areas of public land—approximately 83% of the CCE's land area is publicly owned and managed (Waldt 2004). Public land use and management presents more challenges than using and managing private land. For example, market prices for consumer goods and services influence the development and use of private land, but market prices either do not exist for ecosystem services from public land, or have limited effects on public land use and management. Most decisions about public land are based on statutory authority, regulations, public policy, and public opinion.

Classifying, Using, and Valuing CCE Landscapes

The CCE consists of a mosaic of natural, rural, built-up, and cultural landscapes. Natural landscapes are endowed with natural resources that are managed for a wide range of ecosystem goods and services. This type of management results in environmental amenities; supports economic growth and development; and sustains soil, air, fish, wildlife, and plants. National, state, and provincial parks and other protected natural landscapes in the CCE are managed primarily for public enjoyment and natural resource protection. Protected natural landscapes are generally off limits to residential and commercial development, and in some cases, resource extraction. National, state, and provincial forests are managed for multiple uses, such as recreation, water supply, logging, and fish and wildlife habitats. Most of the CCE's natural landscapes are publicly owned and managed.

Rural landscapes support consumptive and nonconsumptive land uses. Consumptive land uses include crop and livestock production, energy and mineral development, logging, fishing, and hunting. Nonconsumptive land uses encompass tourism, outdoor recreation, low-density residential housing, and commerce. Built-up landscapes, such as cities and towns, are devoted primarily to homes, businesses, government offices, roads, utilities, urban parks, and other land-intensive human activities. Cultural landscapes, such as community parks, scenic highways, battlefields, and historic structures, contain a mix of natural and cultural resources.

Biophysical attributes, land prices, and institutional arrangements strongly influence human and nonhuman landscape uses. Soil type, topography, climate, vegetation, water availability, and land ownership (public versus private) are biophysical attributes that have high spatial correlation. For example, mountainous areas are at higher elevations and contain headwater areas. Compared to valleys, the climate in these areas is more severe (snow and ice), the soil is less productive, and the slopes are steeper, which can lead to debilitating erosion and runoff. Compared to mountainous areas, valleys (in nonarid areas) usually contain valuable water resources (large rivers and lakes) and have more moderate climate, productive soils, and biodiversity. Mountainous areas, then, are more suitable for recreation, tourism, logging, livestock grazing, hydropower development, and natural resource protection. Valleys, on the other hand, are better suited to farming, housing, commerce, public facilities, and transportation.

Land prices have a major influence on private landscape use and management. For example, land used for residential housing generally has a much higher price than land used for agricultural production, especially in rapidly growing areas. This price differential acts as a strong economic incentive for landowners to sell agricultural land to developers or farm and ranch brokers.

Land use and management are influenced by individual and organizational behavior, which is affected by the incentives (rewards) and disincentives (penalties) created by institutional arrangements. Formal institutional arrangements affecting land use and management include laws, zoning and subdivision regulations, covenants, environmental regulations, and property rights. For instance, the organic act establishing a national park is a formal institutional arrangement that determines a park's size, location, and management objectives. Informal institutional arrangements consist of unwritten rules that encompass cultural norms, social conventions, mores, beliefs, etiquette, traditions, and taboos (James 2001). The desire of homeowners to maintain the appearance of the grounds surrounding their homes is an example of an informal institution.

Land ownership influences landscape values. Private land values are determined by the land's location, regional zoning regulations, and market prices of goods and services. Agricultural land is valued according to its productivity, which is determined by soil conditions and water availability. In crop production, for example, land values are determined by crop prices and yields. Private land generally tends to be allocated to the highest valued use. If agricultural land will fetch a higher price for use in a housing development than for farming or ranching, the landowner has a strong incentive to sell that property to a developer or broker.

Public land values are based on the ecosystem goods and services they generate. Market prices of ecosystem goods, along with pertinent statutory requirements and regulations, guide the management of multiple-use public lands. Because ecosystem services do not have market prices, their values are determined in other ways. Generally, the use and management of public land is not sensitive to land prices.

People and communities assign use and non-use values to natural landscapes, be they private or public. Use value can be direct or indirect. When people use a landscape for hiking, camping, and nature viewing, it has use value. Indirect use value arises from the ecosystem services afforded by natural landscapes, such as air and water purification; pollination; and cycling of nutrients, chemicals, and water. Although people do not consume ecosystem services directly, the services have indirect use value because they are essential for human existence (Prato 1998). A landscape also has intrinsic value, which is above and beyond its worth to humans (Capra 1995). Intrinsic value implies that all living things have equal rights regardless of their value to humans.

Landscape and human values are related. For example, the values of natural landscapes in recreation and tourism depend on the consumer goods and services provided by built-up landscapes. In particular, the potential recreation and tourism values of a protected landscape are influenced by the infrastructure of gateway communities, which includes restaurants, motels, service stations, roads, and other goods, along with law enforcement and other services. Conversely, infrastructure availability in gateway communities depends on visitor use in nearby protected landscapes. Similarly, rural landscapes would not be attractive places to live and work without the consumer goods and services supplied by nearby built-up landscapes.

Institutional arrangements—the rules people use to organize repetitive activities with outcomes that affect them and possibly others—have an impact on the relationships among landscape values. These arrangements influence the use, management, and values of land. Subdivision regulations, for example, influence the location and density of residential housing, which in turn affect habitat loss and fragmentation, non-native species, wildlife mortality, and environmental amenities.

Public policies for national parks and forests influence the type, timing, extent, and value of outdoor recreation. As an example, the decision to impose road restrictions and decommissioning in Flathead National Forest in the northern U.S. portion of the CCE results in more secure habitats for grizzly bear, but it eliminates motorized recreation in affected areas (Flathead National Forest 2002a, b). In another example, Canadian land-use regulations permit open-pit coal mining and drilling for coal-bed methane in the upper watershed of the North Fork of the Flathead River in southeastern British Columbia. The North Fork forms the western boundary of Glacier National Park, and U.S. environmental interests oppose such energy developments on the grounds that they are likely to pollute the North Fork and Flathead Lake, harming transboundary populations of bull trout and other wildlife populations (National Parks Conservation Association 2002).

Environmental conditions outside the ecosystem can also affect landscape conditions and values. Under certain meteorological conditions, for example, soil from the Gobi Desert is blown into the CCE. Global climate change can have negative effects on the CCE's biodiversity, water resources, agriculture, and human life.

Managing the CCE Ecosystem

Sustaining landscapes in the CCE is challenging because the region is large and complex, scientific knowledge about the ecosystem is incomplete and data are insufficient, and human and financial resources are limited. Liszewski (2004, 6) points out that "The complex nature of ecosystems, and the increasingly complex nature of human stresses and demands on ecosystems, means that simple and narrowly focused approaches are not sufficient to penetrate modern environmental problems." Traditional, commodity-based resource management is an example of a narrowly focused approach. Ecosystem management, which aims to sustain ecosystem integrity by working with larger spatial scales, longer time periods, and more variables (Franklin 1997), is an alternative to commodity-based land management. To implement ecosystem management, we must learn to work with complex ecosocial systems in a way that meets human needs while preserving ecosystem integrity.

An ecosystem approach to landscape management requires (1) deciding whether the ecosystem's current state is sustainable; (2) selecting management actions for achieving sustainability if the current state is not sustainable; (3) incorporating scientific knowledge into landscape-management evaluations and decisions; and (4) making data and findings accessible and usable to stakeholders and decisionmakers.

To move toward sustainable ecosystem management, we must make two types of decisions. First, we must decide whether the ecosystem's current state is sustainable, which requires defining sustainability. An ecosystem is considered *sustainable* when the social and economic needs for consumer goods and services are met without permanently impairing the ecosystem's long-term capacity to deliver ecosystem goods and services (Franklin 1997; Prato 2000). This decision is subject to two kinds of errors: (1) deciding that the ecosystem is sustainable when it is not, possibly resulting in taking no management actions even though they are warranted; and (2) deciding that the ecosystem is unsustainable when it is actually sustainable, possibly leading to the implementation of management actions when they are unnecessary.

If we decide that the ecosystem is not sustainable, we must then select management action(s) that have the greatest likelihood of achieving sustainability. Here, we could make an error by selecting ineffective management actions that do not achieve sustainability.

Both decision errors arise from uncertainty—first, about the ecosystem's current state, and second, about whether the chosen management actions achieve sustainability. Adaptive management is a promising approach for handling both sources of uncertainty (Holling 1978). This type of management presumes that "because human understanding of nature is imperfect, human interactions with nature should be experimental" (Lee 1993, 53). To adaptively manage an ecosystem, we must work cooperatively with multiple stakeholders (academia, government, the private sector, and the public) to select indicators of ecosystem sustainability; measure the indicators; use those measurements to test alternative hypotheses about the current state of the ecosystem and its responses to alternative management actions; and, if warranted, alter management of the ecosystem based on test results. A limitation of adaptive management is that several of the prerequisites for successful implementation are generally not satisfied (Prato 2004). For this reason, implementing adaptive management involves compromise (see Chapter 16 for a more complete explanation of adaptive management).

To assess ecosystem sustainability and make appropriate landscape-management decisions, we must integrate scientific knowledge, data, and methods that pertain to the social, economic, and ecological processes operating in an ecosystem. Such integration requires communication and cooperation among natural, social, and behavioral scientists, which may be difficult to achieve because of differences in training, background, and vocabulary.

One obstacle to sustainable landscape management is that stakeholders and decisionmakers may not understand or have access to the scientific knowledge, data, and results needed to practice sustainable landscape management. Developing a Web-based spatial decision support tool (SDST) could help. In general, an SDST gives users new insights into the structure of spatial decision problems and helps them generate new alternatives and strategies in a problem-solving context. More specifically, an SDST (1) facilitates natural resource assessments and sustainable landscape management by integrating models, spatial data, and user-provided information; (2) generates graphical and tabular reports of results; (3) does not require the user to collect spatial data or purchase the geographic information system software required for interactive mapping; (4) makes it easier to update data and procedures, and (5) does not require one-on-one communication with individual users. The last two advantages reduce the costs of providing spatial data and analysis capability.

Interactive mapping is a useful feature of many Web-based SDSTs. This capability allows users to identify and assess landscape attributes for a region (e.g., the properties of a soil type or the size and perimeter of a lake), perform queries on those attributes, create maps, apply interpretive functions to maps (e.g., zooming in and out, determining geographic coordinates of a point, and clipping data layers), and print maps.

Achieving Sustainability in the CCE

Ideally, ecosystem sustainability and sustainable landscape management should be evaluated at the ecosystem scale. Such evaluations are particularly difficult, however, in large ecosystems such as the CCE. Difficulties arise from the massive amount of data required; inadequate or missing data; the diversity of social, cultural, political, economic, and biophysical processes that need to be modeled; and differences in resource management goals and objectives. Evaluating ecosystem sustainability and sustainable landscape management at the local/regional scale and then aggregating those evaluations to the ecosystem scale is a more practical and manageable approach. We use three examples to illustrate sustainable landscape-management practices at the local level.

The Nyack Floodplain

The first example illustrates sustainable livestock and natural resource management in the Nyack floodplain. This floodplain is located on the Middle Fork of the Flathead River about 11 km upstream from West Glacier, Montana. It provides important habitats for elk, moose, deer, mountain lion, black bear, cutthroat trout, bald eagle, and Canada geese, and critical habitats for harlequin ducks, grizzly bear, bull trout, and boreal toads. Elk in the region use the floodplain in winter and spring. As long-time floodplain residents, the Dalimata family learned how to make ranching compatible with sustaining wildlife species and their habitats. On their small ranch, the family supports 150 mother cows, produces hay and grain for winter feed, selectively harvests mature timber, and harvests and mills salvage logs from the river corridor on its property. Wood products from the family's sawmill are used to manufacture prefabricated cabins and other value-added products. In addition, the Dalimatas have a trout-farming operation that uses adjudicated state water rights. Stanford (2000, 6) points out, "The Dalimatas have not received, nor have they asked for, compensation for avoiding elk calving areas during the spring, for allowing elk to graze in their hay fields year-round or for hazing the occasional grizzly away from their cattle. These and other activities by ranchers do have real monetary value, however, which figure into the decision by families to sell out to development or hang in there with traditional land-use activities that secondarily foster maintenance of natural attributes of the landscape."

The Rocky Mountain Front

The second example comes from the Rocky Mountain Front, which forms the eastern boundary of the CCE. Stretching from southern Alberta to northern Montana, the front is 81 km wide and 322 km long, and some consider it to be one of the richest and largely unfragmented landscapes on Earth. It offers prime habitats for grizzly bear, black bear, wolf, cougar, lynx, wolverine, elk, deer, and moose, and is one of the few places where grizzly bear habitat extends onto the plains. The outstanding native prairie in the Rocky Mountain Front makes it ideal for ranching.

Unlike many other areas in the Rocky Mountain West, sustainable ranching is being practiced in the Rocky Mountain Front at The Nature Conservancy's (TNC) Pine Butte Swamp Preserve. This 7,287-ha preserve is being operated as a working ranch. Management goals are to support biodiversity and provide seasonal habitats for wildlife, especially prairie habitat for grizzly bear. TNC leases grazing rights on the preserve to local ranchers. Leaseholders are required to use grazing systems that "mimic the buffalo's seasonally intensive use of grass and ... integrate into the rancher's agricultural operations" (TNC 2006, *1*). Using sustainable grazing systems in the Rocky Mountain Front has local economic benefits (livestock grazing) and improves the health of native grasslands. A local rancher, whose land is under a TNC conservation easement, employs sustainable agricultural production practices that are compatible with TNC's conservation objectives.

The Blackfeet Trust

The third example is the Blackfeet Trust, a private, nonprofit land trust focusing on the Blackfeet Indian Reservation, which is located on the northeast side of the U.S. portion of the CCE. The goal of the trust is to "...preserve native plants and animals and perpetuate respect for the land consistent with the culture and heritage of the Blackfeet Nation/people." To move toward this goal, the trust works collaboratively with TNC to reclaim and protect nontribal lands once under tribal ownership, and to prevent development in high-quality prairie foothill, prairie pothole, and wetland ecosystems (TNC 2004).

Sustainable landscape management is more likely to be practiced in the CCE if landowners adopt a stewardship ethic (as do the Dalimatas, TNC, and the Blackfeet Nation), become knowledgeable about sustainable landscape-management practices, and demonstrate the will and ability to implement those practices. A stewardship ethic has a greater chance of developing if the economic and environmental benefits of sustainable landscape management exceed the environmental costs of unsustainable landscape management. An SDST could be useful for helping landowners and managers learn about the benefits and costs of sustainable landscape management. Landowners and managers are more willing and able to adopt sustainable management practices when they receive technical and financial assistance for implementing those practices. The U.S. Department of Agriculture (USDA) and state agencies offer such assistance through their conservation programs. The USDA sponsors programs in the following areas:

- Conservation reserves

- Conservation security
- Environmental quality incentives
- Farm and ranchland protection
- Grasslands reserves
- Resource conservation and development
- Wildlife habitat incentives
- Wetlands reserves.

Land trusts also offer financial incentives for conservation. In particular, land trusts can purchase conservation easements from willing landowners, preventing land from being developed.

References

Beyers, W.B., and P.B. Nelson. 2000. Contemporary Development Forces in the Nonmetropolitan West: New Insights from Rapidly Growing Communities. *Journal of Rural Studies* 16: 459–74.

Capra, F. 1995. Deep Ecology: A New Paradigm. In *Deep Ecology for the 21st Century*, edited by G. Sessions. Boston, MA: Shambhala, 19–25.

Costanza, R., R. d'Arge, R. de Groot, S. Farber, M. Grasso, B. Hannon, and K. Limburg, et al. 1997. The Value of the World's Ecosystem Services and Natural Capital. *Nature* 387: 253–60.

Daily, G.C. 1997. Introduction: What Are Ecosystem Services? In *Nature's Services: Societal Dependence on Natural Ecosystems*, edited by G.C. Daily. Washington, DC: Island Press, 3–6.

Davis, G. 2000. Personal communication with the authors, July 24.

Fagre, D.B. 2000. Changing Mountain Landscapes in a Changing Climate: Looking into the Future. *Changing Landscapes* 2(1): 31–39.

Flathead National Forest. 2002a. Moose Post-Fire Project FEIS [Final Environmental Impact Statement]. Chapter 3—Wildlife Introduction. http://www.fs.fed.us/r1/flathead/nepa/moose/ch3/wildlife.pdf (accessed October 24, 2005).

Flathead National Forest. 2002b. Moose Post-Fire Project, Record of Decision. http://www.fs.fed.us/r1/flathead/nepa/moose/decision.pdf (accessed November 4, 2005).

Flathead on the Move. 2004. Summary of the Flathead on the Move Forum. http://www.crmw.org/MontanaOnTheMove/Data/Flathead%209-22-04%20Forum.pdf (accessed October 18, 2006).

Franklin, J.F. 1997. Ecosystem Management: An Overview. In *Ecosystem Management: Applications for Sustainable Forest and Wildlife Resources*, edited by M.S. Boyce and A. Haney. New Haven, CT: Yale University Press, 21–53.

Hansen, A.J., A. Gallant, J.R. Rotella, and D. Brown. 1999. Natural and Human Drivers of Biodiversity in the Greater Yellowstone Ecosystem. In *Land Use History of North America: Providing a Context for Understanding Environmental Change*, edited by T. Sisk. Washington, DC: U.S. Geological Survey, 61–70.

Holling, C.S. 1978. *Adaptive Environmental Assessment and Management.* Chichester, UK: John Wiley & Sons.

James, S.J. 2001. An Institutional Approach to Protected Area Management Performance. In *The Politics and Economics of Park Management,* edited by T.L. Anderson and A. James. Lanham, MD: Rowman & Littlefield, 3–27.

Lee, K.N. 1993. *Compass and Gyroscope: Integrating Science and Politics for the Environment.* Washington, DC: Island Press.

Liszewski, M.J. 2004. The Glen Canyon Dam Adaptive Management Program. *Water Resources Impact* 6: 10–13.

Mansfield, D. 2002. Smokies the Nation's Most Polluted Park, Says Study. Associated Press release, September 24, Environmental News Network. http://enn.com/news-/wire-stories/2002/09/09242002/ap_48504.asp (accessed August 25, 2005; site no longer active).

National Parks Conservation Association. 2002. State of the Parks, A Resource Assessment: Waterton Glacier International Peace Park. http://www.npca.org/across_the_nation/park_pulse/glacier/glacier.pdf (accessed August 25, 2005; site no longer active).

Power, T., and R. Barrett. 2001. *Post-Cowboy Economics: Pay and Prosperity in the New American West.* Washington, DC: Island Press.

Prato, T. 1998. *Natural Resource and Environmental Economics.* Ankeny, IA: Iowa State Press.

———. 2000. Multiple Attribute Evaluation of Landscape Management. *Journal of Environmental Management* 60: 325–37.

———. 2004. Alleviating Multiple Threats to Protected Areas with Adaptive Ecosystem Management: Case of Waterton Glacier International Peace Park. *George Wright Forum* 20: 41–52.

Rasker, R., B. Alexander, J. van den Noort, and R. Carter. 2004. *Prosperity in the 21st Century West: The Role of Protected Public Lands.* Bozeman, MT: Sonoran Institute.

Riebsame, W.E., H. Gosnell, and D.M. Theobald. 1996. Land Use and Landscape Change in the U.S. Rocky Mountains. I. Theory, Scale, and Pattern. *Mountain Research and Development* 16: 395–405.

Riebsame, W.E., H. Gosnell, and D.M. Theobald (eds.). 1997. *Atlas of the New West.* New York: W.W. Norton.

Stanford, J.A. 2000. Family Stewardship of a River Floodplain: A Demonstration Project. Unpublished paper. Missoula, MT: Flathead Lake Biological Station, The University of Montana, February.

TNC (The Nature Conservancy). 2004. The Rocky Mountain Front. http://nature.org/wherewework/northamerica/states/montana/preserves/art6027.html (accessed June 28, 2004).

———. 2006. Land Stewardship on the Front. http://www.nature.org/wherewework/northamerica/states/montana/science/art15847.html (accessed October 19, 2006).

Travis, W.R., D.M. Theobald, and D. Fagre. 2002. Transforming the Rockies: Human Forces, Settlement Patterns, and Ecosystem Effects. In *Rocky Mountain Futures: An Ecological Perspective,* edited by J.S. Baron. Washington, DC: Island Press, 1–26.

U.S. Census Bureau. 2006. State and County Quick Facts. http://quickfacts.census.gov/qfd/states/30000.html (accessed October 15, 2005).

U.S. Department of Commerce. 2001. *Regional Economic Information System.* Washington, DC: Bureau of Economic Analysis.

Waldt, R. 2004. *Crown of the Continent: The Last Great Wilderness of the Rocky Mountains.* Helena, MT: Riverbend Publishing.

2

The Crown of the Continent Ecosystem
Profile of a Treasured Landscape

Ben Long

In this chapter, I provide a succinct, nontechnical profile of the CCE's history, geology, climate, water resources, plants and animals, landownership patterns, protected areas, Indian lands, and management challenges. The CCE offers visitors, recreationists, and residents alike a sense of freedom and wilderness solitude, spectacularly rich and beautiful scenery, and friendly communities. Its outstanding biophysical features are important scientifically and as a source of environmental amenities. Geologically, the CCE is significant for its ancient rocks and glacially carved mountains. Biologically, the region is home to a full suite of native wildlife, including rare species such as grizzly bear (*Ursus arctos*), wolf (*Canis lupus*), and lynx (*Lynx canadensis*).

Where Is the CCE?

The CCE is an area about twice the size of Massachusetts, where the Continental Divide crosses the 49th parallel (see CCE map on p. xviii). It is split north to south by the Continental Divide. The northernmost boundaries are the headwaters of the Elk River in British Columbia and the Highwood River in Alberta. The southernmost boundary is Montana's Blackfoot River Valley, and the eastern periphery extends slightly into the Great Plains. Other mountain ranges of Montana and British Columbia, such as the Salish Mountains, make up the western fringe of the ecosystem.

Folded within the CCE are the Macdonald, Lewis, Clark, Whitefish, Galton, Lizard, Swan, Mission, Flathead, and Livingston mountain ranges. Between these ranges are the Elk River, the three forks of the Flathead River, and the Swan River valleys. More than 80% of the CCE is public property, and about 32% is protected by law in a pristine state.

This chapter is a condensation of Long, B. 2002. *Crown of the Continent, Profile of a Treasured Landscape*. Helena, MT: Scott Publishing.

The CCE includes a World Heritage Site (the Waterton Glacier International Peace Park), recognized globally alongside the Galapagos Islands, the Serengeti Plain, and the Great Barrier Reef. The World Heritage Site designation recognizes cultural and natural heritage sites that have outstanding global value. Within the CCE, the physical components of the ecosystem—geological chemistry, rainfall and snowmelt, and climate—interact with the living plants and animals with minimal local human disturbance. In short, the CCE is a vibrant, functional ecosystem (see Canada and the United States of America 1995; Rockwell 1995; Chadwick 2000; Gadd 2000; Grinder et al. 2000).

A Rich Native History

From the lakeshores and prairies to the rocky peaks of the Continental Divide, the CCE has been home to humans for at least 10,000 years. Many cultural connections between the land and local tribal members are intact today, and for many, the CCE remains "sacred geography."

People arrived shortly after the Ice Age. The mysterious Clovis people roamed the prairies on the east side of the CCE more than 10,000 years ago, using thrusting spears to hunt prehistoric mammoth and horse.

When glaciers finally melted out of the high passes, people crossed the mountains from the west as well. Bighorn sheep (*Ovis canadensis*) and bison (*Bison bison*) were important prey 5,000 years ago. People ambushed their prey, or drove it into traps or over cliffs. Some cliff traps, called *pishkuns*, were used for thousands of years. Lake outlets were popular places to live, catch and prepare fish, dry meat, and gather plants. While tribes elsewhere in North America developed agriculture, the CCE's climate was more conducive to hunting, fishing, and gathering. Camas (*Zigadenus elegans*), biscuitroot (*Lomatium* spp.), and balsamroot (*Balsamorhiza* spp.) were staples. Huckleberries (*Vaccinium* spp.), Saskatoon berry (*Amelanchier alnifolia*), and other fruit were also important.

Today, the CCE is most closely associated with two tribes—the K'tunaxa bands (the Kootenai in Montana or Kootenay) on the west side, and the Piikanii tribe of the Nitsitappi (the Blackfeet Nation) on the east side of the mountains. Although tribes intermingled socially and for trade and warfare, the contact was not sufficient to blend their cultures. See Chapter 3 for more details about native peoples.

The Modern Era Begins

Even before people of European descent saw the CCE, they triggered changes in the region. Horses, firearms, and smallpox were entering the region well before Anglo-Americans arrived toward the end of the 1700s. The first Anglo to explore the CCE was probably Peter Fidler of the Hudson's Bay Company, and in 1806, U.S. Army Captain Meriwether Lewis (of the Lewis and Clark Expedition) visited the CCE. These journeys opened the region to fur traders. Christian missionaries like Father Pierre Jean De Smet later set up a network of missions throughout the wilderness. Gold prospectors explored the CCE as early as 1850.

Ranchers herded cattle and sheep into the CCE in the 1860s and 1870s. Partly to make way for cattle, and also to obtain meat and hides, the bison were systematically eradicated. The slaughter of the bison was disastrous for native people and wildlife alike. By 1883, only a few hundred bison remained of the millions that once roamed the Great Plains; native people were devastated by their loss. The harsh winter of 1886–1887 killed huge numbers of the white ranchers' cattle and triggered great hardship for native peoples.

Perhaps no one person changed the CCE as much as Canadian-American James J. Hill, who helped finance and build the Great Northern Railroad and Canadian National Railway. The Great Northern crested Marias Pass in 1891, and the Canadian Pacific Railway crossed Crowsnest Pass in 1898. The completion of the railroads on both sides of the international border, along with homesteading legislation and the distribution of land through massive railroad land grants, opened the CCE to waves of settlers. Farmers homesteaded the Elk, Old Man, Swan, Flathead, and Clark Fork valleys.

The railroad also brought tourists. One sightseer was George Bird Grinnell, a well-connected and wealthy New Yorker. A hunter, conservationist, and the editor of *Field and Forest* magazine, Grinnell coined the "Crown of the Continent" phrase. His conservation efforts culminated in the creation of Glacier National Park in 1910. North of the border, Fredrick Godsal had done the same for Waterton Lakes National Park. In 1895, the Canadian federal government set aside a 140-km^2 area as the Dominion Forest Park.

Today, the CCE is maturing into a modern blend of traditional timber, mining, and agricultural industries, along with light industry. In addition, professionals, retirees, and entrepreneurs move to the area to enjoy the high quality of life (Flores 1995).

The Geologic Foundation: The Story in Stone

If any one feature defines the CCE, it is mountains. Geologists and astronomers calculate that the earth formed about 4.6 billion years ago as one of eight planets in our solar system. About 1.5 billion years ago, a huge basin that was largely devoid of terrestrial life occupied a large part of the present-day CCE. The basin, roughly the size of Arizona, comprised rolling uplands that surrounded a saltwater bay, somewhat similar to today's Gulf of Mexico. Because there were no land plants, rainfall resulted in high rates of erosion, which washed silt and sand into the shallow sea.

Sediments gradually filled the basin, stacking up like layers in a cake, and over hundreds of millions of years, these layers hardened into rock. Geologists identify distinct layers of these ancient, hardened sediments that were deposited between 900 million and 1.5 billion years ago. Collectively, they are called the Belt/Purcell Supergroup, a formation that is about 15,000 m thick. The supergroup and the Mississippi Limestone formation are the two dominant geologic formations in the CCE.

Mountains began to form during the Mesozoic, 225–265 million years ago. During that period, North America moved westward, separating from the European tectonic plate and forming the Atlantic Ocean. For the next 100 million years, the western edge of North America buckled and was uplifted. This phenomenon, accompanied by earthquakes and volcanic activity, created the Rocky Mountains. During much of this period, lowland portions east of the rising mountains were swamps, the source of some of today's shallow coal fields. These layers also contain the remarkable dinosaur fossils found in Montana and Alberta.

About 75 million years ago, compressive forces in the earth's crust moved great masses of rock, pushing the old "basement" rocks over much younger material. The mountains formed by this overthrusting process make up the transboundary region of the CCE.

These forces ended about 40–50 million years ago, and portions of the high, mountainous landscape reacted by dropping to form relatively broad valleys, such as the Swan, Mission, and Flathead. The most dramatic is the Rocky Mountain Trench, the long valley that runs north from St. Ignatius to the Yukon.

Next came the great carver of this landscape—ice. Starting about 2 million years ago, the global temperature dropped and ice covered 30% of Earth. Ice was thousands of meters thick over most valleys and hundreds of meters thick over most mountains in the CCE. With so much frozen water, ocean levels dropped and the land bridge connecting Siberia and Alaska was exposed. This land bridge allowed plant and animal species, including humans, to pass between North America and Asia. The last great ice sheet retreated about 10,000 years ago.

As the ice retreated, it left behind a dramatic landscape formed by glacial sculpting of the underlying rock, namely U-shaped valleys, knife-like ridges (*arêtes*), hanging valleys, waterfalls, cirques, cliffs, and fjord-like lakes.

The CCE is a very active landscape. Sometimes, massive amounts of earth move overnight. In 1903, a huge portion of Turtle Mountain—northwest of Waterton Lakes National Park— collapsed, burying the mining town of Frank and creating the Frank Slide.

Because of cool temperatures, lack of available water, and steep slopes, soils in the CCE tend to be immature or poorly developed. Much of the alpine is devoid of soil altogether, comprising only bare rock. Soils are deepest and most fertile in low, flat valleys (see Rockwell 1995; Elias 1996; Blood 1997; Achuff et al. 1997; Blood 2004; Flathead National Forest n.d.).

Climate: A Fortress of Snow and Ice

Climate and weather vary widely in the CCE for three key reasons: (1) it is located within the planet's mid-latitudes, the zone on the globe that is subject to wide seasonal shifts; (2) its position in North America along the Continental Divide is the general boundary between maritime and continental conditions; and (3) the rugged topography and wide ranges in elevation create localized microclimates.

The CCE has long winters and short summers. Indeed, the coldest temperature ever recorded in the contiguous United States, $-57°C$, was measured at Montana's Roger's Pass at the southern end of the CCE on January 20, 1954. Precipitation varies greatly across the CCE, particularly from west to east, but most of the precipitation comes as winter snow. Summer droughts are common. The CCE is in the crosshairs of two major, hemispheric weather patterns: maritime and continental. Maritime air masses, generally moist and of moderate temperature, blow in from the Pacific Ocean. Continental air masses from northern Canada tend to be cold and dry. Although maritime air masses generally produce wet, moderate weather, the continental air masses tend to produce more extreme weather conditions—hot in the summer and very cold in the winter. The collision of maritime and continental air masses produces dramatic storms. (See Chapter 12 for more details about climate and climate change.)

Fire in the Crown: Born to Burn

Fires are inevitable in the CCE whenever summer drought, electrical storms, and wind coincide. For thousands of years, aboriginal humans left their mark on this landscape by setting fires, and lightning served as a natural source of ignition. But in recent decades, people have left their mark on this landscape by suppressing fires.

Modern humans tend to see wildfire as a destructive force, threatening property, endangering lives, and causing air pollution. To our sensibilities, a green forest is a thing of beauty, whereas a blackened one represents death and ugliness. Nature reacts differently for the following reasons:

- Fire is a major determinant of the structure and composition of forests and grasslands.
- By killing trees, fire provides food for insects and the birds that eat those insects.
- Fire kills trees that infringe on grasslands, keeping the grasses healthy.

Plants and animals of the CCE demonstrate countless adaptations to fire. The thick, corklike bark of ponderosa pine (*Pinus ponderosa*) reduces its vulnerability to small fires. The plumage of the black-backed woodpecker (*Picoides arcticus*) is a nearly perfect camouflage against the burned tree trunks where it forages for insects. Prairie grasses evolved with the beneficial recycling effects of periodic fire. Shrubs, such as redstem ceanothus (*Ceanothus sanguineus*; a favorite food of deer and elk) and huckleberry (a favorite of black and grizzly bears), are also regenerated by fire. Lodgepole pine (*Pinus contorta*) cones pop open under hot fires to disperse their seeds, and whitebark pine (*Pinus albicaulis*) requires fire to clear mountain slopes of competing trees, preparing potential seedbeds for a new generation of saplings.

Fire, however, poses challenges for the CCE's human inhabitants. Rural landowners and representatives of timber interests are justifiably concerned about losing property and wasting timber, and smoke is unpleasant and unhealthy. So far, taxpayers have been willing to spend millions of dollars protecting homes and stopping fires. The future challenge is to maximize the benefits of fire while minimizing the risks to safety, health, and property. See Chapter 13 for more details about fire and fire management (see also Rockwell 1995; Gadd 2000).

Headwaters of North America

The CCE has some of the cleanest water in the world and an aquatic environment that is treasured for its beauty, relatively pristine condition, and ecological integrity. The CCE includes a rare "triple divide," where precipitation drains into the Pacific Ocean, the Arctic Ocean, and the Atlantic Ocean, and Hudson Bay.

The CCE includes the upper reaches of 19 major rivers, many of which are free-flowing; a rarity in modern North America. These rivers include the following:

- Blackfoot, Clearwater, Clark Fork, Flathead, Tobacco, Swan, and Stillwater drainages of the Columbia River Basin in Montana
- Elk and Wigwam rivers, and the upper reaches of the Flathead (North Fork of the Flathead) rivers in British Columbia
- Milk, Dearborn, Sun, Teton, and Marias rivers in the Missouri River Basin of Montana (The Milk River also flows through southern Alberta.)

- St. Mary, Waterton, and Belly, which start in Montana and flow into Alberta, and the Old Man and Castle rivers of Alberta. These flow into the Saskatchewan River and Hudson's Bay.

Another striking feature of the CCE's watersheds is its glacial lakes, which are generally deep, cold, and remarkably clean. Flathead Lake is the largest natural freshwater lake in the western United States and one of the 300 largest lakes in the world. It is 47 km long and 27 km wide with a surface area of 50,400 ha.

At 147 m, Upper Waterton Lake is the deepest lake in the Rockies and in Alberta, and is the largest in a chain of lakes that are the focal point of Waterton Lakes National Park. In all, the park is home to 80 lakes and ponds. Glacier National Park has 750 lakes, of which 131 are named. See Chapter 9 for more details about water resources (see also U.S. EPA 1983; Rockwell 1995; Flathead Basin Commission 1997–1998).

Plants

Plant communities change dramatically across the CCE's landscapes, ranging from prairie grasslands to alpine meadows with a patchwork forest in between. Waterton Lakes National Park has about 971 vascular plants on record; Glacier hosts 1,258 plant species; and the Flathead National Forest contains 1,026 species. Many of the same plant species are found in all three locations. Although none of these plant lists represent total numbers for the ecosystem, they help illustrate the extent of the ecosystem's plant diversity.

Plant communities in the CCE can be broken into several zones that reflect differences in climate. From low to high elevation, these zones are as follows:

- Grasslands are common in the valley bottoms and prairie edges of the CCE. Because grasslands are prime locations for agriculture, they have been altered across much of North America. Local grassland is typified by rough fescue (*Festuca* spp.), bluebunch wheatgrass (*Agropyron spicatum*), oat grass (*Danthoina* and *Trisetum* spp.), and june grass (*Koeleria cristata)*, punctuated with Rocky Mountain juniper (*Juniperus scopulorum*), limber pine (*Pinus flexilis*), and a variety of shrubs.
- Foothills and parklands include the aspen stands (*Populus tremuloides*) and open limber pine forests of the Rocky Mountain Front.
- Montane includes low- to mid-elevation forests, dominated by Douglas fir (*Pseudotsuga menziesii*), lodgepole pine, western larch (*Larix occidentalis*), and other coniferous tree species. Ninebark (*Physocarpus* spp.), mountain maple (*Acer* spp.), Saskatoon berry, and other shrubs grow beneath the trees.
- Subalpine occurs at higher elevations where the climate becomes colder and the montane forest fades out to subalpine forest dominated by lodgepole

pine, subalpine fir (*Abies lasiocarpa*), whitebark pine, and alpine larch (*Larix lyallii*).
- Alpine—the zone above treeline—generally begins at about 2,100 m on the west side and 1,800 m on the east side of the divide. It is known for its lush carpet of forbs and wildflowers.

Exotic weeds represent the largest recent change in plant communities. About 10% of the flora of the CCE is non-native, and about 1% of the total flora is categorized as noxious weeds. These weeds displace natives, often with repercussions throughout the native ecosystem. Even small exotics can have wide-ranging effects. Fungus blister rust has devastated both whitebark pine and western white pine (*Pinus monticola*) in the CCE. See Chapters 6, 7, and 17 for details on a variety of plant-related topics (see also Arno and Hammerly 1984; Maser 1988; Whitney 1990; Rockwell 1995; Kershaw et al. 1998; Mantas 1999; Gadd 2000).

Mammals

As does the Greater Yellowstone Ecosystem, the CCE includes one of the most intact assemblages of mammals of any region in southern Canada or the contiguous United States. About 65 species of mammals are native to the CCE. The smallest is the pygmy shrew (*Sorex hoyi*), which is just 8 cm long and a few grams in weight; the largest is the bison at 800 kg. The CCE serves as a reservoir for mammals repopulating other areas and is a link for populations on both sides of the international border.

The CCE is home to both black and grizzly bears. Although black bears (*Ursus americanus*) live across North America, grizzlies now live only in protected places, in sparsely inhabited areas of Canada, and in five distinct recovery areas in the continental United States (refer to Figure15-2).

After being driven to near extinction in the western United States and much of southern Canada, wolves have begun to rebound in the CCE, and have recolonized in the region without direct human intervention. The CCE is also home to coyotes (*Canis latrans*), river otters (*Lontra canadensis*), pine marten (*Martes americana*), bobcats (*Lynx rufus*), cougars (*Puma concolor*), striped skunks (*Mephitis mephitis*), raccoons (*Procyon lotor*), long-tailed weasels (*Mustela frenata*), minks (*Mustela vison*), red foxes (*Vulpes vulpes*), and badgers (*Taxidea taxus*), as well as more rarely seen wolverines (*Gulo gulo*), lynxes, and fishers (*Martes pennanti*).

The CCE has the most diverse assortment of ungulates in North America, including mountain goats (*Oreamnos americanus*), moose (*Alces alces*), elk (*Cervus elaphus*), mule deer (*Odocoileus hemionus*), white-tailed deer (*Odocoileus virginianus*), bison, pronghorn (*Antilocapra americana*), and bighorn sheep.

The CCE is home to many smaller mammals, including deer mice (*Peromyscus* spp.), western jumping mice (*Zapus princeps*), red squirrels (*Tamiasciurus hudsonicus*), northern flying squirrels (*Glaucomys sabrinus*), three species of chipmunks (*Tamias* spp.), porcupines (*Erethizon dorsatum*), and packrats (*Neotoma cinerea*). Several varieties of voles also make their home in the region. Golden-mantled ground squirrels (*Spermophilus lateralis*) are common, as are hoary marmots (*Marmota caligata*). Burrowing rodents include Columbian ground squirrels (*Spermophilus columbianus*), Richardson ground squirrels (*Spermophilus richardsonii*), thirteen-lined ground squirrels (*Spermophilus tridecemlineatus*), and northern pocket gophers (*Thomomys talpoides*). Beavers (*Castor canadensis*) and muskrats (*Ondatra zibethicus*) are abundant semiaquatic rodents.

Hares include the snowshoe hare (*Lepus americanus*) and the white-tailed jackrabbit (*Lepus townsendii*). The pika (*Ochotona princeps*) lives in boulder fields. Six species of bats live in the CCE, eating insects in summer and migrating or hibernating in winter. There are three species of shrew, which are primitive insectivores even smaller than mice.

Of the species that dwelled in the region 200 years ago, only two large mammals are missing from the wild: the bison and the mountain caribou (*Rangifer tarandus caribou*).

The CCE historically had abundant wildlife; wildlife is still abundant today. When the railroads crossed the CCE, some North American wildlife species were annihilated in market-driven slaughters. Market hunters killed uncounted thousands of elk, bison, deer, and bighorn sheep to feed hungry railroad crews and miners. Around 1900, conservationists such as Theodore Roosevelt and George Bird Grinnell began to turn the tide by establishing public ownership of wildlife and key wildlife habitat.

Several areas in the CCE played key roles in wildlife recovery in the twentieth century, particularly the National Bison Range, the Sun River Wildlife Management Area, and the CCE's two national parks. Animals from these places colonized habitats across North America where wildlife was depleted. See Chapters 15 and 17 for more details about mammals (see also Schmidt 1993; Rockwell 1995; Van Tighem 1999; Webster 1999; Gadd 2000).

Aquatic Ecosystems

Waters of the CCE host unique and important aquatic ecosystems. For example, virtually all the genetically pure westslope cutthroat trout (*Oncorhynchus clarki lewisi*) in the world live in the CCE. Microscopic plants (phytoplankton such as algae and many others) float in the water, fed by sunlight. Flathead Lake, for example, has more than 500 species. Although phytoplankton and zooplankton are microscopic, they are the foundation of the aquatic food chain.

The Flathead Basin alone has at least 300 species of aquatic insects. Waterton and Glacier National Parks have a greater diversity of aquatic invertebrates than either Banff or Jasper National Parks to the north and Yellowstone National Park to the south.

The mix of fish in the CCE includes about 27 native species. Because the area has only recently (in geologic time) been unlocked from ice, fish have had less time and opportunity to colonize these waters. Many isolated drainages naturally had no fish. During the first half of the twentieth century, people stocked almost every body of water with a variety of non-native fish, raising havoc with native fisheries.

Because fish such as the bull trout (*Salvelinus confluentus*) and the westslope cutthroat trout need clean, clear, and cold water to spawn, they are indicators of high water quality. Native fishes of the CCE include westslope cutthroat trout, bull trout, lake trout (*Salvelinus namaycush*), lake whitefish (*Coregonus clupeaformis*), and northern pike (*Esox lucius*). Not all those species are native to every lake or stream in the CCE. The northern pike is the namesake native fish of Maskinonge Lake in Waterton Lakes National Park, but is a disruptive exotic in Flathead Lake. These aquatic ecosystems are sensitive, and even relatively minor habitat degradation can have profound negative impacts on species. See Chapter 8 for more details about aquatic ecosystems (see also Rockwell 1995; Deleray et al. 1999; Gadd 2000).

Birds

Birds are the most abundant and diverse vertebrate animals in the CCE. Roughly 270 species of birds call the region home, compared to about 65 mammals, 25 native fish, and 12 reptiles and amphibians. The CCE contains important habitat for resident birds, as well as key migration routes for many thousands of birds every spring and fall. The CCE supports a wide variety of birds because it contains a variety of special habitat niches.

Of the bird species in Glacier National Park, about half breed there. The rest are classified as "accidental" (birds that show up rarely) or as birds that migrate through the park without nesting, although they may spend days resting in the area. Because the CCE is one of the narrowest points in the Rocky Mountains, it forms a funnel that compresses and concentrates the movement of birds, making for some spectacular bird watching. For example, thousands of golden eagles (*Aquila chrysaetos*) glide north and south along this invisible funnel of air. On one fall day at the southern end of the Livingstone Range, biologists counted 1,000 golden eagles.

Conserving migratory birds is challenging because such birds require enormous areas. Destruction of forests in Mexico or Guatemala, for example, could eventually reduce the number of songbirds nesting in Canada during the summer. Swainson's hawks (*Buteo swainsoni*), which summer on the Alberta prairie, may be killed by pesticides used to control grasshoppers in Argentina where the birds winter. Conversely, events in the CCE can have repercussions in other nations. For example, the destruction of native grasslands in the CCE can influence the breeding success of savannah sparrows (*Passerculus sandwichensis*), which migrate between the CCE and South America. See Chapter 15 for more details about native birds (see also Rockwell 1995; Gadd 2000; Holroyd n.d.).

Reptiles and Amphibians (Herptiles)

Because of the CCE's harsh climate, the number of cold-blooded creatures and herptile species is quite small. Glacier National Park, for example, includes only one salamander species, three frogs, and one toad on its list of amphibians. One of the most primitive frog species in the world—the Rocky Mountain tailed frog (*Ascaphus montanus*)—lives in the CCE. It requires clean, fast-running streams. Other resident amphibians include the spotted frog (*Rana luteiventris*), the pacific chorus frog (*Pseudacris regilla*), and the western leopard frog (*Rana pipiens*). The CCE is also habitat for the western toad (*Bufo boreas*) and the long-toed salamander (*Ambystoma macrodactylum*).

Reptiles living in the CCE include the western, Great Plains, and common garter snakes (*Thamnophis* spp.), as well as the rubber boa (*Charina bottae*). Western rattlesnakes (*Crotalus viridis*) exist on the prairie adjacent to the Rocky Mountain Front and in the Mission Valley. The painted turtle (*Chrysemys picta*) is common in wetlands.

Around the globe, experts are increasingly concerned about amphibians because they are disappearing at an unusually fast rate. Because amphibians breathe through their skin, they are particularly sensitive to water pollution and ozone depletion. Wilderness waters can be important laboratories for testing the effects of ozone depletion or other global problems (Rockwell 1995; Gadd 2000).

Insects and Other Arthropods

Insects and other arthropods are among organisms sometimes called "the little things that run the world." Insects of the CCE make up for their small size with sheer numbers. In addition, insects influence their ecosystems to an extent that is disproportionate to their size. For example, since 1980, about 80% of the lodgepole pine in Waterton Lakes National Park has been killed by mountain pine beetles (*Dendroctonus ponderosa*)—insects that are only 1 cm long.

Insects are an important agent for converting plant and animal matter to soil. Some insects, such as ladybug beetles (*Coccinella novemnotata*), congregate in great numbers and are an important source of grizzly bear food in the Mission Mountains and elsewhere. Army cutworm moths (*Euxoa auxiliaris*) retreat to the cool mountain boulder fields during hot summers where grizzlies congregate to feed on them.

Of the estimated 100,000 arthropod species in North America, perhaps 20,000 exist in the Canadian Rockies. This ballpark estimate is probably low (Rockwell 1995; Gadd 2000).

Landownership and Management

Centuries ago, no one owned the CCE. The idea of owning titled land came with European settlement. Today, about 17% of the CCE is in private hands. Most of the land is publicly owned by the people of the United States and Canada and by Indian tribes, and is managed by their governments for the public good. The map of the CCE seems a confusing maze of different owners and agencies with different mandates. Understanding the distinction between private land and public landownership is crucial in understanding the economic and social forces that shape the CCE.

Private Land

The single largest private landowner in the CCE is Plum Creek Timber Company, with about 180,000 ha of land in the Swan, Blackfoot, and Stillwater valleys and the Garnet Range. The Sieben Ranch covers about 15,200 ha at the headwaters of the Blackfoot River. Smaller timber companies own land in the U.S. portion of the CCE. For example, F.H. Stoltze Land and Lumber Company owns land in the northern and western edges of the Flathead Valley. Smaller, nonindustrial, private timberlands can be found in Montana, particularly west of the Continental Divide.

Most flatland in the CCE—including the eastern prairie and the flatlands of the Mission, Flathead, and other broad valleys—is privately owned. Narrow stringers of private land follow the Elk River, the Middle and North forks of the Flathead, and other rivers, as well as highways. East of the Rocky Mountain Front are thousands of hectares of private ranch and farmland. A chain of private land crosses over Crowsnest Pass. Private landownership in Alberta ranges from large ranches of several thousand hectares to hobby farms of less than two hectares.

As the popularity of the CCE as a place to live has increased, property values have risen, giving rural landowners a powerful incentive to sell or develop their property. This is particularly true of land near parks, lakes, ski areas, and other attractions.

Owners of Conservation Land and Conservation Easements

Some private landowners have purchased land for conservation purposes. The Nature Conservancy (TNC) and Nature Conservancy Canada, for example, purchase land to conserve biodiversity. Pine Butte Swamp Preserve, a 7,200-ha ranch on the Rocky Mountain Front near Choteau, Montana, is a flagship preserve for TNC. The preserve is located on a fen that contains a rich array of native plants and animals. On the west side of the CCE, TNC owns the 272-ha Dancing Prairie Preserve, which protects native grassland near Eureka, Montana. In addition, TNC owns the 53-ha Safe Harbor Marsh along Flathead Lake, and the 157-ha Swan River Ox Bow Preserve.

The Boone and Crockett Club owns the 2,400-ha Theodore Roosevelt Memorial Ranch near Dupuyer, Montana, which is operated as a conservation laboratory.

In Canada, the Prairie Crocus Ranching Coalition and the Southern Alberta Land Trust Society have been established to help preserve ranchland and open space in southwestern Alberta. Extensive private conservation easements have been placed on land abutting Waterton Lakes National Park in Alberta and in the Elk River Valley in British Columbia.

State Lands and Parks

When Montana achieved statehood, the federal government granted 2.6-km^2 sections of land to the state. State forests include the Coal Creek State Forest (6,000 ha) on the North Fork of the Flathead, the Stillwater State Forest (36,000 ha) near Olney, the Swan River State Forest (6,000 ha) in the Swan River Valley, and the Clearwater State Forest (96,000 ha). The Kalispell office of the Montana Department of Natural Resources & Conservation also manages 114,400 ha in western Montana.

Montanans and visitors enjoy several state parks in the CCE. Flathead Lake State Park includes six units around the lake. The largest of these is 840-ha Wild Horse Island State Park. State parks are also located at Lake Mary Ronan, Whitefish Lake, and Placid and Salmon lakes.

Montana Fish, Wildlife & Parks operates wildlife management areas that protect deer and elk winter range; wetlands for waterfowl; and other habitat for deer, elk, and other hunted species. The state owns some of these areas outright; others are privately owned but protected by conservation easement. These wildlife management areas include the 560-ha Woods Ranch near Eureka, the 640-ha Khuns area near Kalispell, and the 1,386-ha Ninepipes Waterfowl Production Area in the Mission Valley.

Deer and elk that summer in the CCE spend winters on the Blackfoot/Clearwater Game Range (about 26,000 ha) near Ovando. Several important winter ranges are located on the Rocky Mountain Front, including Bay Ranch (1,548 ha), Blackleaf (4,160 ha), Ear Mountain (4,160 ha), and Sun River (16,574 ha). Freezeout Lakes (5,353 ha) is an important wetland complex near Choteau.

Provincial Parks

Although there are no national parks in British Columbia's portion of the CCE, three provincial parks are striking. In the uppermost reaches of the Elk River drainage is Elk Lakes Provincial Park, which encompasses 17,325 ha. The Akamina-Kishinena Provincial Park is along the Continental Divide in the southeasternmost corner of British Columbia. Akamina-Kishinena, which abuts Waterton Lakes and Glacier national parks, spans 10,922 ha and is managed for backcountry recreation and wildlife habitat. Near Fernie, the 259-ha Mt. Fernie Provincial Park offers developed camping and recreation.

In 1999, the Alberta Ministry of Environmental Protection gave limited protection to 28,000 ha in the Whaleback Ridge/Livingstone Range. These include the 20,000-ha Bob Creek Wildland Park and the 7,000-ha Black Creek Rangeland. Black Creek is the first provincially protected rangeland in Alberta. Other provincial parks along the northeastern boundary of the CCE (in Alberta) include Beauvais Lake (1,149 ha), Police Outpost (220 ha), and Chain Lakes (404 ha). Southwestern Alberta also contains several protected natural areas, including Beehive (5,560 ha), Mt. Livingstone (540 ha), Outpost Wetlands (71 ha), and Plateau Mountain (2,296 ha). These areas are managed by Alberta Community Development, Parks and Protected Areas.

British Columbia

The province of British Columbia is two-thirds unreserved public or "Crown land" (in this context, "Crown" has no relation to the CCE phrase). Unlike provincial parks, Crown land is managed for sustained, multiple uses—including timber, mining, oil and gas development, grazing, recreation, wildlife, and watershed protection—under the Forest Practices Code of British Columbia Act (2005).

Alberta

Most public land in Alberta is provincial and includes both forests and grasslands. Alberta Sustainable Resource Development (ASRD) oversees the management of provincial Crown land. Divisions within the agencies include the Forest Service, which manages forests, and Public Lands, which manages grasslands. Other ASRD divisions include Alberta Fish and Wildlife, which manages wildlife populations, and Forest Protection, which fights forest fires. The Rocky Mountain Forest Reserve is a large portion of provincial Crown land in the mountains north of Waterton Lakes National Park. One component of that reserve is the 970-km^2 Castle Special Management Area, which is just north of Waterton Lakes. In addition to land managed by the ASRD, the province of Alberta has set aside places with special recreational or cultural values. These areas, including Bob Creek Wildland Park and Black Creek Heritage Ranchlands, are managed by Alberta Community Development, Parks and Protected Areas.

The foothills and prairies of the Cardston and Pincher Creek municipal districts are largely under private ownership. Ranching and farming are the dominant land uses, although large areas are being developed for recreational homes and acreages. Traditional ranches in southern Alberta, such as the Palmer Ranch a few kilometers east of Waterton, tend to be large (up to 25,000 ha or more).

Glacier National Park

Founded in 1910, Glacier National Park is among the oldest and largest units in the U.S. National Park System. The park, which is 410,178 ha in size, is managed to preserve natural values and foster public enjoyment. Most of the park is managed as wilderness, except along road corridors and in villages. The National Park Service in the U.S. Department of the Interior (DOI) manages Glacier National Park, which hosts between 1.5 and 2 million visitors annually, mostly from June through September. Within the park's boundaries are 169 ha of in-holdings (private parcels), which were in private hands before the park was created. See Chapter 17 for more information about the management of Glacier National Park.

Waterton Lakes National Park

Founded in 1895, Waterton Lakes National Park encompasses 52,617 ha. Parks Canada in the Department of Canadian Heritage manages the park for preservation and recreation. Unlike Glacier National Park, Waterton contains a town within its borders and is bisected by a four-lane highway.

The two national parks comprise the Waterton Glacier International Peace Park. At 4,667 km^2, the peace park is larger than the state of Delaware and is the transboundary wilderness core of the CCE. The governments of Canada and the United States of America established the peace park in May 1932. In 1976, the United Nations Education, Scientific and Cultural Organization (UNESCO) designated Glacier National Park as a Biosphere Reserve under the Man and Biosphere Program. Waterton Lakes National Park received the same designation in 1979. In 1995, UNESCO named Waterton Glacier International Peace

Park a World Heritage Site, which signifies "outstanding universal values." Chapter 18 discusses some of the challenges of protecting and managing the transboundary area of the CCE.

U.S. National Forests and Other Public Lands

The U.S. portion of the CCE includes segments of Flathead, Lolo, Lewis and Clark, and Helena national forests. About 1.4 million ha of U.S. national forest are contained within the CCE. The Bureau of Land Management in the DOI manages about 5,200 ha along the Rocky Mountain Front for multiple uses. The Bureau of Reclamation (also in the DOI) manages two small parcels near the Willow Creek and Pishkun reservoirs. For a discussion of wildland recreational issues, see Chapter 5.

American Indian Reservations

Two major American Indian reservations overlap the CCE in the United States—the Blackfeet Indian Reservation (also called the Blackfeet Nation), just east of Glacier National Park, and the Confederated Salish-Kootenai Tribes Reservation (also called the Flathead Reservation) in the Mission Valley.

The 600,000-ha Blackfeet Indian Reservation is located in Glacier County, Montana, east of Glacier National Park. Most of the reservation is grassland with some forest on the western edge. The reservation is home to 9,000 people, of whom 7,500 are tribal members. The U.S. Blackfeet Tribe has a total of 14,700 registered members.

The northern boundary of the Flathead Indian Reservation bisects Flathead Lake. The reservation encompasses 520,000 ha. The Salish and Kootenai tribes include about 7,000 members, with about half living on the reservation. Tribal members make up only about 30% of the people living on the reservation. About 51% of the Flathead Indian Reservation is owned by the tribe. About 30% of the land area is forested, including the southwestern portion of the Mission Mountain Range.

Canadian Indian Reservations

Lands owned by the Kainaiwa First Nation lie east of the CCE in southern Alberta, but the tribe's timber reserve lies adjacent to Waterton Lakes National Park. At roughly 1,560 km^2, the area occupied by the Kainaiwa First Nation is the largest area owned by an Indian nation in Canada. Of approximately 8,600 members in the Kainaiwa First Nation, about 6,800 live on the reserve. Other Indian lands include those owned by the Piikani First Nation (Peigan Reserve), located east of Pincher Creek, and the K'tunaxa-Kinbasket First Nations, located in the Kootenay Valley and in or immediately adjacent to the CCE.

National Wildlife Refuges

National wildlife refuges are managed by the U.S. Fish and Wildlife Service (FWS) in the DOI. Typically, no recreational activities are allowed on national wildlife refuges unless specifically authorized; hunting and fishing is automatically allowed on federal waterfowl production areas. National wildlife refuges in the CCE include the National Bison Range (7,410 ha), Pablo (960 ha), and Ninepipes (825 ha) in the Mission Valley. Swan River Refuge (627 ha) is at the head of Swan Lake. The FWS manages 2,295 ha of waterfowl production areas in 13 parcels in the CCE. In addition, 2,440 ha in 24 parcels of private land are in conservation easements (Ferguson 1990; The Nature Conservancy of Montana 1994; BC Parks and BC Ministry of Environment, Lands and Parks 2000; Montana Fish, Wildlife & Parks 2000).

The Challenge of Conservation

History offers several examples of how humankind's unsustainable use of natural resources has led to the demise of societies, ecosystems, and entire species. The CCE faces a major challenge—how to accommodate economic development while sustaining the ecosystem's integrity. The region offers an opportunity for the current generation to demonstrate how to meet this challenge, leaving future generations a natural legacy as rich as the one we inherited (Salwassar n.d.; University of Montana School of Forestry 1994; Pelletier and the Swan Valley Linkage Zone Working Group 1995; USDA Forest Service 1995; The Nature Conservancy of Montana and The Nature Conservancy of Canada 2001; Diamond 2005).

References

Achuff, P.L., et al. 1997. *Ecological Land Classifications of Waterton Lakes National Park, Alberta: Soil and Vegetation Resources.* Alberta: Waterton Lakes National Park.
Arno, S., and R. Hammerly. 1984. *Timberline: Mountain and Forest Frontiers.* Seattle, WA: The Mountaineers.
BC Parks and BC Ministry of Environment, Lands and Parks. 2000. *Kootenay Visitor Guide, 2000.* Fernie, BC: Elk Valley Publishing, Ltd.
Blood, L. 1997. *The Story Behind the Landscape: A Geologic Guide to the Flathead Valley.* Kalispell, MT: Flathead Valley Community College.
———. 2004. Personal communication with the author, July 20.
Canada and the United States of America. 1995. Crown of the Continent World Heritage List Nomination. *Waterton/Glacier International Peace Park Report to the World Heritage Committee.* Calgary, AB: Canada and the United States of America.
Chadwick, D. 2000. *National Geographic Destinations: Yellowstone to Yukon.* Washington, DC: National Geographic Society.

Deleray, M., et al. 1999. *Flathead Lake and River System Fisheries Status Report.* Kalispell, MT: Montana Fish, Wildlife & Parks.

Diamond, J. 2005. *Collapse: How Societies Choose to Fail or Succeed.* New York: Penguin Group.

Elias, S.A. 1996. *The Ice Age History of the National Parks of the Rocky Mountains.* Washington, DC: Smithsonian Institution.

Ferguson, G. 1990. *Montana National Forests.* Helena, MT: Falcon Press.

Flathead Basin Commission. 1997–1998. *Biennial Report.* Kalispell, MT: Flathead Basin Commission.

Flathead National Forest. n.d. *Geology of the Flathead Forest Region.* Kalispell, MT: Flathead National Forest.

Flores, D. 1995. The Rocky Mountain West: Fragile Space, Diverse Place. *Montana: The Magazine of Western History* 45(1): 46–56.

Forest Practices Code of British Columbia Act. 2005. [RSBC 1996]. Chapter 159. Victoria, BC: Queen's Printer. http://www.qp.gov.bc.ca/statreg/stat/F/96159_00.htm (accessed October 3, 2006).

Gadd, B. 2000. *Handbook of the Canadian Rockies.* Jasper, AB: Corax Press.

Grinder, B., V. Haig-Brown, and K. Van Tighem (eds.). 2000. *Voices in the Wind: A Waterton-Glacier Anthology.* Waterton Lakes, AB: Waterton Natural History Association.

Holroyd, G. n.d. Bird Conservation in the Yellowstone to Yukon Ecoregion. In *A Sense of Place.* Environment Canada, Ottawa, ON: Canadian Wildlife Service.

Kershaw, L.J., J. Pojar, and A. MacKinnon. 1998. *Plants of the Rocky Mountains.* Edmonton, AB: Lone Pine Publishing.

Mantas, M. 1999. *Vascular Plant Checklist for the Flathead National Forest.* Kalispell, MT: Flathead National Forest.

Maser, C. 1988. Ancient Forests, Priceless Treasures. *Mushroom, the Journal of Wild Mushrooming* 21: 8–18.

Montana Fish, Wildlife & Parks. 2000. *FWP 2000 Land Inventory.* Helena, MT: Montana Fish, Wildlife & Parks.

Pelletier, K., and the Swan Valley Linkage Zone Working Group. 1995. *Managing Private Land in the Swan Valley Linkage Zones for Grizzly Bears and Other Wildlife.* Missoula, MT: U.S. Fish and Wildlife Service Forestry Sciences Laboratory.

Rockwell, D. 1995. *Glacier National Park: A Natural History Guide.* New York: Houghton Mifflin Co.

Salwasser, H. n.d. Conservation of Biological Diversity in the United States. In *Our Living Legacy: Proceedings of a Symposium on Biological Diversity.* Washington, DC: U.S. Department of Agriculture (USDA) Forest Service.

Schmidt, K.J. 1993. *Biodiversity in Glacier.* West Glacier, MT: Glacier Natural History Association.

The Nature Conservancy of Montana. 1994. *Five-year Strategic Plan for the North Fork of the Flathead River.* Helena, MT: The Nature Conservancy of Montana.

The Nature Conservancy of Montana and The Nature Conservancy of Canada. 2001. *The Crown of the Continent: Too Precious to Lose.* Helena, MT, and Toronto, ON: The Nature Conservancy of Montana and The Nature Conservancy of Canada.

University of Montana School of Forestry. 1994. *Considerations for Sustainability in the*

Crown of the Continent Ecosystem. Missoula, MT: University of Montana School of Forestry Boone and Crockett Wildlife Conservation Program.

USDA Forest Service. 1995. *Fish, Wildlife and Habitat Management Framework for the Bob Marshall Wilderness Complex.* Missoula, MT: USDA Forest Service.

U.S. EPA (U.S. Environmental Protection Agency). 1983. *Flathead River Basin Environmental Impact Study Final Report.* Washington, DC: USEPA.

Van Tighem, K. 1999. *Wild Animals of Western Canada.* Canmore, AB: Altitude Publishing.

Webster, D. 1999. Walking Down Grizzlies' Road, Yellowstone to Yukon. *Smithsonian* (November): 58–72.

Whitney, S. 1990. *Western Forests: The Audubon Society Nature Guides.* New York: Alfred A. Knopf.

PART II
Social Dimensions

3

Native Peoples and Archaeology of Waterton Glacier International Peace Park

Brian O.K. Reeves

Native peoples have inhabited the CCE for the last 10,000 years. In this chapter, I describe the activities of native peoples in the portion of the CCE that became Waterton Glacier International Peace Park. I also discuss the area's cultural–historical record, patterns of resource harvesting and occupancy, relationships to past climates and environments, and changes in cultural patterns (particularly during the last 1,500 years). All these historical factors have implications for understanding the Peace Park of today. I draw heavily from technical reports written about the ethnology (Reeves and Peacock 2002) and archaeology (Reeves 2003) of Glacier National Park (GNP).

The Waterton Glacier International Peace Park is a primary protected area in the CCE, and contains the greatest variety of plants and animals found in the northern Rocky Mountains. The areas east of the Continental Divide feature large and productive montane, foothills, and grassland valleys, as well as fruitful lakes and marshes. Herds of mountain bison (shortened to "bison" or buffalo in the rest of this chapter), bighorn sheep, mule and white-tailed deer, and elk, coexisting with numerous edible berries and roots, made the area particularly attractive to Native Americans, who have been associated with these lands for more than 1,000 years. The native peoples were also drawn seasonally to the subalpine and alpine grasslands, which were easily accessed from the valley floors. At times, the banks of large mountain valley lakes west of the Continental Divide that flow into the generally heavily forested North Fork of the Flathead River Valley also invited settlement (Reeves and Peacock 2002; Reeves 2003).

The region that became the Peace Park was known to the *Piikáni* (one of the three tribes of the *Nitsitapii* or *The People*; commonly referred to in the United States as the Blackfeet) as *Miistakis*, "The Backbone of the World." It remains an area of profound importance to Native Americans, particularly the Piikáni, who resided along the eastern slopes of the mountains and adjacent plains, and the *K'tunaxa*, who lived west of the mountains in the Flathead and Kootenai River valleys and traveled east to hunt buffalo and other game. For these peoples, the mountains are their most sacred place.

The GNP evokes strong feelings in Native Americans, specifically the Blackfeet. The Peace Park is considered to be one of the last remnants of primeval wilderness in the U.S. Rocky Mountains. Many believe that Native Americans played little or no role in what came to be known as the Peace Park, or that they did not occupy it until very recently and then at best only peripherally (Sheire 1970; Bucholtz 1976; Newell et al. 1980). This perception prevails today despite the fact that, for the last 30 some years, most archaeologists have known that Native Americans seasonally frequented the northern Rockies for the last 10,000 years (see, for example, Reeves 1974).

Historical Native Tribes

The K'tunaxa

K'tunaxa bands occupied the upper reaches of the Kootenai and Columbia rivers, the Columbia Lakes region, and valleys in the Rocky Mountain to the east. Some K'tunaxa bands resided west of the Continental Divide year-round; others moved east to hunt buffalo in the foothills and plains on a seasonal basis. Traditional K'tunaxa territory included the eastern slopes of the Rockies from the valley of the North Saskatchewan River south to the Peace Park region. The three bands traditionally associated with the region were the *Gakawakamitukinik* (Michel Prairie), *Akanahonek* (Tobacco Plains), and *Akiyinik* (Jennings). In the historical and anthropological literature, the bands are typically known by the place names given in parentheses. The Gakawakamitukinik's traditional lands included Crowsnest Pass and the adjacent Rocky Mountains. Smallpox obliterated this band in the 1730s (Schaeffer 1982).

The Akanahonek traveled east over the passes during the fall, winter, and spring to hunt buffalo. The people conducted the winter hunt, which dates back thousands of years, on foot. After acquiring horses in the early 1700s, they hunted on horseback in the spring and fall. Prior to this period, the Akanahonek did not go east during the warm-weather months. The families left the Tobacco Plains in January and traveled by snowshoe east over passes such as the Buffalo Cow Trail (commonly known as South Kootenay Pass) in Waterton Lakes National Park (WLNP); an unnamed pass at the head of Logging Lake; Kootenai;[1] Swiftcurrent; and *Packs-Pulled-Up* (Logan) in the GNP. The Akanahonek hunted sheep and other game along the way, rejoining as a band on the eastern slopes to hunt buffalo. In late March, they would return westward with heavy packs of dried buffalo meat. The Akanahonek continued to hunt and trap on the eastern slopes of the Peace Park region into the late 1800s. Today, most descendants of the Akanahonek reside on reserves in British Columbia.

Although the Akiyinik may have joined the Akanahonek in the winter buffalo hunts before they acquired horses (as they did later), it was after they obtained the horse and moved their winter camps from the Kootenai River to the head of the Flathead Valley in the late 1700s that they began to go east in large numbers for the spring and fall hunts. The Akiyinik used the Cut Bank and Marias passes. Marias Pass was approached from the west side of the Continental Divide by a route up the South Fork. They continued hunting on the east side of the divide into the late 1800s. The Akiyinik were removed to the Flathead Reservation on the west side of Flathead Lake in the mid to late 1800s.

The Nitsitapii

Three tribes comprise the *Nitsitapii*: the *Kainaiwa* (Many Chiefs, often referred to as the Bloods) who have resided on the Blood Reserve in southern Alberta since 1883; the *Siksiká* (Blackfoot) who resided on the Siksiká Nation east of Calgary; and the Piikáni (Peigan). The Piikáni consist of two divisions—the North Piikáni who reside on the Peigan Nation in southwest Alberta and the South Piikáni who reside on the Blackfeet Indian Reservation in Montana and officially call themselves the Blackfeet.

Five hundred years ago, the Nitsitapii's traditional territory extended from the North Saskatchewan River to the Missouri River and east from the Rocky Mountains to today's province of Saskatchewan. Kainaiwa and Siksiká traditional lands centered on the plains and parklands to the east and north of Piikáni lands. The Piikáni traditionally ranged along the eastern slopes and adjacent western plains from the Bow River south to the Missouri River, and east as far as the Sweetgrass Hills. Favored wintering grounds were toward the mountains in the big wooded bottoms on the Marias, Oldman, and Bow rivers. Some Piikáni bands traditionally spent their summers in the eastern slopes and adjacent foothills of the Peace Park, journeying east only to join friends and relatives for the annual tribal Sundance at the Sweetgrass Hills.

The Lame Bull Treaty of 1855 set aside the lands north of the Missouri River, east of the Continental Divide, for the Blackfeet (Piikáni) and other tribes. The size of the reservation was reduced a number of times in the late 1800s. In 1892, the Blackfeet ceded the mountain portion of the reservation. They were guaranteed the right to hunt, fish, and cut timber as long as the lands known as the *Ceded Strip* remained public. The political pressures behind this cession related to the discovery of purported commercially developable deposits of silver and other metals in the region north of Marias Pass. In 1898, the Ceded Strip became part of the Lewis and Clark National Forest, and in 1910, the lands north of Marias Pass became the GNP. In the view of the U.S. government, the land occupied by the park was no longer considered multiple-use public land (like national forests) and the Blackfeet lost or had their traditional rights in the park severely curtailed. Today, many Blackfeet still contend that they never gave up these rights to the area.

The Piikáni's relationship with the Peace Park land is not only long-standing, stretching back more than 2,000 years, but it is a fundamental part of their traditional religion and way of life. Many traditional accounts and stories are associated with Miistakis, and it is highly sacred to the Piikáni elders. They believe that the place has great power and that many sacred "doings" took place there and continue today. Sacred rituals and materials (such as paints, animals, and plants) are believed to have come from the land, which is also used for vision quests.

Ninastakis (The Chief Mountain) continues to be the ongoing focal place for Nitsitapii vision questing and other religious activities. The most powerful of the *Up-Above-People* live on the mountain, including Thunder Bird who gave the first medicine pipe to the people in the long-ago time. South Piikáni, Kainaiwa, and Siksiká traditionalists continue to visit the area, as do traditional people from other tribes, including the Cree and *Tsuu T'ina*. The Ninastakis straddles the Blackfeet Indian Reservation and the GNP, with approximately half the mountain in each locale. In recent years, increasing numbers of nonnatives visited the base as well as the summit of the mountain, taking sacred offerings and disturbing ceremonies. In addition, the Blackfeet Tribal Business Council continues to log the forested north slopes. A slide on the mountain's north face in 1992 temporarily halted these activities, which, if allowed to continue, will ultimately result in the spiritual degradation of this sacred place.

The Piikáni have a rich and extensive ethnobotanical knowledge. More than 80 species of plants have been or still are collected within the Peace Park. The Piikáni believe that plants in the Peace Park are bigger and have more power than those outside the park because the area is not overgrazed by cattle. A number of very important ranges for species on slopes east of the Continental Divide coincide with the Peace Park and do not occur north of the Carbondale River–Crowsnest Pass area or south of the Birch Creek–Teton River area.

The Piikáni's well-developed set of plant management techniques ensures a continual supply of plants for food, medicinal, and spiritual purposes. These techniques continue to be culturally significant to the people. Although the elders recognize and appreciate the role that the Peace Park plays in protecting the plants, many are frustrated about having to "sneak in" to collect plants. The elders are also concerned that no one is looking after the plants and that they are being crowded out.

Precontact Cultural History

Native cultural history in the northwest plains and Rocky Mountains before the Europeans arrived (known as *precontact times*) is usually divided into three periods: early, middle, and late. These periods are based on the major technological changes in projectile-point types that accompanied significant advances in the weapon systems used in hunting and warfare (see Figure 3-1).

Early Precontact Period

This period (ca. 10,500–7,750 radiocarbon years[2] ago) is characterized by four archaeological complexes identified by stone projectile points presumably used with throwing and stabbing spears. The earliest recorded site recovered a Clovis Complex point along the Belly River in the GNP. The Clovis Complex is the oldest easily identifiable American culture. Paleoenvironmental data from the GNP and vicinity (Carrara 1989) indicate that plants and animals recolonized the valley floors and the alpine zone 11,000–10,000 radiocarbon years ago, creating a generally hospitable environment for the peoples of the Clovis Complex to recolonize the eastern valleys. The late glacial environment was cold and dry, and treeline was as much as 500 m lower than it is today (Carrara 1989). The lower eastern-slope valley floors were populated by extensive grasslands with restricted open lodgepole-pine-dominated forests at higher elevations toward the Continental Divide. Today these valley floors and mountainsides are dominated by old-growth, subalpine, spruce-dominated forests. Lodgepole-pine forests were established in the Kootenay Lakes area at the head of Upper Waterton Lake 10,000 radiocarbon years ago (Carrara 1989).

Reforestation of the west-side valleys was also complete by this time. Lodgepole-pine forests may have been established as early as 11,000 radiocarbon years ago. Spruce or larch was also present in the North Fork of the Flathead River Valley by at least 10,600 radiocarbon years ago. The Hungry Horse Reservoir area of the Middle Fork of the Flathead River was colonized by a sagebrush–juniper steppe by 12,300 radiocarbon years ago. It was later replaced by an open spruce and whitebark and/or limber- and lodgepole-pine forest with a sagebrush/grass understory. By 10,500 radiocarbon years ago, an almost pure ponderosa-pine woodland community had replaced these pioneer communities.

A Late Ice Age suite of big game roamed the Peace Park's eastern slopes. Bison and bighorn sheep were probably common, and lake and river fisheries were present. Migratory waterfowl moving along the Rocky Mountain Flyway used the lakes as staging areas.

The Ice Age abruptly ended 9,900 radiocarbon years ago. In a matter of four to five years, the regional climates of the Northern Hemisphere warmed significantly and the dry, cold climate was replaced with a dry, warm climate. Because the inclination of the earth's axis changed, summer solar radiation increased, which resulted in very warm summers in northern latitudes. These changes had significant short- and long-term impacts on regional vegetation, big game, and seasonally resident Clovis-related descendant hunters in the areas east of the

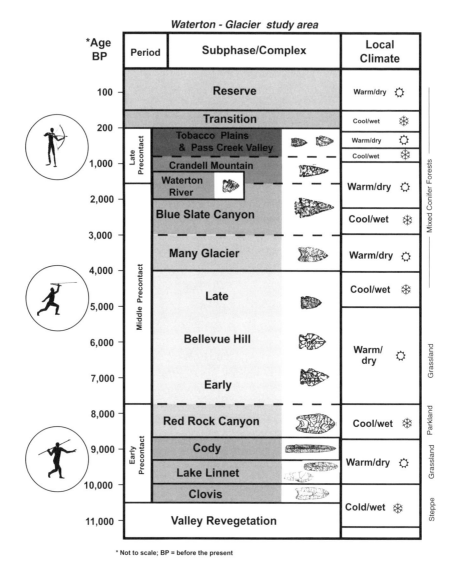

FIGURE 3-1. Native Culture History of Waterton-Glacier

Continental Divide and northwest plains. The changes can explain, on a regional level, the apparent lack of cultural continuity between the early fluted-point hunters and later cultural complexes.

The first-defined occupation is the Lake Linnet Complex (10,000–9,500 radiocarbon years ago). Early resource harvesting in the alpine zone consisted of quarrying highly siliceous metamorphosed green argillite at specific locales along the Continental Divide in the GNP. It appears that by approximately 9,500 radiocarbon years ago, the native people had established basic patterns of seasonal camping within the eastern-slope valleys, traveling back and forth over the Continental Divide passes, and seasonal occupancy of the alpine zone. Styles of spear points, often made of basalt obtained from quarries in the Intermountain West, indicate that most Lake Linnet cultures lived west of the mountains. These first people reoccupied the western slopes after the Cordilleran Ice Cap disintegrated and the large proglacial lakes that occupied the mountain valleys drained. The dry, warm climate and open forests and grasslands facilitated movement of these people both north into the Cordillera and east across the Continental Divide passes onto the eastern slopes. These people were hunters, trappers, gatherers, and fishers, and were perhaps better adapted to the warm and dry climate than the peoples of the earlier Clovis Complex.

Around 9,300 radiocarbon years ago, the regional climate became drier, resulting in the development of xerophytic grasslands in the adjacent foothills. This general climate type, known as the Altithermal, persisted in the foothills until approximately 5,000 radiocarbon years ago. Regional, negative climatic fluctuations, which are not necessarily recognized in the pollen records taken to date, occurred during this early–mid Holocene, warm, dry interval. Around 8,400 radiocarbon years ago, Hudson Bay experienced a massive catastrophic flood (Barber et al. 1999). This event, combined with the change in the earth's axis to its present inclination and the associated changes in solar radiation, resulted in an abrupt cold snap and a cooler and wetter climate in the eastern slopes, which in turn caused a treeline descent of about 100 m (Reeves and Dormaar 1972). Pine parklands expanded into the grasslands of the valley floors and foothills (Vivian et al. 1998; Barber et al. 1999). Around 8,000 radiocarbon years ago, this climatic episode ended with a return to warm conditions.

A new cultural complex known as Cody appeared in the northwest plains and Rocky Mountains approximately 9,500 radiocarbon years ago. Cody people were the first complex of communal bison hunters who expanded north from the southern plains and the Rocky Mountains, adapting to a variety of environments along the way. Cody Complex points have been found in the Peace Park, but have not as yet been recovered from excavated contexts. In the Central Rockies, Cody appears to have a different settlement pattern than other cultural complexes; these peoples often summered in the high country, hunting a variety of wildlife and using watercrafts (Johnson et al. 2004).

Red Rock Canyon (ca. 9,500–7,750 radiocarbon years ago) is the last and best-defined early-period cultural complex. Basic resource harvesting and occupancy patterns established during this period include seasonal camping, fishing, and hunting for bison and bighorn sheep in the valleys east of the Continental Divide; summer hunting and quarrying in the alpine zone; camping at the foot of the lakes west of the Continental Divide (to quarry a silicified marlstone known as Bowman chert); and traveling back and forth across the mountain passes during snow-free months. It is not uncommon to find tool stones from central and southern Montana Rocky Mountain sources that indicate trade and exchange with culturally related groups to the south. This continues the pattern that Cody Complex peoples had established earlier.

The Middle Precontact Period

This period (ca. 7,750–1,600 radiocarbon years ago) begins with the appearance of notched dart points used with the atlatl or throwing stick, along with new cultural complexes and technologies. The earliest culture, Bellevue Hill, dates 7,750–3,500 to 4,000 radiocarbon years ago. Developing out of the earlier Red Rock Canyon, people of this culture were long-standing residents of the region who intensified earlier resource harvesting and occupancy patterns. This culture seasonally occupied both the eastern valley floors and the alpine zone during the height of the mid-Holocene Altithermal, which ended about 5,000 years ago, and during early Late Holocene times until approximately 4,000–3,500 years ago when the climate again became cooler and wetter. Alpine glaciers re-formed in some areas of the northern Rockies at the beginning of the Late Holocene. The warmer, drier conditions characterizing most of the Bellevue Hill times extended the season in the alpine zone, reduced snowpacks, and probably increased favorable forage for grazers. These factors led to increased travel between the west and east sides of the Continental Divide and more extensive seasonal resource harvesting and occupancy of the alpine zone by the Bellevue Hill group (Figure 3-2).

In the eastern valley floors, bison were communally hunted using natural traps, such as ice-block depression, wetlands, and constructed corrals (Reeves 1978). Buffalo jumps, such as *Head-Smashed-In,* were used. Bison and bighorn sheep were also hunted on the bedrock ridges at the lower valley fronts. Judging from the increase in depth of the archaeological deposits at the Narrows site between the Upper and Middle Waterton Lakes, fishing also intensified. Toolstone use patterns focused on locally and regionally available materials. Tool stones from central and southern Montana quarries are much less common than they were in earlier times, reflecting a change in north–south trading and exchange systems. This is possibly the result of increased summer–fall aridity on the slopes and plains east of the Continental Divide, which restricted travel south to the quarries. It could also reflect an expansion of local habitats along the favored foothills and slopes east of the divide, which resulted in the focusing and localizing of activities.

During the mid-Holocene Altithermal, the treeline, which probably descended during the late Early Holocene, cool, wet interval, ascended again. The forests became more open and the grasslands of the montane region more expansive. Effective precipitation was probably reduced by more than 40% from today's values (Reeves and Dormaar 1972). More frequent and violent chinooks, which reduced winter snowpacks, expanded and enhanced the

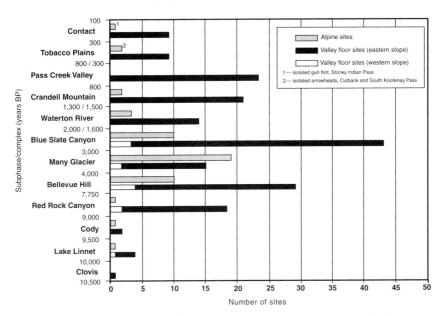

FIGURE 3-2. Waterton-Glacier Valley Floor and Alpine Culturally Assignable Sites

fall–spring bison ranges in the eastern valley floors, possibly resulting in larger local herds of bison.

Pollen records from Lake Linnet in the Waterton Valley (Christensen and Hills 1985) show a marked increase in charcoal with the onset of Late Holocene conditions. These data and the pollen spectra suggest that at this time, native people starting setting fire to the montane forests and grasslands of the eastern valley floors to enhance hunting conditions and maintain grassland productivity. Prescribed burning continued for the next 5,000 years in many valleys of the northern Rockies until smallpox reached epidemic levels in the 1700s and the natives were relocated to the reserves and reservations in the mid to late 1800s.

Along the South Fork of the Flathead River, pines were replaced by juniper during the height of the dry period (ca. 7,200 radiocarbon years ago). Available moisture was at least 50% below today's level. Moister conditions and pine woodlands began to return approximately 4,700 radiocarbon years ago. The forest closed and modern conditions prevailed by 2,500 radiocarbon years ago.

The Many Glacier Subphase (ca. 4,000–3,000 radiocarbon years ago) followed Bellevue Hill, locally representing an intrusive culture from the south that originated in the northern Great Basin. These peoples brought new resource harvesting and food processing patterns, such as sheep trapping, stone boiling, and pemmican preparation, to the region. Many Glacier occupancy patterns differ from other cultures. They are short-term occupations in specific locales. The culture appears to represent minor occupants of this region who spread to the north during a warm and dry interval following the first of the Late Holocene cold and wet episodes. Tool stones from the Greater Yellowstone region are more common, indicating more extensive trading and exchange patterns than in Bellevue Hill.

The most extensive and intensive occupation of the valleys and alpine zone areas occurred between about 3,000 and 1,600 years ago during the Blue Slate Canyon Subphase (see Figure 3-2). The climate was wetter and cooler during early occupancy (3,000–2,500 years ago). This was a time of major intensification and extensification throughout the northern plains and the adjacent Rocky Mountains. Large communal bison corrals and jumps developed, and the discovery of net sinkers indicates that the native people were fishing with weighted, slack-water nets on the slopes east of the Continental Divide. An extensive trading and exchange system of chert from quarries in central and southern Montana and obsidian from the Greater Yellowstone region developed.

New cooking technologies—hot-rock roasting using pit ovens—also appeared in the northern Rockies and adjacent plains around this time, and the use of stone boiling intensified. In addition to cooking large cuts of meat, fish, and various green vegetables, the native peoples baked inulin-rich roots, such as balsamroot, various lomatiums, and blue camas, in these ovens. The ovens have been found in considerable abundance in some of the valleys east and west of the Continental Divide. Blue camas was most probably introduced in areas east of the divide as part of this new food preparation complex, which originated on the Columbia Plateau.

Blue Slate Canyon peoples in the Peace Park mined chert intensively at the Bowman quarry complex; hunted and camped throughout the alpine zone; and drove bison in spring, fall, and mid-winter in the valleys east of the Continental Divide. They left behind thick deposits of cultural refuse at large base camps located at the bottom of major east- and west-side valley lakes. In the Peace Park, Blue Slate Canyon most likely represents both ancestral Blackfoot- and Kutenai-speaking peoples.

Another locally intrusive culture into the Peace Park area at the end of the middle period was the Waterton River Subphase, the local representative of the Northern Plains Besant Phase (ca. 2,000–1,400 radiocarbon years ago). The Peace Park was on the western edge of the territory of these plains-oriented, communal, Arapaho-speaking bison hunters. Their presence is limited in contrast to the earlier and contemporary Blue Slate Canyon Subphase with whom they sometimes camped and hunted in the valley floors and alpine zone (Figure 3-2).

Late Precontact and Postcontact Periods

The record of this period, the last 1,600–125 years, is marked by the introduction of the bow and arrow, pottery, and a major elaboration on bison driving in the adjacent plains. These and other cultural changes resulted in significant shifts in precontact resource harvesting and occupancy patterns in areas east of the Continental Divide. Seasonally resident groups' summer–fall resource harvesting and occupancy patterns increasingly focused on the adjacent foothills and plains.

Resource harvesting and occupancy in the Peace Park first became less intensive and more seasonally structured during the Crandell Mountain Subphase, approximately 1,600–1,000 to 800 years ago. Summer seasonal occupancy of the alpine zone and slopes west of the Continental Divide was also markedly reduced (Figure 3-2).

Patterns of resource harvesting and occupancy within valley floors on the eastern slope of Crandell Mountain continue, in part, from those followed in the preceding Blue Slate Canyon Subphase from which it developed. Communal bison hunting continued, and roots were unearthed and roasted in season. Focal settlement locales at the bottom of the Inside Lakes continued to be occupied. Use of Bowman chert continued, indicating the persistence of seasonal camping, at least for quarrying at the Bowman Lake/Creek locale. Use of the lake and stream fisheries on the east side of the Continental Divide appears to have diminished markedly or to have been abandoned. Occupation seems less extensive, more geographically focused, and less intensive than Blue Slate Canyon, particularly in the alpine region (Figure 3-2).

Seasonal data from bison kills in the WLNP area for both Crandell Mountain and Pass Creek Valley indicate a late winter–spring pattern as opposed to the earlier Blue Slate Canyon, which included both spring and fall kills. Fall bison kills appear to be absent, suggesting that Crandell Mountain groups, who were seasonally resident in spring and summer and hunted in the alpine zone, left the valleys to join the large fall communal bison hunts. This accounts for the abandonment of the fisheries.

These trends intensify in the last two precontact cultures (ca. 1,000–800 to 200 years ago): Pass Creek Valley, the local representative of the Old Women's Phase of the northwest plains and slopes east of the Continental Divide, representing the precontact Piikáni, and the Tobacco Plains Phase of the northern Rocky Mountains/Rocky Mountain Trench, representing the precontact K'tunaxa. Warm and dry conditions during the Crandell Mountain Period changed to cooler and wetter conditions for a brief two centuries or so about 1,000 years ago, then changed back to warmer and drier (the Medieval Warm Period). Next came the cooler and wetter conditions of the Little Ice Age, which peaked about 300 years ago.

When the Bowman quarries and the lake and river net fisheries on the east side of the Continental Divide were abandoned, people no longer hunted and camped in the alpine zone along the Continental Divide in summer, nor to any extent at the bottom of the western valley lakes or along the North and South forks of the Flathead River. Camps were smaller and more seasonally limited in the eastern valley floors and bison driving was more seasonally restricted than in earlier times. The precontact Piikáni were seasonally resident primarily in the cool to cold weather months. Like those of most earlier cultures, their tool stones continued to show a strong orientation toward the plains and the south-central Montana mountain quarries. In contrast, the precontact K'tunaxa tool-stone pattern in the Peace Park is typical of that found in the Crowsnest Pass and middle Kootenay River to the north–northwest, dominated by the use of Top-of-the-World chert (a quarry source located in the Rocky Mountains in southeast British Columbia northwest of Fernie).

Tobacco Plains residency in the Peace Park appears to have been primarily restricted to mid-winter bison hunting and processing camps similar to some of the earlier Crandell Mountain mid-winter Subphase components in the Peace Park. These camps are characterized by large amounts of Top-of-the-World chert and no doubt represent camps associated with the Akanahonek's traditional mid-winter bison hunt. The only two alpine occurrences are isolated arrow points found at the summits of the Cut Bank and South Kootenay passes (Figure 3-2).

Radical changes occurred at the end of precontact times. Beginning with the postcontact phase in the early 1700s, the native peoples acquired horses, guns, knives, kettles, and other traded goods. Smallpox, which decimated local populations, spread unchecked throughout the region in the early 1730s. These changes resulted not only in a major decline in native populations during the 1700s, but also in radical changes in resource harvesting, occupancy, material culture, territory, and warfare. As a result, archaeological sites of this age are uncommon in the Peace Park (Figure 3-2). Native campsites in the Peace Park dating to the mid-1800s indicate that the bison herds were essentially wiped out by the 1860s.

Climate Change or Culture Change?

Did climate change precipitate the significant cultural changes that occurred during the last 1,600 radiocarbon years? It seems unlikely. Evidence from the archaeological sites in the adjacent foothills and plains shows continued seasonal native occupation tied to the movements of bison between summer and winter ranges. Fluctuations in temperature and effective precipitation did occur during these and earlier times (McKinnon 1986), reflecting the historic cycles of wet and dry periods and the larger episodic climatic cycle. These events did not result in any discernible period of overall abandonment by animals or people during the Late Precontact Period. Indeed, during this time communal bison driving intensified, regional populations grew, and cultures enlarged (Reeves 1989).

Certainly the yearly and longer-term cycles in precipitation and temperature affected the local abundance of game and edible plants in the Peace Park, as did species-specific cycles and the cyclical appearance of animal and plant diseases.

These cycles were small in scale and resulted in localized impacts on precontact native resource harvesting and occupancy.

Climatic changes during the last 1,000 radiocarbon years ended in the Little Ice Age when most glaciers advanced to their maximum position since the end of the Wisconsin Ice Age. In the Peace Park, this event and the climatic change that preceded it resulted in increased winter precipitation, cooler and cloudier climate, and shorter summers, all of which were likely to have affected flora and fauna. If there were such impacts, they do not seem to have had an appreciable impact on native resource-harvesting and occupancy patterns in the eastern valley floors during this time period or in the alpine zone during the Little Ice Age.

Despite the impacts of the smallpox epidemic in the late eighteenth century, the Akanahonek and Akiyinik continued to cross the passes to hunt bison in valleys east of the Continental Divide during the mid-winter, to hunt deer and elk in the valley bottoms on the west side of the divide and pursue bighorn sheep in the alpine zone. They also crossed over the passes on horses for the spring and fall bison hunts, hunting extensively in valleys on the east side of the divide in the nineteenth century, as did the Piikáni.

After the initial smallpox epidemics and the arrival of the horse in the early 1700s, other peoples arrived from the east, north, and west to hunt in the areas east of the Continental Divide. These arrivals included Crow, Shoshone, Assiniboine and Cree, Upper Kalispel, and Salish, along with the occasional arrival of mounted tribes from farther west. The Peace Park remained very attractive to native tribes during the Little Ice Age of the last two centuries.

When George Bird Grinnell and James Willard Schultz, the first white sportsman hunters to leave a published record of their hunting trips, began to hunt in the St. Mary and Many Glacier country in the1880s, they traveled with Piikáni guides and found abundant game. They also met mounted Akanahonek hunting parties in the St. Mary area, and Kainaiwa hunting parties in the Swiftcurrent/Many Glacier region.

All these activities occurred from the height of the Little Ice Age to its end. It appears that there was a considerable amount of game in the alpine zone, which was almost eradicated by prospectors, trappers, and hunters before the turn of the century and the establishment of the Kootenay (Canada) and Flathead (United States) Forest Reserves in 1898, and the GNP and the WLNP in 1910 and 1895, respectively.

In contrast, the west side of the GNP, and the North Fork of the Flathead River Valley appear not to have had much native use in the last few centuries. When Euro-American settlers first came into the North Fork Valley in the 1880s, the first natives they saw were Stoney Indians from the north and, later, Montana Cree who had come into the area to hunt in the summer. The North Fork Valley was not part of the traditional subsistence territory of the Akiyinik. They did not camp, hunt, fish, or collect plants there, which is not surprising because it is quite unproductive compared to the Flathead Valley to the south.

Conclusions

Climatic change does not appear to explain the variability in native resource-harvesting and occupancy patterns that occurred in the Peace Park during the last 1,600 radiocarbon years. Large-scale cultural changes that occurred at the regional level appear to be largely related to the introduction of the bow and arrow and its impact on communal bison hunting among the ancestral Nitsitapii, Arapaho, and other tribes. Populations grew and became more socially complex. Military societies developed and new religious rituals appeared that caused changes in the traditional resource-harvesting and occupancy patterns associated with the Peace Park area.

Furthermore, the early post-Columbian smallpox epidemics (Campbell 1989) appear to have swept the Plateau, the Southwest, and the Southeast culture areas of North America 400 to 500 years ago. If these epidemics had penetrated the headwater regions of the Missouri, Columbia, and Saskatchewan basins, there would have been far-reaching consequences for native occupancy and resource harvesting in the Peace Park, as well as for the plant and animal species. Perhaps the abundance of big game and fish when first observed, hunted, and fished by Grinnell and others in the late 1800s partly reflects the impacts of smallpox epidemics as well as earlier cultural changes.

It appears, for example, that the superabundance of large-sized lake trout netted by Kootenai Brown and other earlier settlers from the Middle and Upper Waterton lakes in the late 1800s reflects in part the abandonment of the native fishery several centuries before. Similarly, the largely forested habitat of the North Fork River in the Flathead Valley in the late 1800s might well reflect the abandonment and the cessation of prescribed burning in that area by natives centuries earlier. Today, valleys east of the Continental Divide are becoming increasingly choked with aspen, which reflects the long-term impacts of the extinction of the bison herds and the removal of the natives and their prescribed burning since the mid-1800s. These cultural changes; the loss of traditional animal species (bison) and native practices (i.e., prescribed burning, selective hunting, and plant harvesting); and natural and human-induced climate change will cause the Peace Park to bear little resemblance to the region over the past 10,000 years.

Notes

1. Spelled Kootenay in Canada and Kootenai in the United States.

2. The term radiocarbon years means "calendar years before the present." By convention, the "present" begins at 1950 ad, the standard baseline for radiocarbon dating.

References

Barber, D.C., et al. 1999. Forcing of the Cold Event of 8,200 Years Ago by Catastrophic Drainage of Laurentide Lakes. *Nature* 400: 344–51.

Bucholtz, C.W. 1976. *Man in Glacier*. West Glacier, MT: Glacier Natural History Association.

Campbell, S.K. 1989. Postcolumbian Culture History of the Columbia Plateau: AD 1500–1900. PhD dissertation. Seattle, WA: University of Washington Department of Anthropology.

Carrara, P.E. 1989. *Late Quaternary Glacial and Vegetative History of the Glacier National Park Region, Montana.* Bulletin No. 1902. Washington, DC: U.S. Geological Survey.

Christensen, O.A., and L.V. Hills. 1985. Part 1: Palynologic and Paleoclimatic Interpretation of Holocene Sediments, Waterton Lakes National Park, AB. In *Climatic Change in Canada:* National Museum of Natural Sciences Project on Climatic Change in Canada during the Past 20,000 Years, edited by C.R. Harrington and G. Rice. *Syllogis* 55: 345–96. Ottawa, ON: National Museum of Natural Sciences.

Johnson, A.M., B.O.K. Reeves, and M.W. Shortt. 2004. Osprey Beach: A Cody Complex Occupation on Yellowstone Beach. Final report on file. Mammoth Hot Springs, Yellowstone National Park: Yellowstone Center for Resources.

McKinnon, N.A. 1986. Paleoenvironments and Cultural Dynamics at Head-Smashed-In Buffalo Jump, Alberta: The Carbon Isotope Record. MA thesis. Calgary, AB: University of Calgary Department of Archaeology.

Newell, A.S., D.W. Walter, and J.R. McDonald. 1980. *Historic Resources Study, Glacier National Park and Historic Structures Survey.* Report on file. West Glacier, MT: Glacier National Park Library.

Reeves, B.O.K. 1974. Prehistoric Archaeological Research on the Eastern Slopes of the Canadian Rocky Mountains 1967–71. *Canadian Archaeological Association Bulletin* 6: 1–31.

———. 1978. Bison Killing in the Southwestern Alberta Rockies. In Bison Procurement and Utilization: A Symposium, edited by L. B. Davis and M.C. Wilson. *Plains Anthropologist* Memoir No. 14., 23(82): 63–78.

———. 1989. Communal Bison Hunters of the Northern Plains. In *Hunters of the Recent Past*, edited by L.B. Davis and B.O.K. Reeves. London: Unwin and Hyman, 169–94.

———. 2003. *Miistákis: The Archaeology of Waterton-Glacier International Peace Park.* Final report on file. West Glacier, MT: Glacier National Park Library.

Reeves, B.O.K., and J.F. Dormaar. 1972. A Partial Holocene Pedological and Archaeological Record from the Southern Alberta Rocky Mountains. *Alpine and Arctic Research* 4: 325–36.

Reeves, B.O.K., and S. Peacock. 2002. *Our Mountains Are Our Pillows: Ethnographic Overview—Glacier National Park.* Final report on file. Glacier National Park, MT: National Park Service.

Schaeffer, C.E. 1982. Plains Kutenai: An Ethnological Evaluation. *Alberta History* 30: 1–9.

Sheire, J.W. 1970. *Glacier National Park Historic Resource Study.* Washington, DC: National Park Service Office of History and Historic Architecture, Eastern Service Center.

Vivian, B.C., K. Bosch, and B.O.K. Reeves. 1998. *Historical Resource Conservation Excavations.* Consultant's report (Lifeways of Canada Limited) to the City of Calgary, Archaeological Survey of Alberta, Edmonton, AB.

4
Economic Growth and Landscape Change

Tony Prato, Dan Fagre, and Ramanathan Sugumaran

Human-induced landscape change is impairing the capacity of the CCE to provide environmental amenities and sustain ecological processes (Turner and Meyer 1994; Prato 2005). In this chapter, we discuss the impacts of economic growth and development on landscape change and wildlife habitat in the CCE.

Landscape change involves changes in land cover and land use. Land cover refers to the physical, chemical, or biological features of a landscape. Agricultural land, grassland, and forestland are land covers. Land use refers to the uses humans make of a particular land cover. For example, agricultural land can be used for food and fiber production, cattle grazing, timber harvesting, and recreation.

Human disturbances in the form of subdivisions, shopping malls, roads, and other land uses have more significant and permanent effects on the ecological integrity of landscapes than natural disturbances such as drought, insect epidemics, and fire (Vitousek 1994). Landscape change alters the amount of energy, water, and nutrients available to people, plants, and animals; increases the spread of exotic (non-native) species; accelerates natural processes of ecosystem change; and diminishes system services (Vitousek et al. 1997). System services include air and water purification, provision of water supplies, mitigation of floods and drought, detoxification and decomposition of wastes, generation and renewal of soil, maintenance of biodiversity, partial stabilization of climate, nutrient cycling and services, and pollination, among others (Costanza et al. 1997). On the positive side, growth and development increases jobs, income, and sociocultural amenities.

Economic growth and development in the CCE boosts income, employment, and population, and accelerates the conversion of agricultural land and/or forestland to residential and commercial developments. Most old-growth forests that once existed outside national park and wilderness areas have been harvested, rivers have been altered by hydroelectric power development, lakes and streams have been polluted by agricultural and urban runoff, fish and wildlife habitat has been eliminated or degraded, large areas have been invaded by non-native species, and air pollution has increased. Development on private lands surrounding Glacier National Park, which is part of the CCE, continues to pose threats to the park's natural resources (Keiter 1985; Prato 2004). Overall, the ecological integrity of natural landscapes in the CCE has diminished as a result of growth and development.

Several studies have demonstrated that environmental degradation contributes to lower personal incomes, impaired human health, and depressed economic conditions (see, for example, Meyer 1993; Templet and Farber 1994). Many stakeholders (resource planners and managers, scientists, citizens, special interest groups, and policy actors) in the CCE are seriously concerned about how to sustain the environmental amenities and quality of life supplied by the region's natural landscapes under continued pressure from rapid growth and development. Here we outline an integrated framework that CCE stakeholders can use to assess and compare the economic, landscape, and wildlife-habitat impacts of alternative future growth and development scenarios.

An Integrated Ecosystem Model

This type of model specifies the relationships among key ecosystem processes and variables, estimates those relationships using economic and ecological data, and simulates economic and ecological responses to growth and development. Developing an integrated ecosystem model to assess and compare economic and ecological impacts of scenarios for future growth and development requires that concepts and methods from the social and natural sciences be synthesized. Social science concepts and methods useful for this purpose include (1) remote-sensing data and land-cover classification methods for creating land-cover maps; (2) geographic information systems (GIS) for developing baseline land-use maps; (3) economic impact assessment models for simulating the impacts of growth on regional output and employment; (4) residential and commercial development models for determining the demand for land for these uses; (5) multiple-attribute evaluations for ranking the suitability of developable parcels for residential and commercial uses; and (6) efficiency analyses for eliminating inefficient growth and development futures. Natural science concepts and methods useful in developing integrated ecosystem models include landscape ecology for measuring changes in landscape patterns and habitat suitability criteria for evaluating the effects of changes in landscape patterns on wildlife habitat.

Scale Issues

Integrated ecosystem models are developed for a particular ecosystem, referred to as the study area. In selecting a study area, we must consider three factors:

Economic Growth and Landscape Change 57

1. Landscape patterns/processes and wildlife behavior are often scale-dependent, meaning that they vary with the size of the study area (Turner et al. 2001).
2. Larger study areas require more data and present greater modeling challenges than smaller study areas.
3. The county is the smallest geographic unit in the United States for which economic impact assessment data and models are available.

An example of the first factor is the extensive home ranges of large carnivores. If the study area is too small, it is unlikely to contain the range of habitat types that large carnivores require. As a result, its value in comprehensive evaluations of the impacts of landscape change on large carnivores will be limited. In terms of the second factor, the fact that the CCE is large (44,000 km^2) and spans an international border makes obtaining consistent data problematic. Data consistency is also an issue in the CCE because several Canadian data sets are proprietary. In addition, capturing the range of ecological and economic complexity of large ecosystems is difficult. Simulating economic, landscape, and wildlife impacts for the entire CCE is not feasible. Finally, the third factor implies that economic impacts of growth can be assessed only for study areas that are approximately conterminous with counties, at least in the U.S. portion of the CCE.

Flathead County Model

In this section, we describe an integrated ecosystem model for assessing the economic, landscape, and wildlife impacts of growth and development in Montana's Flathead County from 2002 to 2014 and from 2002 to 2024 (CARES 2006a). Flathead County is located in the northern U.S. portion of the CCE. The integrated ecosystem model for Flathead County is called the Flathead Landscape Ecosystem Model (FLEM; see Figure 4-1). Each box in the figure represents an element, designated by a lowercase letter, of the model. The arrows indicate how the elements influence each other. FLEM is suitable for modeling wildlife and economic impacts of landscape change in other areas of the CCE, as well as in other ecosystems. Space limitations do not allow us to describe the functional relationships being used in the model.

Land Cover

FLEM simulates future changes in land cover based on historical changes in land cover. To create land-cover maps that identify the location and amounts of land in various cover classes (i.e., agricultural, coniferous forest, deciduous forest, built-up area, water, and snow/rock/ice) land-cover classification methods were applied to multiseason, multiperiod (1982 and 2002) Landsat™

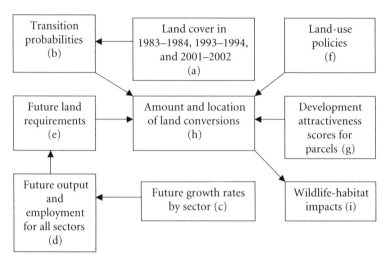

FIGURE 4-1. Schematic of Flathead Landscape Ecosystem Model

satellite images for Flathead County (Figure 4-1a; CARES 2006b). The images have a spatial resolution of 30 m, which means that landscape features as small as 30 m can be detected.

Transition Probabilities

FLEM models the conversion of agricultural and forested lands to built-up land by combining transition probabilities (Figure 4-1b; see also Dale et al. 2004) with economic growth scenarios. A transition probability is a number between 0 and 1 that indicates the likelihood that land converts from one cover class to another in a particular time interval. Transition probabilities are estimated based on acreages in various land-cover classes in two years. Suppose the transition probabilities for land-cover changes between two historical years that are 10 years apart are 0.75 for the conversion of agricultural land to built-up land and 0.25 for the conversion of forested land to built-up land. If the economic growth scenarios indicate that an additional 405 ha of built-up land are needed in the next 10 years, 304 ha of agricultural land (0.75 x 405) and 101 ha of forested land (0.25 x 405) must be converted to built-up land. This step in the procedure determines only the total acreage converted from agricultural and forested lands to built-up land, not the locations of the conversions on the landscape.

Alternative Futures Analysis

Uncertainty about the rate of future economic growth and land-use policies is a major challenge in using integrated ecosystem models to simulate the landscape and wildlife impacts of future economic growth and development. FLEM handles this type of uncertainty using alternative futures analysis (ALFA), also known as *scenario planning* (Figure 4-1c, 4-1f; see also Peterson et al. 2003). In the Flathead County study, stakeholders first developed nine alternative futures embodying a range of future economic growth rates and land-use policies. Next, the research team used FLEM to simulate economic, landscape, and wildlife impacts of the nine futures. ALFA has been used in other areas, including the Willamette River Basin in Oregon (Hulse et al. 2000; Baker et al. 2004) and the Upper San Pedro River Basin in Arizona and Sonora, Mexico (Steinitz et al. 2003).

The alternative futures selected for Flathead County represent combinations of three sets of future growth rates (low, moderate, and high) for outputs in 11 primary industries, along with three land-use policies (unrestrictive, moderately restrictive, and highly restrictive). Growth rates cover the periods from 2000 to 2014 and from 2014 to 2024. The 11 primary industries are (1) services; (2) retail trade; (3) manufacturing; (4) government; (5) construction; (6) finance, insurance, and real estate; (7) transportation and public utilities; (8) wholesale trade; (9) agricultural, forestry, and fishery services; (10) farming and ranching; and (11) mining. Land-use policies define (1) the mix of new home types; (2) densities for new homes; (3) setbacks of new homes and commercial structures from wetlands and water bodies; (4) new residential and commercial developments relative to wetlands, streams, rivers, lakes, ponds, and shallow aquifers; (5) expansion of infrastructure (e.g., roads, sewer, power, and water) relative to future development; and (6) new conservation easements for agricultural land, forestland, and open space.

Future Output and Employment

FLEM simulates economic output in the county for 2014 and 2024 by applying the three sets of growth rates specified in the alternative futures (Figure 4-1c) to economic output in 11 industries in 2000 (Figure 4-1d). The model determines future growth in industry outputs by subtracting outputs in 2002 from estimated outputs in 2014 and 2024. Employment levels in 2014 and 2024 are estimated by entering estimate future economic outputs for all industries in IMPLAN®, a menu-driven software program developed by the U.S. Department of Agriculture (USDA) Forest Service. IMPLAN permits nonsurvey regional input–output analysis of any county or combination of counties in the United States (Lindall and Olson 1993).

Future Land Requirements

FLEM simulates future increases in land required for new residential housing (Figure 4-1e), based on simulated increases in employment (additional employment increases the demand for housing) derived from IMPLAN for each growth scenario. Users can specify values for the unemployment rate, housing vacancy rate, number of resident workers per housing unit, percentage of new jobs filled by those commuting into the county, percentage of the resident labor force that fills jobs, average yearly migration outflow (population), labor-force-to-population ratio, and percentage of new construction for nonpermanent residents. FLEM simulates future increases in county land required for new commercial development by multiplying the average acreage in commercial establishments per worker in the base period (2002) by the estimated number of workers (i.e., employment) for each alternative future.

Locations of Future Land Conversions

The model simulates locations of future land conversions to residential and commercial developments by combining (1) land-use policies (Figure 4-1f), which influence the amount and location of land available for residential and commercial developments (Figure 4-1h); and (2) the relative attractiveness of developable agricultural and forested parcels for residential and commercial developments (Figure 4-1g). Land-use policies are specified in the alternative futures. The model determines the relative attractiveness of developable parcels for residential or commercial development by ranking those parcels from most to least attractive based on their development-attractiveness scores. A developable parcel has restrictions based on ownership (e.g., national park and national forestland cannot be developed) and on land-use policies.

Development-attractiveness scores are estimated using a multiple-attribute evaluation method that selects attributes for parcels and assigns weights to those attributes (Prato 1999). Developable parcels are converted to residential and commercial developments in the order of their development-attractiveness scores, subject to any restrictions that the land-use policies impose on the location of new developments. Attributes used to calculate the scores for residential-use parcels include (1) location with respect to growth boundaries that establish where sewer service is available; (2) maximum acceptable distances from a major highway, the edge of town, and amenities (i.e., mountains, lake, river, preserve, golf course, ski resort, park, forest, and the elevation difference between the valley floor and the parcel); and (3) minimum acceptable distances from disamenities (i.e., industrial facility or park, mining facility, mobile-home park, edge of town, railroad tracks, and airport). Separate development-attractiveness scores are determined for high-density (2.8 units per ha), urban (2.2 units per ha), suburban (0.8 units per ha), rural (1 unit per 2.5 ha), exurban (1 unit per 3 ha), and agricultural homes (1 unit per 19 ha), and for tract homes versus custom homes.

Development-attractiveness scores of parcels for commercial development are based on two attributes—maximum acceptable distances from major highways and population centers. These scores are calculated by combining attribute values for developable parcels (determined using a GIS) and the relative importance or weights of attributes (using an additive utility function).

Wildlife-Habitat Impacts

Wildlife-habitat impacts of an alternative future are simulated based on the amount and location of future land conversions for that future and on wildlife-habitat suitability criteria for different species (Figure 4-1i). Specifically, FLEM simulates impacts of alternative futures on habitat for avian, aquatic, and terrestrial species using a three-step procedure: (1) land-cover maps are simulated for 2014 and 2024 for the nine alternative futures; (2) landscape patterns are determined for the nine land-cover maps; and (3) potential wildlife-habitat implications of those landscape patterns are assessed. For each alternative future, changes in landscape patterns resulting from land conversions between 2002 and 2014 and between 2002 and 2024 are assessed by applying APACK to the 2002 land-cover map and the simulated land-cover maps for 2014 and 2024 (Mlandenoff and Dezonia 1997). APACK, developed by the University of Wisconsin Department of Forest Ecology and Management, calculates landscape metrics for a particular land cover (Turner et al. 2001). Two landscape metrics are especially useful in evaluating impacts of landscape change on wildlife habitat: (1) the frequency of patch characteristics (e.g., number of patches in a specific size class and diversity of patch types); and (2) the spatial relationships among different objects, such as interpatch distance (Griffiths et al. 1993).

FLEM assesses the probable wildlife impacts of simulated changes in landscape patterns based on established relationships among those patterns and the quantity and quality of wildlife habitat. Examples of such relationships follow.

- Increases in the number of small patches influence the movement and persistence of organisms and the redistribution of matter and nutrients.
- Matrix porosity, which measures patch density, reflects the degree of species isolation and potential genetic variability present within animal and plant populations.
- Increases in patch size appear to be positively correlated with species, habitat diversity, or both.
- Patch area is a good indicator of the number of tree and bird species present.
- The amount of edge and interior spaces influences the abundance and diversity of organisms and wetland functioning.
- A lower edge-to-interior ratio favors rare species (Geoghegan et al. 1997).

Identifying Efficient Alternative Futures

Once the economic and wildlife-habitat impacts of the nine alternative futures have been simulated, we can assess the efficiency of those futures. Figure 4-2

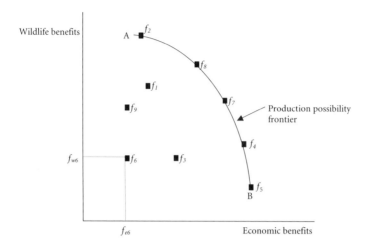

FIGURE 4-2. Hypothetical Wildlife and Economic Benefits for Nine Alternative Futures

illustrates hypothetical trade-offs between wildlife and economic benefits for the nine hypothetical futures. Wildlife benefits are measured using a composite index of wildlife-habitat suitability. The index integrates the landscape patterns for each alternative future determined using APACK and the effects of those patterns on habitat determined using established relationships between landscape patterns and habitat suitability (see Prato [2005] for an example of such an index). Wildlife benefits and wildlife-habitat impacts are opposites— wildlife benefits decrease (increase) when wildlife habitat is degraded (enhanced). Economic benefits are measured by estimated total economic output in the county for each alternative future as determined using the IMPLAN model. Each point in the diagram represents the combination of wildlife benefits and economic benefits for a particular alternative future. For example, alternative future f_6 offers wildlife benefits of f_{w6} and economic benefits of f_{e6}. The nine alternative futures depicted in Figure 4-2 are not intended to correspond to the nine alternative futures discussed earlier.

Alternative futures that yield greater wildlife and economic benefits are preferred. For instance, f_8 is preferred to f_1. Alternative futures f_1, f_3, f_6, and f_9 are inefficient because they result in fewer wildlife and/or economic benefits than alternative futures f_2, f_4, f_5, f_7, and f_8. The five alternative futures on Line AB (f_2, f_4, f_5, f_7, and f_8), then, are preferable to the four alternative futures (f_1, f_3, f_6 and f_9) below Line AB. The line is known as the *production possibility frontier*. If this frontier has the shape illustrated in Figure 4-2, alternative futures involve trade-offs between economic and wildlife benefits. In other words, moving between any two alternative futures on the production possibility frontier causes economic benefits to increase and wildlife benefits to decrease, and vice versa.

Ranking the efficient alternative futures (those on the production possibility frontier) is not possible without information about stakeholder preferences for wildlife and economic benefits. Multiple-attribute evaluation can be used to score and rank efficient alternative futures for Flathead County. In the case of two attributes, this procedure requires that stakeholders assign weights to wildlife and economic benefits, respectively (e.g., 0.6 to economic benefits and 0.4 to wildlife benefits where the weights must sum to one). Because stakeholders are likely to have different preferences for wildlife benefits and for economic benefits, different stakeholders are likely to assign different weights to the attributes. Differences in weights could result in different rankings of efficient alternative futures.

Wildlife benefits, trade-offs between wildlife and economic benefits, the extents of those trade-offs, and the efficiency of alternative futures are all likely to vary across species. For example, the shape of the production possibility frontier in Figure 4-2 is appropriate for generalist species such as grizzly bear, whose habitat is adversely affected by landscape fragmentation. In contrast, white-tailed deer seem to thrive in fragmented landscapes, meaning that there may be few, if any, trade-offs between wildlife and economic benefits for white-tailed deer.

A Decision-Support Tool

FLEM quantifies wildlife and economic impacts of nine specific alternative futures, but it has a limitation—it cannot necessarily be used to assess the impacts of other alternative futures. For example, suppose a stakeholder wants to evaluate an alternative future that differs from the nine alternative futures. The best outcome in this case is that two of the nine alternative futures bracket the new alternative future in terms of economic growth rates and land-use policies. The wildlife and economic benefits of the new alternative future would fall somewhere in between the corresponding benefits for the two alternative futures. On the other hand, if none of the nine alternative futures brackets the new alternative future, the nine alternative futures are of limited value. This limitation can be alleviated by embedding FLEM in a Web-based, interactive, decision-support tool that allows stakeholders to define their own alternative futures, using FLEM to simulate and rank the wildlife and economic benefits of those futures. Such an interactive decision-support tool extends the usefulness and accessibility of FLEM and similar models to stakeholders (Prato et al. 1995; Sugumaran et al. 2000).

Conclusions

Evaluating the potential landscape, economic, and wildlife impacts of future economic growth and development in the CCE is challenging because (1) the large and complex nature of the ecosystem makes integrated ecosystem modeling very demanding; (2) considerable uncertainty exists about future economic growth and land-use policies; and (3) a large amount of compatible data are needed to implement the model. The development and application of integrated ecosystem models for the CCE can be simplified by selecting a study area that is substantially smaller in size than the CCE. The integrated ecosystem model described here (FLEM) is being applied to Flathead County, Montana, in the northern U.S. portion of the CCE (see CARES [2006a, 2006b] for more details).

Stakeholders and analysts can use the FLEM model to

- Identify historical changes in land cover using remote-sensing data
- Estimate transition probabilities for future land conversions among cover classes based on historical land-cover changes
- Simulate future outputs and employment in primary industries
- Determine total acreage required for residential and commercial developments
- Identify developable parcels that are best suited for conversion to residential and commercial developments
- Asses the probable impacts of alternative futures on economic output, employment, landscape patterns, and wildlife habitat
- Eliminate alternative futures that yield undesirable (inefficient) combinations of wildlife and economic benefits
- Identify and rank desirable alternative futures.

Stakeholder accessibility to and use of FLEM can be increased by incorporating the model in a Web-based, interactive decision-support tool that allows stakeholders to create and evaluate alternative futures of their own choosing. This capability, which has the potential to alleviate adverse ecological impacts of land-use change caused by economic growth and development, is helpful in land-use planning.

References

Baker, J.P., D.W. Hulse, S.V. Gregory, D. White, J. Van Sickle, P.A. Berger, D. Dole, and N.H. Schumaker. 2004. Alternative Futures for the Willamette River Basin. *Ecological Applications* 14: 313–25.

CARES (Center for Agricultural, Resource and Environmental Systems). 2006a. *Assessing Ecological Economic Impacts of Landscape Change in Montana's Flathead County.* http://cares.missouri.edu/montana/ (accessed November 6, 2006).

———. 2006b. *Assessing Ecological Economic Impacts of Landscape Change in Montana's Flathead County.* http://cares.missouri.edu/montana/land_cover.html (accessed November 6, 2006).

Costanza, R., R. d'Arge, R. de Groot, S. Farber. M. Grasso, B. Hannon, K. Limburg, et al. 1997. The Value of the World's Ecosystem Services and Natural Capital. *Nature* 387: 253–60.

Dale, V.H., D.T. Fortes, and T.L. Ashwood. 2004. A Landscape-Transition Matrix Approach for Land Management. In *Integrating Landscape Ecology into Natural Resource Management*, edited by J. Liu and W.W. Taylor. Cambridge, UK: Cambridge University Press, 265–93.

Geoghegan, J., L.A. Wainger, and N.E. Bocksteal. 1997. Spatial Landscape Indices in a Hedonic Framework: An Ecological Economics Analysis Using GIS. *Ecological Economics* 23: 251–64.

Griffiths, G.H., J.M. Smith, N. Veitch, and R. Aspinall. 1993. The Ecological Interpretation of Satellite Imagery with Special Reference to Bird Habitats. In *Landscape Ecology and Geographic Information Systems*, edited by R. Haines-Young, D.R. Green, and S. Cousins. Bristol, PA: Taylor and Francis, Inc., 253–72.

Hulse, D., J.E. Eilers, K.E. Freemark, D. White, and C. Hummon. 2000. Planning Alternative Future Landscapes in Oregon: Evaluating Effects on Water Quality and Biodiversity. *Landscape Journal* 19: 1–19.

Keiter, R.B. 1985. On Protecting the National Parks from the External Threats Dilemma. *Land and Water Law Review* 20: 355–420.

Lindall, S., and D. Olson. 1993. *MICRO IMPLAN 1990/1985 Database Documentation*. St. Paul, MN: Minnesota IMPLAN Group.

Meyer, S.M. 1993. *Environmentalism and Economic Prosperity: Testing the Environmental Impact Hypothesis. Project on Environmental Politics and Policy*. Boston, MA: Massachusetts Institute of Technology.

Mlandenoff, D.J., and B. Dezonia. 1997. *APACK 2.0 User's Guide.* Madison, WI: University of Wisconsin Department of Forest Ecology and Management.

Peterson, G.D., G.S. Cumming, and S.R. Carpenter. 2003. Scenario Planning: A Tool for Conservation in an Uncertain World. *Conservation Biology* 17: 358–66.

Prato, T. 1999. Multiple Attribute Decision Analysis for Ecosystem Management. *Ecological Economics* 30: 207–22.

———. 2004. Alleviating Multiple Threats to Protected Areas with Adaptive Ecosystem Management: Case of Waterton-Glacier International Peace Park. *George Wright Forum* 20: 41–52.

———. 2005. Modeling Ecological Impacts of Landscape Change. *Environmental Modelling & Software* 20: 1359–63.

Prato, T., C. Fulcher, and F. Xu. 1995. Decision Support System for Total Watershed Management. In *Animal Waste and the Land-Water Interface*, edited by K. Steele. New York: Lewis Publishers, 333–42.

Steinitz, C. et al. 2003. *Alternative Futures for Changing Landscapes: The Upper San Pedro River Basin in Arizona and Sonora*. Washington, DC: Island Press.

Sugumaran, R., C. Davis, J. Meyer, T. Prato, and C. Fulcher. 2000. Web-Based Decision Support Tool for Floodplain Management Using High Resolution DEM. *Photogrammetric Engineering and Remote Sensing* 66: 1261–65.

Templet, P.H., and S. Farber. 1994. The Complementarity between Environmental and Economic Risk: An Empirical Analysis. *Ecological Economics* 9: 153–65.

Turner, B.L., and W.B. Meyer. 1994. Global Land-Use and Land-Cover Change: An Overview. In *Changes in Land Use and Land Cover: A Global Perspective*, edited by W.B. Meyer and B.L. Turner. Cambridge, UK: Cambridge University Press, 3–10.

Turner, M.G., R. H. Gardner, and R.V. O'Neill (eds.). 2001. *Landscape Ecology in Theory and Practice: Pattern and Process*. New York: Springer-Verlag.

Vitousek, P.M. 1994. Beyond Global Warming: Ecology and Global Change. *Ecology* 75: 1861–76.

Vitousek, P.M., H.A. Mooney, J. Lubchenco, and J.M. Melillo. 1997. Human Domination of the Earth's Ecosystems. *Science* 277: 494–99.

5

Sustaining Wildland Recreation in the CCE
Issues, Challenges, and Opportunities

Stephen F. McCool and John C. Adams

The CCE is a spectacular mountain setting for remote, nature-based recreation, a type of recreation that relies on undeveloped settings, few roads, diverse wildlife and fish populations, unmanaged natural processes, and relatively few other visitors. It is a place where one can find solitude and serenity, remoteness and isolation, and escape from the everyday demands of contemporary society. In this chapter, we reflect on the growing importance and management of wildland recreation and tourism in the CCE, development and patterns of recreational use, and positive and negative consequences of recreation and tourism development.

Recreation and tourism development opportunities of the type found in the CCE are declining significantly in North America, and the CCE remains one of the few regional-scale areas in the United States and southern Canada where these opportunities are both relatively abundant and readily accessible. Each year, hundreds of thousands of nonlocal visitors experience the scale, grandeur, scenery, adventure, and wildlife of the CCE. Sustaining these opportunities in the face of such use requires wrestling with contentious and complex issues and challenges to stewardship. Addressing these issues and meeting these challenges with proper and appropriate management can lead to long-term educational, recreational, and economic development benefits.

Measured in terms of public land use and resort and second-home development on private lands, recreation is a dominant land use in the CCE. Recreation—including developments, transportation and access systems, information and education, and rules and regulations—is managed within a context of increasing population; structural shifts in recreation demand; changing expectations of what values public lands sustain; federal, state, and provincial legislative complexities; and policy turbulence. Outdoor recreation is critical to the quality of life of CCE residents and tourism is essential to the region's economic vitality. One of the greatest challenges facing CCE communities and public land-management agencies today is how to accommodate population and tourism growth while sustaining the high-quality environment on which both outdoor recreation and tourism depend.

Despite the outstanding quality and relative abundance of nature-dependent recreational opportunities, the social and economic significance of recreation to both countries, and the complexity and contentious nature of its management, relatively little critical, scientifically based discussion about recreation and tourism issues, challenges, and opportunities has taken place. Here we address four issues that influence the future of recreation and tourism development in the CCE: (1) the complex contextualizing forces that affect the capability to manage outdoor recreation and tourism development; (2) the regional patterns of recreational use; (3) the positive and negative consequences of recreation and tourism development; and (4) the issues, challenges, and opportunities confronting the area.

Context

As interest in CCE stewardship mounts, five particular recreation-related changes are significant: (1) population change and associated demographic shifts; (2) changes in values and orientations toward the environment and public lands; (3) economic restructuring in western communities; (4) reduced capability to manage public lands, and (5) technological change.

Population on the Rise

Since World War II, the U.S. population has increased by about 50 million people every 20 years. Concurrently, rapid advances in agricultural technology have stimulated rural to urban migration, with occasional periods of brisk reversals. As a result, the public land states of the West have grown extraordinarily fast, particularly in the last 30 years. In Flathead County, Montana, for example, net in-migration has accounted for 75% of the population growth since the early 1990s. Counties that contain relatively large amounts of public land, particularly national parks and wilderness, have seen relatively fast rates of growth as many economically "footloose" households have chosen to live in amenity-rich areas (Rasker et al. 2004). Much of this growth has occurred outside municipal boundaries, often in the wildland–urban interface. Amenity-driven population growth and migration has not been as dramatic in the Canadian portion of the CCE. Most population growth in Canada is occurring north of the CCE in the Bow Valley, including Calgary, Banff, and Canmore.

Population growth has been accompanied by demographic changes. Today's population is younger and more homogenous, with a more equitable distribution of income than that of the older and more ethnically diverse population with more concentrated income and wealth. These changes are leading to major transformations in demand for outdoor recreational activities, primarily from consumptive to nonconsumptive, extractive to appreciative, and high impact to low impact (McCool 1996a).

Changing Values and Orientations

Demographic changes in population have been accompanied by equally telling shifts in social values and orientations toward public land. Generally, the population is not only more sensitive to environmental issues, but also much more active in outdoor recreational pursuits, with significant differences observed between old-timers and newcomers. Increases in population and per capita recreational participation rates have increased the demand for recreational opportunities. As a result, increasing pressure has been placed on public land managers to offer more opportunities for different styles of recreation on a greater area of the land base.

Other changing values, such as a growing appreciation for the symbolic or nonutilitarian values of the environment and shifts in public expectations about land products and services cause the government to act more like a private business and exert more pressure on the stable land base that agencies administer. In-migration, largely from urban areas along the Pacific Coast and in the Midwest, brings urban values and different social constructions of the natural environment. These changes give rise to conflict over the acceptability of natural resource-management practices.

Economic Restructuring

Since the 1980s, the economic bases of rural communities in the North American West, including those in the CCE, have been dramatically restructured. The economic significance of resource commodity development—minerals, grazing, and wood fiber—has declined sharply as a result of policy changes, shifting values, and corporate decisions. Many communities that once depended heavily on logging and sawmills now have no wood-processing facilities. Many once-prosperous communities have turned to tourism to replace the jobs and incomes lost from sawmill closures. Although international tourism is expected to more than double between 2002 and 2020 (World Tourism Organization 2001), it remains a highly, perhaps globally, competitive industry. Successful tourism development requires community, business, and individual skill sets that differ markedly from those required by manufacturing industries. In addition, the fast-growing segments of tourism that are important in the CCE depend on the region's natural and cultural heritage values.

Economic restructuring poses two significant questions for communities and their interactions with public land-management agencies: (1) How do communities capitalize on the outstanding natural wealth—defined primarily in terms of amenities—located on public land without unacceptably altering the resources on which that wealth is based? and (2) What is the capacity of communities to deal with these changes in terms of the financial and social capital needed to develop a sustainable tourism industry?

Stagnant Management Capabilities

Public agencies play a dominant role in this changing environment. In addition to administering many of the amenities important to both residents and tourists, the agencies are also responsible for being good stewards of the resources on which the amenities are based. Yet agencies operate in a stagnant institutional environment because the laws governing management are increasingly viewed as obstacles to good stewardship. In addition, on the American side of the international border, the financial resources available for good stewardship are limited by an increasingly conservative political ideology and a staggering U.S. federal deficit. This environment is challenging the agencies' traditional conservation leadership roles and politicizing resource-management decisions.

Changing Technology

The technological revolution of the last 25 years, particularly in the information-dissemination and transportation arenas, has dramatically affected how people recreate in the CCE. Technologies such as the Internet have made information about recreational opportunities in the CCE, much of which had been stored and available only in adjacent communities, readily available to a vast number of people around the world. Internet searches on the phrases Bob Marshall Wilderness, Glacier National Park (GNP), and Waterton Lakes National Park (WLNP), for example, return literally tens of thousands of hits. The vast number of Web sites that describe these areas, local lodging and resorts, and outfitting opportunities often include debates about their management policies. This information influences expectations about an anticipated recreational engagement, broadens the range of choices considered, and amplifies the global context for local tourism development. As a result, managers face increasing numbers of visitors seeking opportunities that may or may not be appropriate for the conditions and objectives established for a particular area.

In the transportation arena, advances in aircraft and off-road vehicle technologies have increased the demand for recreation in the CCE. First, air transport has made the CCE more accessible from places far from the Rocky Mountains—the region is now just a few hours away from Canadian and American population centers. Second, changes in technology for all terrain vehicles and snowmobiles have redistributed use spatially; places previously accessible only by foot and horse are now penetrated more readily and frequently. Although much of the CCE is in legislative reserves (which we discuss later) that prohibit off-road mechanized travel, much of the area can be accessed by mechanized means.

Recreational and Tourism Resources and Trends

The CCE encompasses a rich resource base for nature-based tourism and outdoor recreation, ranging from school-trust-fund land (in the United States) and Crown land (in Canada; used primarily by locals), to the grandeur and international tourism magnet of the Waterton Glacier International Peace Park. In addition to the 457 thousand ha in the Peace Park and the 682 thousand ha in the Bob Marshall Wilderness Complex, the CCE includes the Mission Mountains Wilderness, the Rattlesnake Wilderness, the Ten Lakes Wilderness Study Area, the Flathead Wild and Scenic River, Flathead Lake, and numerous state and provincial parks on both sides of the border. About one-third of the land area in the CCE is under formal protected status (Table 5-1). Although Crown land owned by the Canadian government, school-trust-fund land, and national forests are generally governed by a multiple-use philosophy, protected areas are managed under some type of federal, state, or provincial mandate. Typically, these mandates restrict the level, location, and type of recreational use and development. Additionally, the region includes or abuts four downhill ski areas, one of which, Big Mountain (Whitefish, Montana), is an international destination resort.

Tourism patterns in the CCE vary spatially. For example, visits to the GNP have stagnated (National Park Service 2004) and visits to the WLNP have declined over the past five years (Parks Canada 2004). Data from provincial parks in Alberta indicate steady visitor patterns (Alberta Parks and Protected Areas Division 2002). Although southern Alberta and southeastern British Columbia are experiencing stable and declining visitation, respectively (Alberta Economic Development 2004; Tourism British Columbia 2004), tourist visitation in Montana rose 15.4% from 1993 to 2003 (Wilton 2004a). Flights at the two busiest regional airports are up dramatically. The annual number of passengers departing the Missoula and Kalispell airports increased 86% and 148%, respectively, between 1989 and 2003.

Although the term *Crown of the Continent* is beginning to gain currency, the extent to which tourists overlook the international boundary and view the area as a single region is unclear. We note that since 9-11, increasing security concerns have made it more difficult to cross the international border in the Peace Park between the United States and Canada. Some data indicate that the two provinces and Montana are visited jointly. For example, of the Montana non-

TABLE 5-1. Areas of CCE in Protected Status, 2005

Area	Total area (hectares)	Protected area (hectares)	Percentage
British Columbia	821,802	46,791	5.7
Alberta	536,620	103,166	19.2
Montana	2,778,195	1,160,000	41.8
Total	4,136,618	1,309,957	31.7

resident visitors who used a rental car, 3% rented in either Alberta or British Columbia (Nickerson et al. 2002). Nearly 1 in 10 visitors to Flathead County (the heart of the CCE in Montana) are from Alberta, second only to California among states and provinces (Wilton 2004b). This indicates that at least some cross-boundary interregional tourism occurs. Many visitors to the WLNP visit the GNP, and vice versa.

Nearly all tourism in the region is nature-based, meaning that it is driven by quality outdoor environments. In this regard, the Peace Park is the flagship recreational destination in the region. Nearly one in five nonresidents who vacation in Montana cite the GNP as their primary reason for visiting (Nickerson et al. 2002). Sixty percent of summer visitors to Flathead County list the GNP as the primary amenity that attracted them to the area, whereas less than 1% indicate that national forests are the primary draw (Dillon and Praytor 2002). The top six recreational activities in which Flathead County visitors engage are wildlife watching (53%), nature photography (44%), day hiking (43%), recreational shopping (34%), camping in developed areas (29%), and picnicking (29%) (Dillon and Praytor 2002).

Despite the relative lack of drawing power, national forests host a great deal of recreation. The Flathead National Forest, which is almost entirely within the CCE, was visited 1.4 million times in 2000. The Lewis and Clark National Forest, the most popular portions of which are within the CCE, welcomed 476,000 visitors in 2001.

Resident Recreation

Obtaining accurate data for outdoor recreational participation and demand is difficult. Nonetheless, anecdotal evidence and personal experience suggest that recreational use of nonpark public land, such as national forests, is increasing significantly. For example, in Montana, the number of fishing days increased from 2.6 million in 1996 to 4.1 million in 2001. Recreational growth in the CCE is likely to be greater than it is in the state as a whole.

Outdoor recreation is extremely important to CCE residents. In many respects, participation in outdoor recreation is part of the regional identity— even those who do not hike or hunt are likely, in our experience, to consider the opportunity to do so an important element of community life. In Montana,

52% of households have a member who watches wildlife as a recreational activity, 37% have a member who day hikes, and 29% have a member who photographs wildlife (Ellard et al. 1999). Nearly 24% of state residents fished and nearly 19% hunted in 2001 (U.S. Fish and Wildlife Service and U.S. Census Bureau 2002). At least 59% of visitors to Flathead National Forest are from the CCE.

Tourism and the Regional Economy

Tourism and outdoor recreation make significant contributions to the economic base of the CCE. The nonresident travel industry is responsible for 7% of Montana's total employment, similar to employment percentages in construction, agriculture, and finance/real estate industries (Wilton 2004a). We believe the percentage is similar for the CCE. Eight percent of all jobs in the Missoula Bureau of Economic Analysis Area, which encompasses a region larger than but including much of the U.S. portion of the CCE, is directly attributable to outdoor recreation (Crone and Haynes 1999). Nonresident visitors to Flathead County spend approximately $146 million per year (Nickerson and Wilton 2002). GNP representatives estimate that expenditures by park visitors generate nearly 1,500 regional jobs. According to Carol Edgar, executive director of the Flathead Convention and Visitor Bureau, "You can't measure the mark Glacier Park has made on this community. The whole economy is tied to the park" (National Parks and Conservation Association 2003).

More than 90% of nonresidents who stay overnight in Flathead County list vacationing as one of the purposes of their visit (Nickerson and Wilton 2002). Three-quarters of nonresident visitors to Montana have been to the state before; more than three-quarters of these visitors travel as a couple or family. Groups who visit Montana primarily for pleasure typically spend $110 per day (McMahon and Cheek 1998). The top six categories of expenditures made by visitors to Flathead County are, in descending order, retail sales, restaurants and bars, gas and oil, groceries and snacks, auto rental and transportation, and lodging (Dillon and Praytor 2002).

Consequences and Management Response

The combination of a growing population, rising income, and intensifying tourism is increasing the demand for nature-based recreational experiences in the CCE. Although specific projections do not exist, the region can expect to continually host greater numbers of visitors with a broader array of expectations and demands. Growth in visitation and population combined with land-tenure and land-use designations have significant consequences, as we discuss in the next section.

Institutional Limitations on Accommodating Recreation and Tourism

The fixed development footprint in the premier national parks (about 5% of the total area for the GNP and the WLNP) limits the ability to accommodate visitation growth within the parks. For example, there has been little if any change in the capability of the two national parks to accommodate more developed camping in the last 30 years. For a variety of social and political reasons, it is unlikely that U.S. federally administered land will have the additional capacity to handle greater numbers of visitors in the near future. Of course, such limits protect the values on which much of the tourism industry depends.

On the U.S. side of the border, growth in population and visitation in Flathead County is the dominant influence. Because about 87% of the land in the county is managed by a federal or state agency, population growth must be accommodated on a relatively small area of private land. This means that facilities for tourism must compete with other land uses, such as transportation and utility corridors, retail and commercial services, and manufacturing. These other land uses can affect the quality of the environment that is needed to support upscale, nature-dependent resorts and tourist activity. The Blackfeet Indian Reservation comprises most of Glacier County, which lies on the east side of the Continental Divide in Montana. Despite the fact that tourism in the GNP began in East Glacier, Glacier County is largely undeveloped for tourism with the notable exceptions of East Glacier, St. Mary, and Babb.

On the Canadian side of the CCE, much of the development is east of the Continental Divide and concentrated at the Waterton townsite. The west side of the Continental Divide, especially the Flathead River drainage, is wild and remote. This area, however, contains mineral resources, which if extracted could cause conflicts with recreation and tourism development. Chapter 18 discusses these conflicts in more detail.

In addition to the limited areas of private land for development, much of the public land in the CCE is under a variety of congressional, parliamentary, and administrative mandates, which severely restrict agency discretion in developing new tourism and recreational opportunities. For example, the U.S. portion of the CCE contains approximately 1.2 million ha of existing or proposed congressionally designated wilderness areas. Roads, mechanized vehicles, commercial timber harvesting, and other developments are prohibited in wilderness areas; nonmechanized forms of backcountry recreation are allowed. These constraints reflect the finite amount of land available for development of recreation and tourism opportunities in the CCE. In addition, they focus development of facilities and transportation systems into a relatively small area, thus requiring greater care in design and management.

Negative Consequences of Recreation and Tourism

Recreation, even in its backcountry forms, is not an environmentally benign land use. Environmental impacts from recreation occur at two scales. First, in backcountry areas, impacts occur at a relatively small scale at campsites, along trails, and at trailheads. Research indicates that low levels of backcountry use can have relatively large impacts, but that increases in impacts become disproportionably smaller as use increases (Leung and Marion 2000). Variables such as travel method (e.g., horse, hike, snowmobile, or ski); usage seasons; soils; and behavior (e.g., practicing low-impact behaviors), among others, dramatically affect the biophysical impacts of recreational activities. At this scale, the level of disturbance to wildlife is a particularly important question for which few real answers are clear.

Second, impacts are greater in more-developed front-country areas that are home to resorts and other services. Onsite and offsite impacts result from transportation modes and corridors, power transmission, lodges, ski areas, and waste disposal. As indicated by the Banff–Bow Valley study (Ritchie 1998), such impacts may be difficult to address. Many variables are affected at larger scales, including secondary and tertiary effects on water and air quality, energy consumption, wildlife populations and habitats, recreational opportunities, and quality of life. The challenges of managing the impacts of recreation and tourism development, although they have important technical dimensions, are largely issues of determining the social acceptability of these impacts, monitoring to ensure that they do not exceed acceptable limits, and taking appropriate management action to achieve and sustain acceptable resource conditions (Cole and Stankey 1998).

Increasing Conflicts among Recreational Activities

If anything typifies recreation in the CCE, it is the presence of conflict. Conflict, defined as goal interference, generally arises from differing behaviors, values, and orientations toward recreation. Examples include clashes between snowmobilers and cross-country skiers, hikers and equestrians, and even bank anglers and float anglers. Conflict is the normal outgrowth of a healthy, vibrant, and pluralist society.

Unresolved conflicts, however, tend to escalate in both volume and hyperbole, which leads to polarization of the conflicting groups and often embroils the management agency in the conflict. Advances in public-engagement processes, such as dispute-resolution mechanisms, collaborative approaches, and citizen juries can result in more successful planning efforts (Crowfoot and Wondolleck 1990).

Upcoming Challenges and Opportunities

As demands for recreation, tourism, and other resource uses escalate in the CCE, contentious and often intractable disputes are bound to continue. These conflicts more often reflect divergent values than resource-management methods. In the sections that follow, we discuss four major issues that arise from growing and diversifying demands for recreation and tourism.

Frameworks for Identifying Trade-Offs

Managing recreation and tourism development often involves making trade-offs among competing objectives. In backcountry settings, the trade-off is generally between protecting a pristine environment and allowing recreational access that has environmental impacts (Cole and Stankey 1998). The legislative mandate for the various federal reserves in the CCE is to protect natural resources and environmental quality. Protection is aimed at sustaining ecological integrity in Canada and natural processes in the United States. Allowing recreational access and use compromises this goal. Because relatively low levels of use have disproportionately large impacts, managers are confronted with the question "How much impact is acceptable?" At its core, impact acceptability is a value issue—what is acceptable to one group may not be acceptable to another. This in itself raises other questions, such as what values have important influences on visitor management and tourism development.

Addressing the question of impact acceptability requires a planning framework that identifies objectives, explicates values, identifies trade-offs, and suggests appropriate courses of action. A useful framework is transparent, incorporates public-engagement processes, builds ownership in decisions, requires specific and explicit objectives, and includes monitoring. Fortunately, several frameworks exist that meet these criteria.

Integrated Management

Not only are the demands for land increasing in the CCE, but they are diversifying. Tourism and recreation are significant components of this suite of demands. Land management in the CCE requires knowledge of the processes that link people to the environment. Such processes are complex and dynamic, and resource-management decisions are fraught with uncertainty. The public expects that decisionmakers will consider a wide range of demands, benefits, values, and alternatives when making plans about particular places and programs. Satisfying these demands can be accomplished only within an institutional and administrative setting that encourages integrative thinking and decisionmaking. As Bardwell (1991) and others argue, environmental problems are particularly messy, wicked, and difficult to frame and resolve.

Outdoor recreation and tourism development are no exception; they affect other values, and management for other values often affects the ability of landscapes to furnish certain recreational opportunities. Providing recreational opportunities often involves value judgments (e.g., what opportunities should be available, where, and for whom). Such problems generally cannot be resolved and successfully addressed through disciplinary thinking, but must be attended to by a variety of scientific disciplines, varying forms of knowledge, and spatial and temporal scales. Consequently, planning and decisionmaking processes must be holistic and intrinsically integrative in their design. Unfortunately, much recreation and tourism development employs single-use, reductionist approaches to natural resource management (e.g., timber, wildlife, range, and fisheries).

Thinking and Acting Regionally

Visitors to the CCE bring along a variety of motivations and expectations that include experiencing a nature-dominated environment. They also expect to learn about nature and native culture, and to appreciate scenery, challenge, adventure, solitude, and escape (Miller et al. 1996). Of course, no one site can meet all recreational expectations, and resource conditions and recreational opportunities vary across recreational sites.

In contrast, protected-area managers in the CCE are responsible for good stewardship of the places under their individual jurisdiction, but not of the ecosystem as a whole. Although the site-oriented focus of decisionmaking ensures that recreation-induced impacts are minimized at the site scale, it has potential negative consequences at the regional scale (McCool and Cole 2001). These include homogenization of recreational opportunities, suboptimization of the supply of recreational settings in relation to demand, and displacement of problems from one area to another. These consequences are frequently subtle and incremental, and may often be disguised in larger debates about managing public lands for other values.

These effects are exacerbated by differences in institutional mandates and goals across jurisdictions. For example, management of wildlife in Montana is conducted by the state's Department of Fish, Wildlife & Parks. This agency, which has an interest in maximum sustainable harvest, licenses and regulates the hunting, fishing, and trapping that take place primarily on federal land in the CCE, but exercises no control over land-management decisions in those areas. Threatened and endangered species in terrestrial ecosystems are managed by the U.S. Fish and Wildlife Service (FWS), which functionally exercises veto power—but not actual control—over land-management decisions. Public land-management decisions are driven by a multiple-use philosophy for lands administered by the U.S. Department of Agriculture (USDA) Forest Service, the dual mandate of providing for public enjoyment and preserving natural resources for lands administered by the National Park Service (NPS) and strict preservation for lands managed under the Wilderness Act of 1964. Achieving sustainability for game, nongame, and threatened/endangered species is difficult because jurisdictional goals, mandates, and responsibilities differ. The goal of sustainability requires great cooperation and, ultimately, greater consonance of missions.

Avoiding negative management consequences will require a regional-scale examination of recreational settings, challenges, and responses. This conclusion is not new; many others have called for regional-scale assessment, management, and monitoring of biodiversity, wildlife, fisheries, and water and air quality (see, for example, Lee 1995). To be successful, a regional approach must examine human–natural system interactions, something yet to be done in the CCE. Chapter 17 discusses this approach at greater length.

Commercializing and Privatizing Public Lands

An increasingly conservative national political climate is exerting greater pressure to commercialize and privatize publicly administered recreational opportunities. Commercialization involves public agencies adopting business models as paradigms for recreation management; privatization involves leasing, contracting, or selling traditionally public recreational opportunities to the private sector. Such policies may lead to significant inequities in public access to recreational opportunities, loss of public support for land-management agencies, and a narrowing of the range of opportunities available.

Sustainable Tourism as a New Paradigm

Tourism contributes significantly to local economies in the CCE, particularly those west of the Continental Divide. As the tourism industry continues to expand, questions arise about whether tourism is or should be sustainable, and what it means to be sustainable (McCool 1996b, 2002). The issue of sustainable development was raised in the 1987 final report of the World Commission on Environment and Development (*Our Common Future*; see reference list). The report stimulated global dialogue about integrating economic development, quality of life, and environmental protection objectives, a dialogue that has dramatically changed the discussion about the role of tourism in achieving these three objectives.

Sustainable tourism is a form of development that is generally small in scale, has few negative environmental consequences, and benefits the local resident population. It can be conceived as a "kinder, gentler" form of tourism development. The notion of sustainable tourism has made few inroads in the United States, but is much further along in Canada. For example, the Tourism Industry Association of Canada recently adopted a code of ethics and guidelines for sustainable tourism, which aims to reduce visitor and tourism-industry impacts—both social and environmental—and enhance economic benefits to local residents. Sustainability is grounded in the notion of intergenerational equity, which implies that future generations have the same right to enjoy the benefits of protected areas as the current generation. Achieving intergenerational equity requires diligent stewardship of protected areas and public lands.

The question of what values tourism should sustain is fundamental to integrating tourism into the larger regional economy rather than treating it as a distinctly separate entity. A study of Montana tourism leaders done in the late 1990s suggested that tourism should sustain Montana's natural and cultural heritage, opportunity for economic enhancement, and quality of life (McCool et al. 2001).

Conclusions

Overlapping and partial jurisdictions, along with the lack of definitive, reliable data, hamper our ability to make simple declarative conclusions about tourism and outdoor recreation trends, challenges, consequences, and opportunities in the CCE. Although investigators have conducted considerable site-based research and regional studies of outdoor recreation and tourism, the resulting information has not been synthesized. Such a synthesis might be achieved through a collaborative effort that involves not only scientists and managers, but the public as well. A venue that focuses on how outdoor recreation opportunities could be sustained within environmental constraints and the stagnant organizational capability could improve decisionmakers' understanding of future challenges and opportunities.

It seems clear, however, that outdoor recreation is an important part of local culture and quality of life in the CCE. Additionally, nature-based tourism is one of the largest industries in the region, directly and indirectly employing approximately 8% of the working population. We cannot overemphasize the importance of maintaining high-quality natural environments for outdoor recreation to the region's quality of life and economy. The comparative advantage of the region is not commerce, cultural opportunities, or historical sites; it is plainly and simply a scenically spectacular and relatively pristine environment.

The public agencies responsible for managing natural landscapes in the CCE face many challenges. Yet their ability to develop tourism infrastructure is limited by, for example, (1) stagnant or declining resources for dealing with increasing recreational demands; (2) a dearth of scientific information on the social and environmental impacts of recreation; and (3) unsupportive legislative and administrative mandates for making the political, value-laden decisions that increasingly dominate their work. Although many other individuals and institutions affect the extent to which the CCE is able to develop sustainable tourism and outdoor recreation, regional success will be measured by how land-management agencies overcome these obstacles.

The range of recreation and tourism issues in the CCE is wide. The ecosystem has a substantial, publicly administered wildland and natural resource base that is managed for relatively pristine, nature-based forms of recreation. The myriad issues discussed in this chapter pose significant challenges at the regional scale. The concept of sustainable tourism may be useful in integrating tourism development and environmental protection while ensuring employment and income opportunities for the region's residents. A focused plan of research and technical assistance that involves wildland managers, biophysical and social scientists, and the public could increase the leadership capacity of land-management agencies in terms of recreation and tourism opportunities on public wildlands.

Regional success in establishing sustainable tourism and outdoor recreation also hinges on implementing frameworks for making the trade-offs inherent in recreational management and for integrating interdisciplinary and multiscale planning and decisionmaking. One of the major challenges for CCE managers is thinking and acting at the regional scale to avoid irreversible losses in natural resource values. In addition, political pressure to increase commercialization and privatization of public wildland resources continues. Whether such proposals are in the public interest is an issue in need of further deliberation. Finally, identifying appropriate sustainable tourism goals is fundamental to garnering public consensus and support for further tourism development. Successfully resolving these issues is critical to the people, the economy, and the environment of the CCE.

References

Alberta Economic Development. 2004. *Alberta South TDR: Visitor Statistics 1998–2002.* Edmonton, AB: Alberta Economic Development, Government of Alberta.

Alberta Parks and Protected Areas Division. 2002. *Visitation Statistics: Provincial Parks and Recreation Areas.* Edmonton, AB: Alberta Community Development, Parks and Protected Areas Division, Policy and Program Coordination Branch.

Bardwell, L. 1991. Problem Framing: A Perspective on Environmental Problem-Solving. *Environmental Management* 15: 603–12.

Cole, D.N., and G.H. Stankey. 1998. Historical Development of Limits of Acceptable Change: Conceptual Clarifications and Possible Extensions. In *Limits of Acceptable Change and Related Planning Processes: Progress and Future Directions*, edited by S.F. McCool and D.N. Cole. Missoula, MT: U.S. Department of Agriculture (USDA) Forest Service, 5–9.

Crone, L.K., and R.W. Haynes. 1999. *Revised Estimates for Direct-Effect Recreational*

Jobs in the Interior Columbia River Basin. Portland, OR: U.S. Forest Service Pacific Northwest Research Station.

Crowfoot, J.E., and J.M. Wondolleck. 1990. *Environmental Disputes: Community Involvement in Conflict Resolution.* Washington, DC: Island Press.

Dillon, T., and H. Praytor. 2002. *Exploring Tourism Development Potential: Resident Attitudes in Kalispell, MT.* Missoula, MT: The University of Montana-Missoula Institute for Tourism and Recreation Research (ITRR).

Ellard, J.A., N. Nickerson, and K. McMahon. 1999. *Recreation Participation Patterns by Montana Residents.* Missoula, MT: The University of Montana-Missoula ITRR.

Jacob, F.R., and R. Schreyer. 1980. Conflict in Outdoor Recreation: A Theoretical Perspective. *Journal of Leisure Science* 12: 368–80.

Lee, K.N. 1995. Deliberately Seeking Sustainability in the Columbia River Basin. In *Barriers and Bridges to the Renewal of Ecosystems and Institutions*, edited by L.H. Gunderson, C.S. Holling, and S.S. Light. New York: Columbia University Press, 214–38.

Leung, Y., and J.L. Marion. 2000. Recreation Impacts and Management in Wilderness: A State-of-Knowledge Review. In *Wilderness Science in a Time of Change Conference: Wilderness Ecosystems, Threats and Management*, edited by D.N. Cole, S.F. McCool, W.T. Borrie, and J. O'Loughlin. RMRS-P-15. Ogden, UT: USDA Forest Service Rocky Mountain Research Station, 23–48.

McCool, S.F. 1996a. *Toward Understanding the Social-Economic Context of Glacier National Park.* Missoula, MT: The University of Montana College of Forestry and Conservation.

———. 1996b. *Searching for Sustainability: A Difficult Course; an Uncertain Outcome.* Unpublished technical completion report. Missoula, MT: The University of Montana College of Forestry and Conservation.

———. 2002. Still Seeking Sustainability. In *Proceedings of the 1999 International Symposium on Coastal and Marine Tourism*, edited by M.L. Miller, J. Auyong, and N.P. Hadley. Seattle, WA: University of Washington School of Marine Science, 327–32.

McCool, S.F., and D.N. Cole. 2001. Thinking and Acting Regionally: Toward Better Decisions about Appropriate Conditions, Standards and Restrictions on Recreation Use. *George Wright Forum* 18: 85–98.

McCool, S.F., R.N. Moisey, and N.P. Nickerson. 2001. What Should Tourism Sustain? The Disconnect with Industry Perceptions of Useful Indicators. *Journal of Travel Research* 40: 124–31.

McMahon, K., and K.A. Cheek. 1998. *Expenditure Profiles and Marketing Responsiveness of Nonresident Visitor Groups to Montana.* Missoula, MT: The University of Montana-Missoula ITRR.

Miller, T.A., S.F. McCool, and W.A. Freimund. 1996. *1996 Glacier National Park Visitor Study, Final Report.* Missoula, MT: The University of Montana.

National Park Service. 2004. *Glacier National Park Visitation Data.* Washington, DC: U.S. Department of Interior.

National Parks and Conservation Association. 2003. *Gateway to Glacier: The Emerging Economy of Flathead County.* Washington, DC: National Parks and Conservation Association.

Nickerson, N., B. Sutton, and A. Aronofsky. 2002. *Nonresident All Year and Four Season*

Comparison: Visitor Profile. Missoula, MT: The University of Montana-Missoula ITRR.

Nickerson, N., and J. Wilton. 2002. *Niche News: Flathead County Visitor Characteristics.* Missoula, MT: The University of Montana-Missoula ITRR.

Parks Canada. 2004. *Parks Canada Attendance 1999–00 to 2003–04.* Ottawa, ON: Parks Canada.

Rasker, R., B. Alexander, J. van den Noort, and R. Carter. 2004. *Prosperity in the 21st Century West: The Role of Protected Public Lands.* Bozeman, MT: Sonoran Institute.

Ritchie, J.R.B. 1998. Managing the Human Presence in Ecologically Sensitive Tourism Destinations: Insights from the Banff–Bow Valley Study. *Journal of Sustainable Tourism* 6: 293–313.

Tourism British Columbia. 2004. *2003 in Review.* Victoria, BC: Tourism British Columbia, Government of British Columbia.

U.S. Fish and Wildlife Service and U.S. Census Bureau. 2002. *2001 National Survey of Fishing, Hunting, and Wildlife-Associated Recreation: Montana.* Washington, DC: U.S. Fish and Wildlife Service and U.S. Census Bureau.

Wilton, J. 2004a. *1993–2003 Montana Nonresident Visitation Trends.* Missoula, MT: The University of Montana ITRR.

———. 2004b. *The Economic Review of the Travel Industry in Montana.* Missoula, MT: The University of Montana ITRR.

World Commission on Environment and Development. 1987. *Our Common Future.* Oxford, UK: Oxford University Press.

World Tourism Organization. 2001. *Tourism 2020 Vision—Africa.* Madrid, Spain: World Tourism Organization.

PART III
Biophysical Dimensions

6

Alpine Ecosystem Dynamics and Change
A View from the Heights

George P. Malanson, David R. Butler, and Dan Fagre

The alpine, from the treeline ecotone up to the highest peaks, is the pinnacle of the CCE. This environment is what defines high mountains, and although it is a valuable biological resource, it is also an extraordinary aesthetic resource. In thinking of the alpine, we jump back and forth between the grand views and all they mean in mountain regions, and the subtle, tiny views of life tucked into niches. The two are closely connected. Here we examine how the great extent of high peaks along the Continental Divide conditions and controls where life can emerge and, in doing so, creates the sparse microworlds of the alpine in the CCE.

Alpine Climate

The first thing that comes to mind when we think about alpine climate is the cold. We know that temperatures drop as we go to higher elevations. But if we spend time in the alpine, we quickly realize that we should think of wind first. In general, mountain climates are poorly understood because of the small number of long-term weather-recording stations at high elevations. These stations are few because they are isolated from human habitation (especially airports, where most weather stations are located) and because they are difficult to maintain in mountain weather. We know the broad outlines of climate in the CCE alpine, however—in summer the solar radiation is intense, the winds are high, and the temperatures are fairly warm. Summer storms bring a moderate proportion of the precipitation to the region. In winter, solar radiation and temperatures are low and precipitation as snow has the potential to exceed the input of water from summer storms, but winds are very high and snow is swept about (Finklin 1986). Selkowitz et al. (2002) established the connection between snow and the broader continental setting of the CCE, finding that Pacific Ocean temperatures had a definite influence (a phenomenon called the Pacific Decadal Oscillation [PDO]).

These climate conditions are set within the context of the western cordillera of North America. At a continental scale, the CCE landscape is influenced by its position in the approximate middle of the overall north–south chain of the Rocky Mountains. At the regional scale, the CCE is positioned toward the drier end of a west–to–east moisture gradient, a function of its inland position relative to moisture-bearing air masses that travel inland from the Pacific Ocean. Because the CCE is positioned along the Rocky Mountains, the contrast between the west and east sides of the ecosystem is relatively sharp. The mountains themselves play an important role in separating climates dominated by Pacific maritime air masses in the west from those controlled by continental air masses in the east. This contrast magnifies the normal sharp gradients found in mountains (Fagre et al. 2003).

Alpine Geomorphology and Soils

In the CCE alpine, geomorphic processes exert an all-encompassing influence on soil development. Geomorphology and soils combine to create the landscape on which plants establish and survive in the harsh environment at and above treeline. In some cases, geomorphic processes and soils development assist in creating fine-scale environments amenable to the establishment and survival of tree seedlings. In other cases, these processes completely preclude establishment success.

The geomorphic processes of the alpine environment do not carve landforms separate from atmospheric and tectonic forces, but should instead be considered part of a continuum of processes and spatial and temporal scales (Malanson and Butler 2002). Temporal scales are significant because disequilibrium exists in the CCE alpine (and elsewhere). The impact of the Laramide Orogeny, responsible for the Lewis Overthrust and its attendant mountain-building processes in the CCE, cannot be overstated. In addition, the CCE landscape still bears sharp testament to the widespread, intense effects of Pleistocene glaciation, and is still adjusting to the changing landforms, slopes, and climatic conditions that have characterized the post-Pleistocene period. The Little Ice Age, a relatively cold period that ended around 1850, impressed its own set of geomorphic and climatic characteristics on the CCE alpine environment, and adjustments to that period of glaciation and cold climate continue as well (Butler and Malanson 1990).

Spatial scales are equally significant in considering the CCE alpine. At the mountain-range scale, the moisture and consequent geomorphic gradient across the CCE creates heavily glaciated peaks in the west, such as the sharp glaciated horns of Kinnerly and Kintla Peaks in Glacier National Park (GNP), as well as the less-glaciated rounded uplands in areas such as Scenic Point in the GNP's Two Medicine area. At the scale of individual peaks and valleys, intense frost processes are especially significant on the more gently sloping uplands of the GNP's eastern peaks (Butler et al. 2004).

Past geomorphic processes dramatically influence current processes. Across widespread areas of the eastern uplands of the GNP, relict solifluction topography (Butler et al. 2004) results in sufficiently gentle slopes for soil development. Solifluction occurs in areas where frozen ground beneath the surface (i.e., permafrost) precludes deep downward percolation of moisture, such that an "active layer" above the permafrost becomes saturated and begins slowly oozing downslope. This helps to create terraces that are roughly parallel and laterally extensive. Nonvegetated treads (the flat surface) and vegetated risers (the slope) produce a characteristic turf-banked "staircase." Stony tread surfaces are only a thin (<10 cm) veneer overlying finer-grained sediments similar to those found beneath riser vegetation. The turf-banked terraces in the CCE are relicts from a former period of more intense and colder climate. Although they are undated, it is likely that they were actively produced by the intense cold associated with Pleistocene glaciation. The extent of these slope processes in the CCE and their potential significance at a regional scale have not been examined, although Schmid (2004) does describe tundra and krummholz (the prone form of trees; from the German, meaning *twisted wood*) soils in these eastern uplands. Based on an analysis of profiles (~50 cm, quite deep for this type of soil), Schmid concluded that soils are well developed. In addition, soils beneath krummholz did not differ at depth, indicating that the krummholz advance was recent.

Alpine Plant Communities

The vegetation of the CCE alpine extends from the treeline ecotone up through the sparsest tundra. The ecotone is the zone of transition between the subalpine forest and the tundra, where the forms and species are intermixed. It is usually classified by form, as in forest with alpine meadows; intermixed fingers of tundra, trees, and dwarf trees; and patches of krummholz. Other characterizations that apply to the tundra are dense versus sparse, setting (wet and dry, correlated with dense and sparse), and species composition.

Alpine Wildlife

Although the diversity of wildlife is relatively limited in the alpine, some key interactions underpin its processes. One of the charismatic megafauna of the

CCE, grizzly bear depends on one of the least charismatic creatures—the army cutworm moth. White et al. (1998) documented the importance of these moths in the diet of grizzlies. The moths congregate on rocky slopes in the alpine of the CCE in mid-summer, drawing grizzlies to the mountaintops (White et al. 1998). There, the grizzly also affects alpine geomorphology (Butler 1992). Through extensive digging for roots and bulbs, the bear overturns considerable amounts of alpine tundra vegetation, causing disturbances that increase the diversity of alpine plant species. The disturbances also have profound effects on nutrient availability and soils development in areas such as the wet meadows surrounding Logan Pass, where such activities are spatially concentrated (Tardiff and Stanford 1998). Similar effects of other earthmovers in the alpine, including the burrowing and tunneling activities of Columbian ground squirrels, northern pocket gophers, and hoary marmots are probably also important, but these effects have not been studied in detail in the CCE. Gardner et al. (1983) did, however, examine the effects of ground-squirrel burrowing in terms of the amount of sediment excavated and eroded on slopes in the Mount Rae area of Kananaskis Country.

Other charismatic megafauna, such as mountain goats, have been widely studied in the CCE, and also illustrate linkages between surface processes, landforms, and vegetation. The trampling and wallowing effects of mountain goats have been known for some time and are well documented. Chadwick's work (1983) in the CCE is especially noteworthy. Trampling by goats is of sufficient environmental impact to produce well-marked animal trails on the landscape.

The Tundra

The tundra is the treeless area of the alpine above the treeline ecotone. In terms of flora, the tundra is more diverse than its harsh environment might portend. Bamberg and Major (1968) identified 185 species in the area of Siyeh Pass, including the broad expanse of East Flattop Mountain, in the GNP, Montana. Of these 185 species, they found 62 in their sixteen 50 x 20 cm sampling plots. Choate and Habeck (1967) reported 136 species at Logan Pass, GNP, identifying 87 species in sampling plots. Many of the CCE's alpine tundra species are near the southern limits of fairly extensive Arctic–alpine ranges.

Whereas Bamberg and Major (1968) differentiated eight plant communities at Siyeh Pass and Choate and Habeck (1967) found six at Logan Pass, Lesica (2002) categorizes all of the communities into four topographic types of alpine tundra (Figure 6-1). One of these, talus, supports little vegetation because of its instability, but some species are endemic to the talus within the CCE (e.g., sky pilot [*Polymonium viscosum*]). What Lesica (2002) describes as fell-fields include the broad, solifluction-dominated uplands east of the Continental Divide that are the most extensive type of tundra in the CCE. Because of the

Alpine Ecosystem Dynamics and Change 89

FIGURE 6-1. Four Topographic Types of Alpine Tundra

Notes: (a) scree; (b) fell-fields and solifluction terraces; (c) deeper turf; (d) snowbanks.
Source: Lesica (2002).

solifluction processes at work, the plants occupy the areas where the stones are not at the surface. The result is stripes of vegetation and rock, most of which are perpendicular to the slope and present a stairstep appearance, in which the stony treads alternate with the vegetated risers. The risers are often dominated by alpine dryad (*Dryas octopetala*) and/or kinnikinick (*Arctostaphylos uva-ursi*), but willow (*Salix arctica, S. reticula*) can form somewhat higher risers. Lesica (2002) notes that this vegetation grades into what he calls turf, but at the more extreme portion of the gradient, wet meadow is found. Along this gradient, grasses dominate the relatively dry sites; sedges dominate the wetter sites. The most scenic, aesthetically valued parts of alpine tundra are the wetter sites, such as the extensive meadows at Logan Pass. Finally, Lesica notes that some types of alpine vegetation are associated with permanent or persistent snow.

Damm (2001) has done the most extensive classification and description of alpine vegetation in the CCE. He divided 700 detailed samples into 46 associations (a vegetation classification unit based on species composition), some with further divisions, using the European phytosociological system. Although most of Damm's work was devoted to the details of classifications per se, he also describes geology, soils, topography, and wind exposure/snow cover/moisture conditions (sometimes differentiated) for many associations. These descriptions form a basis for a more systematic study of the relationships between vegetation and environment in the future.

Treeline Ecotones

Treeline ecotones are the zones of transition from forests below to the treeless tundra above. Studies of alpine treeline ecotones around the world tell us that they are affected by climate and that temperature seems to be the most important variable. The global latitudinal gradient in alpine treeline elevation indicates this importance, but does not illuminate just how it works or what causes variability in a locale such as the CCE. A number of investigators have examined the treeline ecotone in the CCE.

Cairns (1995) found a mean treeline elevation of 2,139 m with a standard deviation of 144 m in the GNP. Using topographic maps, Becwar and Burke (1982) estimated that 80% of the transition from forest to tundra in the GNP spans a 550-m range; in contrast, in Rocky Mountain National Park, Colorado, 80% of the transition covers 200 m. The variability in alpine treeline ecotone elevation in the GNP may result from combinations of variability in macroclimate, microclimate, topography, snow and debris avalanches, and competition with tundra (Malanson and Butler 1994). Brown and colleagues (1994) noted that the elevations of treeline in the GNP are not controlled by climate in many locations. In some areas, the steepness of the slope, rising to vertical, precludes trees at higher elevations—sometimes evidenced by the lone tree growing in a crevice high on a cliff. In other areas, slope instability suppresses trees through rock movement. Fire and avalanche have also affected treeline and these factors contribute to the range reported by Becwar and Burke (1982). Malanson and Butler (1994) considered how soil factors might control the elevation of treeline. They found that treeline was lower where at least some nutrients were higher in adjacent tundra. This led them to speculate that in nutrient-rich sites, tundra would have an advantage over tiny tree seedlings, tipping the balance and limiting trees to even lower elevations. Some of the highest elevation patches of dwarf trees and krummholz in the CCE are found where the soils are nearly absent just upslope. In a more spatially detailed study, however, Cairns (1999) reports a positive effect of krummholz on soil nutrients.

Tree species in the ecotone in the CCE are predominately subalpine fir, with Engelmann spruce (*Picea engelmannii*) and whitebark pine being less dominant. Alpine larch, Douglas fir and, rarely, lodgepole pine can also occur. Understory shrubs include willow and gooseberry (*Ribes lacustre*). In one of the most notable species patterns at high elevation, whitebark pine establishes in a sheltered site, then other species establish in its lee (downwind from the whitebark). Whitebark pine, which has a large seed, depends on dispersal by animals, mostly Clark's nutcracker (*Nucifraga columbiana*; Tomback 2001). Because the whitebark pine seed carries a relatively large energy reserve, it can be cached in soil and, later, can establish in what might otherwise be a lethal environment for a tiny seedling. Moreover, it is possible that the small, wind-dispersed seeds of the other tree species are simply blown off the tundra before they can successfully germinate. These species can establish without the shelter of whitebark pine, but typically only in wind-sheltered places or where the smaller seeds have fallen into the interstices of a mat of tundra plants such as *Arctostaphylos* or *Dryas*.

As in other areas of the Rockies, a variety of tree forms appear in the CCE ecotone, but three forms predominate. In the CCE, we first see upright trees with reduced growth rates that result in dwarf forms. Second, we find krummholz, in which the would-be tree grows laterally. In this form, branches may move into the ground, develop root characteristics, and reemerge. Third, resulting from a process called *layering,* we observe upright krummholz, which appears to be a variant of krummholz that consists of upright, short (~1 m) stems. In the dwarf upright forms, the Leaf Area Index (LAI; a sum of all the leaf layers), is on the order of 8; in the latter two forms, the LAI exceeds 12 in a dense canopy approximately 20 cm deep. Cairns (2005) reports that canopy form alters carbon balance, with dwarf trees being more productive and krummholz less so, even in the harshest environments. No understory exists for the krummholz forms. Instead, we often find a layer of imbricated duff, which is composed primarily of fallen needles and some small twigs that apparently are glued together by resinous compounds. The exact nature of this duff has not been investigated, but it appears to be somewhat hydrophobic, meaning that it could affect the hydrology beneath krummholz by altering both infiltration and soil evaporation patterns.

After conducting a series of carbon-balance studies, Cairns and Malanson (1998) arrived at two key conclusions. First, treeline elevations were probably out of equilibrium with recent climate. Second, the carbon balance is sensitive to moisture, which in turn depends in part on soil depth. Their simulations of the soil's role in determining treeline patterns, however, showed that the soil spatial pattern, which itself is random in tundra, is not matched by the vegetation pattern. As a result, they conclude that there is a role for positive feedback by vegetation itself (Malanson et al. 2001).

Spatially, the treeline forms particular types of patterns with tundra and meadow. The pattern made by tundra and tree as binary classes is not random. Brown (1994) found that topography explained a small but significant part of the pattern differences among closed-canopy forest, open-canopy forest, meadow, or unvegetated surfaces in the ecotone area. Brown concluded that topography influences disturbance regimes that affect treeline. In addition to the commonly observed gradient between forest and tundra patches, the CCE has strongly developed linear ecotone features. Ribbon forest, which consists of strips of trees alternating with alpine meadow, appears at the lowest end of the ecotone. In the middle range, fingers of dwarf forest stretch up into the tundra, and among the patches at the higher elevations (which are often of the upright krummholz form), many are elongated. Allen and Walsh (1996) differentiated six types of spatial patterns within the ecotone and found that these were linked to differences in topography, disturbance, and geologic substrate. Cairns and Waldron (2003), however, found that 48–59% of ecotone transitions in the GNP also had a sigmoid wave pattern indicative of climate control.

Ribbon forest is exceptionally well developed and extensive in the CCE. In the classic paper on the topic, Billings (1969) used an aerial photo of West Flattop Mountain in the GNP, Montana, as his primary illustration, although he discussed field observations elsewhere. Billings hypothesized that the pattern was created by an interaction among trees acting as a snow fence and that the effect of snow suppresses trees in the meadow but enhances their establishment a certain distance downwind (in the lee of the snow drift). It is difficult to envision how this develops through time because the height of the trees and the size of the drift keep changing. Looking around the CCE, we can see that such ribbons often have strong curvature, following the strike of the anticlines and synclines in the sedimentary rock. Butler and coworkers (2003) determined that ribbon forest in the GNP was controlled by stratigraphy and glaciation. On Flattop Mountain and elsewhere, glaciers moving parallel to the strike in the rocks carved grooves, which are now occupied by meadow. The alternating ridges are covered by trees. The resulting ribbons affect, but do not originate from, snowdrift patterns.

Alpine Resources

For people, the alpine zone is an important resource in several contexts. Although the most obvious consumable resource is water, perhaps the primary resource is aesthetics, which make the zone a recreational resource. Parts 2 through 4 of this book cover these topics in more detail. Here we note some of the alpine-specific aspects of the CCE resource.

Water

At high elevations, most precipitation falls as snow (Finklin 1986). Because much of the snow above treeline is blown off by wind, the nature of this water resource is distinctly modified. In general, slopes facing west and southwest, especially if they are convex, hold little snow. The opposite slopes, especially if concave, accumulate snow. The loss of snow from large expanses, then, leads to concentration in others, adding to glaciers in particular as well as to summer snowfields. These snow concentrations form a local water source for wet tundra and meadow. Snow from the high alpine is also redistributed locally to the treeline ecotone, where it accumulates in drifts that can completely pack patches of krummholz, substantially increase depth in tree islands and fingers, and fill snow glades in ribbon forest. Regionally, this blown snow can end up as input to the subalpine forest, but given the position of the CCE next to the

Great Plains, much of what leaves the alpine will inevitably end up leaving the ecosystem.

Recreation and Aesthetics

The CCE's alpine is the central focus of the region's aesthetic appeal and recreational attractiveness. Although the mountains as a whole, including the valleys with the lower slopes, lakes, and rivers, meet the nineteenth-century idea of the "sublime," the alpine is most important because it differentiates mountains from hills. With its vast open views of the CCE, the alpine is a key destination point for recreationists. Of particular note is the Going-to-the-Sun Road in the GNP, which brings 475,000 vehicles (probably containing more than 1 million visitors) to Logan Pass each year. It is the crossing of the alpine, reaching such a high point (contrast it to crossing the Continental Divide at Summit, Montana, on U.S. Highway 2, which is surrounded by coniferous forest) that makes this trip so popular. Although farther exploration is relatively limited, hikes to the alpine sites of Hidden Lake Overlook, Grinnell Glacier, and Iceberg Lake in the GNP are all among the most popular day hikes. Additionally, the relatively good chance of encountering mountain goats or bighorn sheep when visiting popular CCE alpine sites ensures high levels of visitor interest, tying recreational values to biodiversity values.

Alpine Change

In this section, we introduce two drivers of change in the CCE alpine, and then concentrate on potential ecological responses.

Drivers: Climate Change and Geochemical Change

Two aspects of global change drive the ecosystem change in the CCE alpine. First, the global climate is changing, primarily warming (IPCC 2001). What this means for the future of the CCE is uncertain because regional projections are less certain than those of global scale. Multiple projections, though, indicate warmer temperatures and slightly higher precipitation, experienced by the region as a whole in the twentieth century (USGCRP 2000). Direct effects in the alpine are likely to be on the plants, which depend on climate factors.

Second, human activities are resulting in a global redistribution of geochemical materials, especially nutrients such as nitrogen and phosphorous. For example, land-use and urban changes in Asia have changed the trans-Pacific movement of dust to North America, and investigators have determined the implications of such changes for the Colorado Front Range (Williams and Tonnessen 2000). Although differences between the geology of Colorado and that of the CCE make transfer of conclusions premature, changes in such inputs should be examined in more detail. In the CCE, ongoing research is documenting the input of several nutrients in snowfall (USGS 2003). This work ties the two drivers together. The geochemical inputs are linked to climate and will change as the climate changes.

Response: Vegetation Change

Recognizing that modifications in vegetation might potentially follow climatic and geochemical change in the CCE alpine, research scientists have developed a global network of monitoring stations called The Global Observation Research Initiative in Alpine Environments (GLORIA; Grabherr et al. 2000). Under the GLORIA program, investigators record tundra vegetation on mountaintops around the world. The program is most intense in Europe, where it began, with at least 18 current sites and with more being added all the time. The CCE hosts the original North American site, which includes the summits of Dancing Lady, Bison, Pitamakin, and Seward mountains in the GNP, Montana. So far, the greatest diversity has been recorded on the northeastern flanks of these summits and the least diversity has been seen on the southwestern exposures. Community structure is related to topography and slope stability, once again illustrating a link between biological diversity and geomorphology in the CCE.

The usefulness of GLORIA for assessing response to climatic change depends on site context. Alpine tundra is isolated much like an island, and the GLORIA sites function as ecological islands within islands. Because species on islands likely undergo some turnover even in the absence of exogenous forces such as climate change, the equilibrium turnover rate must be taken into consideration when examining changes. The isolation of GLORIA sites depends on the extent of tundra within which they are embedded and the spatial relation of that tundra with that on other mountains. Nonetheless, long-term changes can be detected beyond the species turnover that occurs on an ecological island.

Researchers have assessed the potential of the alpine treeline ecotone as an indicator of climate change in some detail. Based on the idea that tree species reach their climatic limits in the alpine and can thus move upslope if the climate warms (Malanson et al. 2004), investigators postulate that ecotones could be used to monitor climate change. As we have seen, the pattern of treeline in the CCE is often controlled by topography, and variations in treeline elevation may not result directly from a temperature gradient. It may be possible, however, to identify some areas where treeline elevation does result directly from a temperature gradient. The implication here is that some areas of tundra are more susceptible than others to invasion by tree species, making these areas preferable as indicator sites.

Bekker (2005) reports a definitive advance of treeline on Lee Ridge in the GNP. This advance, which took place primarily during the nineteenth century, may be linked to climate change. By mapping the current trees and taking cores for tree-ring counts, Bekker was able to document tree-establishment sequences that are spatially associated with previously established trees. This phenomenon, which led to linear fingers of trees extending into the tundra, suggests that sheltering from climatic extremes was still necessary to facilitate seedling establishment even after the Little Ice Age ended around 1850.

Based primarily on repeat photography, Butler and colleagues (1994) and Klasner and Fagre (2002) found minor changes in treeline position or pattern in the GNP during the twentieth century, although patches of krummholz and trees may have increased in density. Butler and DeChano (2001), however, illustrate widespread upslope migration of forests as revealed in repeat photography from fire lookouts in the GNP. Treeline position or pattern changes seem to depend on where we look.

Using tree-ring samples at 28 sites throughout the eastern portion of the GNP, we found that a burst of new establishment within existing stands probably occurred during the 1930s. This evidence is inexact, however, because we were unable to cross date our cores. Evidence for more recent actual advance of trees into tundra comes from an intensively studied site on Lee Ridge. Here, Alftine and colleagues (2003) found evidence for an accelerating advance of trees during the negative phase of the PDO, which ended rather abruptly when the PDO phase switched to positive.

The potential for treeline as an indicator has also been the motivation for some detailed studies of tree responses to climate. Cairns (2001) documented the pattern of winter desiccation within krummholz of subalpine fir. Although the details of elevation, slope, and canopy position are of technical interest, the broader contribution to the indicator concept is that multiple scales of wind—one of the most difficult problems in climate change prediction—are most important.

Treeline may not respond closely to climate change for several reasons. First, establishment may be limited. If whitebark pine is a necessary pioneer and if its overall abundance has been lessened by white pine blister rust (*Cronartium ribicola*), or if it could be lessened in the future by an upslope change in the range of mountain pine beetle (e.g., Logan et al. 2003), the whitebark seed supply will be lower, which could affect the rate of treeline advance (cf. Malanson 1997). Second, although some (e.g., Körner 1998) have suggested that treeline is controlled by the negative feedback of trees that shade and create cooler soil temperatures, limiting the growth of tree forms, it is really the advance of krummholz that threatens the alpine tundra. Ecotones are commonly affected by positive feedback, in which trees or krummholz improve the possibility of establishment and growth, as in the observations of the role of whitebark pine.

The role of positive feedback in this CCE ecotone has been investigated in some depth in computer simulations (e.g., Malanson 2001). Alftine and Malanson (2004) report that including a directional positive feedback (as wind should induce) in a simulation produces the patterns seen on Lee Ridge. In general, positive feedback decouples the pattern of vegetation from the climate through the modification of the microclimate—the temperatures, moisture regimes, humidity, and winds experienced by the vegetation—by the vegetation itself.

There are limits to treeline response beyond the direct effects of climate, however. Given that the degree of soil development underneath both tundra and krummholz is relatively similar, questions arise as to what soil characteristics in the tundra do vary, such that they might account for observed spatial variation in tree-seedling establishment. Malanson and colleagues (2002) illustrated that soil depth beneath solifluction treads and adjacent risers did not vary significantly. In addition, significant variations in surface penetrability were not seen between vegetated risers and adjacent treads (Butler et al. 2004). At sites where the process of turf exfoliation had exposed fine-grained sediment at the base of the riser, however, surface penetrability *did* vary significantly. At those exfoliated sites, surface penetrability was significantly easier, suggesting that such sites may be amenable to tree-seedling establishment. Resler (2006) illustrates that seedlings established in tundra at three GNP study sites were almost exclusively concentrated either at the base of exfoliated turf risers or in the lee of boulders extending above the general tundra surface. In both cases, the seedlings are protected from severe tundra winds and ice-crystal blasting, and fine-grained material (at the base of exfoliated risers, or in the lee eddy behind tundra boulders where loess can be deposited) is available for seedlings to take root.

At a more abstract level, Malanson and Zeng (2004) found evidence that treelines in places such as the broad solifluction-dominated slopes east of the Continental Divide in the CCE may have rates of advance that are dominated by endogenous processes. In other words, the ecological process in which vegetation modifies the climate is more important to seedling establishment than the regional climate. This is because the microclimate experienced by establishing seedlings is significantly altered by the effects of existing vegetation on wind, available water, and canopy and soil temperatures. The alteration is enough to render the regional climate no longer useful as an indicator.

At the lower end, ribbon forests are clearly unstable. Many show evidence of new establishment reaching out into the glades, so that the trees at the edge of the forest sweep from the tall trees in the older central portion down through successively younger trees to new seedlings (Figure 6-2). This advance can be seen in other alpine meadows as well and is likely a response to both climate change and positive feedbacks such as the microclimatic improvement made by existing vegetation as described previously.

Alpine Ecosystem Dynamics and Change 97

FIGURE 6-2. Alpine Glades in Preston Park, Glacier National Park

An Alpine of the Future

Although its areal extent is considerably smaller than other biomes, the alpine is often cited as an example of the ecological impacts of climate change. The reason for the attention is its insularity. The top of each mountain is like an island, and the potential effect of climatic warming is for biomes now on lower slopes to rise, squeezing the alpine biome to a small area or off the mountaintops altogether. We might envision changes from tundra to extensive krummholz and, eventually, to forest. In this case, continued survival of alpine biota would depend on dispersal to mountaintops to the north. This potential survival scenario is complicated, however, given that bare rock characterizes much of the mountain area above the current alpine and at the high elevations to the north of the CCE. Although a few centuries of change may develop soils sufficient to maintain the alpine biome, in the short term it is literally caught between a rock and a hard place.

It does appear that some concern is warranted. Following the Little Ice Age and continuing into recent decades (Butler and DeChano 2001; Alftine et al. 2003; Bekker 2005), trees have advanced upslope into tundra. The evidence of this, however, is not uniform over all slopes (Butler et al. 1994; Klasner and Fagre 2002). We can conclude that the alpine is threatened by continued climate change and possibly by continued geochemical change; the irregularity in response does not result from uncertainty about the climate and should not be thought of as indicative of a lack of significant climatic change. Instead, the differing responses illustrate the complexity (in terms of multiple drivers and limits) of the CCE alpine.

These changes, however complex, are of critical importance to the public land managers of mountain regions such as the CCE. The alpine is of high aesthetic value, even defining mountains for many people. Charismatic wildlife species such as mountain goats are part of the CCE's biodiversity but, through tourism, are part of the economy as well. Understanding the dynamics of the alpine and the threat of climate change, coupled with monitoring of impacts, will bring a stronger focus to development of management strategies designed to minimize unwanted effects.

References

Alftine, K.J., and G.P. Malanson. 2004. Directional Positive Feedback and Pattern at an Alpine Tree Line. *Journal of Vegetation Science* 15: 3–12.

Alftine, K.J., G.P. Malanson, and D.B. Fagre. 2003. Feedback-Driven Response to Multi-Decadal Climatic Variability at an Alpine Forest-Tundra Ecotone. *Physical Geography* 24: 520–35.

Allen, T.R., and S.J. Walsh. 1996. Spatial and Compositional Pattern of Alpine Tree Line, Glacier National Park, Montana. *Photogrammetric Engineering and Remote Sensing* 62: 1261–68.

Bamberg, S.A., and J. Major. 1968. Ecology of the Vegetation and Soils Associated with Calcareous Parent Materials in Three Alpine Regions of Montana. *Ecological Monographs* 38: 127–67.

Becwar, M.R., and M.J. Burke. 1982. Winter Hardiness Limitations and Physiography of Woody Timberline Flora. In *Plant Cold Hardiness and Freezing Stress: Mechanisms and Crop Implications*. Volume 2. Edited by P.H. Li and A. Sakai. New York: Academic Press, 307–23.

Bekker, M.F. 2005. Positive Feedback between Tree Establishment and Patterns of Subalpine Forest Advancement, Glacier National Park, Montana, U.S.A. *Arctic, Antarctic, and Alpine Research* 37: 97–107.

Billings, W. D. 1969. Vegetational Pattern Near Alpine Timberline as Affected by Fire-Snowdrift Interactions. *Vegetation* 19: 192–207.

Brown, D.G. 1994. Predicting Vegetation Types at Tree Line Using Topography and Biophysical Disturbance Variables. *Journal of Vegetation Science* 5: 641–56.

Brown, D.G., D.M. Cairns, G.P. Malanson, S.J. Walsh, and D.R. Butler. 1994. Remote Sensing and GIS Techniques for Spatial and Biophysical Analyses of Alpine Tree Line through Process and Empirical Models. In *Environmental Information Management and Analysis: Ecosystem to Global Scales*. Edited by W.K. Michener, S. Stafford, and J. Brunt. Philadelphia, PA: Taylor and Francis, 453–81.

Butler, D.R. 1992. The Grizzly Bear as an Erosional Agent in Mountainous Terrain. *Zeitschrift für Geomorphologie* 36: 179–89.

Butler, D.R., and G.P. Malanson. 1990. Non-Equilibrium Geomorphic Processes and Patterns on Avalanche Paths in the Northern Rocky Mountains, U.S.A. *Zeitschrift für Geomorphologie* 34: 257–70.

Butler, D.R., and L.M. DeChano. 2001. Environmental Change in Glacier National Park, Montana: An Assessment through Repeat Photography from Fire Lookouts. *Physical Geography* 22: 291–304.

Butler, D.R., G.P. Malanson, and D.M. Cairns. 1994. Stability of Alpine Tree Line in Glacier National Park, Montana, U.S.A. *Phytocoenologia* 22: 485–500.

Butler, D.R., G.P. Malanson, M.F. Bekker, and L.M. Resler. 2003. Lithologic, Structural, and Geomorphic Controls on Ribbon Forest Patterns. *Geomorphology* 55: 203–17.

Butler, D.R., G.P. Malanson, and L.M. Resler. 2004. Turf-Banked Terrace Treads and Risers, Turf Exfoliation, and Possible Relationships with Advancing Tree Line. *Catena*: 259–74.

Cairns, D.M. 1995. Carbon Balance Modeling at the Alpine Treeline Ecotone in Glacier National Park, Montana. PhD dissertation. Iowa City: University of Iowa.

———. 1999. Multi-Scale Analysis of Soil Nutrients at Alpine Tree Line in Glacier National Park, Montana. *Physical Geography* 20: 256–71.

———. 2001. Patterns of Winter Desiccation in Krummholz Forms of *Abies lasiocarpa* at Treeline Sites in Glacier National Park, Montana, USA. *Geografiska Annaler* 83A: 157–68.

———. 2005. Simulating Carbon Balance at Tree Line for Krummholz and Dwarf Tree Growth Forms. *Ecological Modelling* 187:314–28.

Cairns, D.M., and G.P. Malanson. 1998. Environmental Variables Influencing Carbon Balance at the Alpine Treeline Ecotone: A Modeling Approach. *Journal of Vegetation Science* 9: 679–92.

Cairns, D.M., and J.D. Waldron. 2003. Sigmoid Wave Transitions at Alpine Tree Line. *Geografiska Annaler* 85A: 115–26.

Chadwick, D.H. 1983. *A Beast the Color of Winter—The Mountain Goat Observed*. San Francisco, CA: Sierra Club Books.

Choate, C.M., and J.R. Habeck. 1967. Alpine Plant Communities at Logan Pass, Glacier National Park. *Proceedings of the Montana Academy of Sciences* 27: 36–54.

Damm, C. 2001. A Phytosociological Study of Glacier National Park, Montana, U.S.A., with Notes on the Syntaxonomy of Alpine Vegetation in Western North America. PhD dissertation. Göttingen, Germany: Georg-August-Universität Göttingen.

Fagre, D.B., D.L. Peterson, and A.E. Hessl. 2003. Taking the Pulse of Mountains: Ecosystem Responses to Climatic Variability. *Climatic Change* 59: 263–82.

Finklin, A.I. 1986. *A Climatic Handbook for Glacier National Park, with data for Waterton Lakes National Park*. INT-GTR-204. Ogden, UT: U.S. Department of Agriculture (USDA) Forest Service.

Gardner, J.S., D.J. Smith, and J.R. Desloges. 1983. *The Dynamic Geomorphology of the Mt. Rae Area: A High Mountain Region in Southwestern Alberta*. University of Waterloo Department of Geography Publication Series No. 19. Waterloo, ON: University of Waterloo.

Grabherr, G., M. Gottfried, and H. Paull. 2000. GLORIA: A Global Observation Research Initiative in Alpine Environments. *Mountain Research and Development* 20: 190–91.

IPCC (Intergovernmental Panel on Climate Change). 2001. *Climate Change 2001: Impacts, Adaptation, and Vulnerability Technical Summary*. A Report of Working Group II of the Intergovernmental Panel on Climate Change 2001, WMO [World Meteorological Organization] and UNEP [United Nations Environment

Programme]. Cambridge, UK: Cambridge University Press.

Klasner, F.L., and D.B. Fagre. 2002. A Half Century of Change in Alpine Treeline patterns at Glacier National Park, Montana, U.S.A. *Arctic, Antarctic, and Alpine Research* 34: 53–61.

Körner, C., 1998. A Re-Assessment of High Elevation Treeline Positions and their Explanation. *Oecologia* 115: 445–59.

Lesica, P. 2002. *A Flora of Glacier National Park*. Corvallis: Oregon State University Press.

Logan, J.A., J. Regniere, and J.A. Powell. 2003. Assessing the Impacts of Global Warming on Forest Pest Dynamics. *Frontiers in Ecology and the Environment* 1: 130–37.

Malanson, G.P. 1997. Effects of Feedbacks and Seed Rain on Ecotone Patterns. *Landscape Ecology* 12: 27–38.

———. 2001. Complex Responses to Global Change at Alpine Tree Line. *Physical Geography* 22: 333–42.

Malanson, G.P., and D.R. Butler.1994. Tree–Tundra Competitive Hierarchies, Soil Fertility Gradients, and the Elevation of Tree Line in Glacier National Park, Montana. *Physical Geography* 15: 166–80.

———. 2002. The Western Cordillera. In *The Physical Geography of North America*. Edited by A. Orme. Oxford, UK: Oxford University Press, 363–79.

Malanson, G.P., and Y. Zeng. 2004. Uncovering Spatial Feedbacks at Alpine Tree Line Using Spatial Metrics in Evolutionary Simulations. In *GeoDynamics*. Edited by P.M. Atkinson, G. Foody, S. Darby and F. Wu. Boca Raton, FL: CRC Press, 137–50.

Malanson, G.P., N. Xiao, and K.J. Alftine. 2001. A Simulation Test of the Resource Averaging Hypothesis of Ecotone Formation. *Journal of Vegetation Science* 12: 743–48.

Malanson, G.P., D.R. Butler, D.M. Cairns, T.E. Welsh, and L.M. Resler. 2002. Variability in an Edaphic Indicator in Alpine Tundra. *Catena* 49: 203–15.

Malanson, G.P., D.R. Butler, and S.J. Walsh. 2004. Ecological Response to Global Climatic Change. In *WorldMinds*. Edited by D.G. Janelle, B. Warf, and K. Hansen. Dordrecht, Netherlands: Kluwer Academic Publishers, 469–73.

Resler, L.M. 2006. Geomorphic Controls of Spatial Pattern and Process at Alpine Treeline. *Professional Geographer* 58: 124–38.

Schmid, G.L. 2004. The Role of Soils in Recording Environmental Change at Alpine Tree Line. PhD dissertation (unpublished). San Marcos, TX: Texas State University-San Marcos Department of Geography.

Selkowitz, D.J., D.B. Fagre, and B.A. Reardon. 2002. Interannual Variation in Snowpack in the Crown of the Continent Ecosystem. *Hydrological Processes* 16: 3651–65.

Tardiff, S.E., and J.A. Stanford. 1998. Grizzly Bear Digging: Effects on Subalpine Meadow Plants in Relation to Mineral Nitrogen Availability. *Ecology* 79: 2219–28.

Tomback, D.F. 2001. Clark's Nutcracker: Agent of Regeneration. In *Whitebark Pine Communities*. Edited by D.F. Tomback, S.F. Arno, and R.E. Keane. Washington, DC: Island Press.

USGCRP (U.S. Global Change Research Program). 2000. *Climate Change and America: Overview Document. A Report of the National Assessment Synthesis Team*. Washington: USGCRP.

USGS (U.S. Geological Survey). 2003. Snowpack Analysis of Chemical Loading in Glacier National Park. http://www.nrmsc.usgs.gov/research/snowpack_analysis.htm (accessed December 6, 2006).

White, D., K.C. Kendall, and H.D. Picton. 1998. Grizzly Bear Feeding Activity at Alpine Army Cutworm Moth Aggregation Sites in Northwest Montana. *Canadian Journal of Zoology* 76: 221–27.

Williams, M.W., and K.A. Tonnessen. 2000. Critical Loads for Inorganic Nitrogen Deposition in the Colorado Front Range, USA. *Ecological Applications* 10: 1648–65.

7
Conserving Biodiversity
Michael Quinn and Len Broberg

Literature about the CCE abounds with references to its rich biodiversity. In this chapter, we investigate approaches and challenges to conserving native biodiversity or species richness in the CCE. The term "biodiversity" is widely used and often refers primarily to species richness. In this chapter, we take a broader view of biodiversity, in particular of native biodiversity, and its conservation in the CCE.

The term biodiversity was coined at an international conference held in 1985 to address the global state of biological diversity. The conference proceedings were published under the title *Biodiversity* (Wilson 1988) and the environmentally aware public, the popular media, politicians, and the scientific community rapidly adopted the term. The term's use has increased exponentially and the number of new scientific publications on biodiversity currently exceeds 3,000 per year.

Less than four years after the publication of the seminal conference proceedings, the Convention on Biological Diversity (CBD) was signed at the 1992 United Nations Conference on Environment and Development in Rio de Janeiro. Article 2 of the CBD defines biodiversity as "the variability among living organisms from all sources including, inter alia, terrestrial, marine and other aquatic ecosystems and the ecological complexes of which they are part: this includes diversity within species, between species and of ecosystems" (CBD 2001–2005). A review by DeLong (1996) included 85 different definitions of biodiversity that vary in their degree of inclusiveness and concreteness. The CBD, which has now been ratified by 188 countries (Canada has ratified and the United States has signed but not ratified), is a political–economic statement about equitable sharing of the burdens and benefits of biodiversity conservation. The CBD advocates an integrated ecosystem approach to the conservation and sustainable use of biodiversity. Therefore, although the term biodiversity is often defined in relation to its hierarchical ecological elements, conserving biodiversity is a highly complex amalgam of political, legal, economic, cultural, and moral issues. Takacs (1996, *24*) suggests that the "complexity of the biodiversity concept does not only mirror the natural world it supposedly represents; it is that plus the complexity of human interactions with the natural world, the inextricable skein of our values and its value, of our inability to separate our concept of a thing from the thing itself." Despite the difficulty in achieving a unified definition of the term (or possibly because of it), biodiversity has become a central term in the international politics of sustainable development as well as the theory and practice of conservation biology. Perhaps one of the greatest benefits of the biodiversity concept is its potential to promote and advance interdisciplinary approaches to science, policy, and management.

In this chapter, we approach biodiversity conservation as the product of science, policy, law, funding, and human values and attitudes. Conservation of species, communities, ecosystems, or landscapes all depend on the complex interaction of these factors across a variety of ownership, stewardship, and management regimes. Through the lens of individual species or communities, we examine the policy and legal framework within which "on-the-ground" conservation of biodiversity occurs in the United States (Montana) and Canadian (British Columbia and Alberta) portions of the CCE.

Biodiversity in the CCE

The native biodiversity found in the CCE is extensive and relatively intact. The richness and interactions across all levels of biological organization are the result of a geologically complex landscape and orthographic effects in the western interior of North America. The CCE is characterized by a geographically compressed set of elevation and moisture gradients that result in a juxtaposition of ecological systems ranging from prairie to alpine and from subhumid to semiarid.

The CCE is recognized for its species richness and relatively intact assemblage of native species. The region is the most southern part of North America that still possesses the full suite of large mesocarnivores including wolves, coyotes, grizzly bears, black bears, cougars, lynx, bobcats, wolverines, fishers, martens, and river otters (see Chapter 2 for scientific names not given in this chapter). This region hosts source populations of carnivores that are critical to the long-term persistence of species north and south of the international border. Moose, elk, white-tailed deer, mule deer, bighorn sheep, and mountain goats all thrive in the CCE, giving the region the richest assemblage of native ungulates in North America. Wild populations of two species of large mammals, however, have been extirpated from the CCE—bison and woodland caribou (although caribou are found to the immediate north and west of the CCE). Overall, at least 1,200 species of vascular plants, 250 species of birds, and

65 species of mammals are native to the CCE. Herpetofauna is less diverse than in regions further south, but several species do occur in the area, including such rare species as the long-toed salamander and the spotted frog. Native fish species have been dramatically influenced by non-native introductions, but no native species have been extirpated entirely. Although the invertebrate fauna of the region is poorly known, the CCE is recognized as being particularly rich in lepidopterans.

Beyond the species level, the CCE is characterized by significant ecosystem diversity. The details and differences of the classification systems used to delineate and describe the ecosystem are a product of comprehensive collaborative approaches to ecological land classification that are based on biogeoclimatic characteristics of the landscape. It is within the context of the ecoregion that essential processes and flows of matter and energy occur. The area being labeled as the CCE in this book comprises portions of five ecosections (Border Ranges, Crown of the Continent, East Front Ranges, Porcupine Hills, and Swan-Mission Ranges).

Protected Areas in the CCE

Conserving landscapes by designating protected areas has a long and illustrious history in the CCE. Through the iconic efforts of individuals such as John George "Kootenai" Brown on the Canadian side and George Bird Grinnell on the American side, Waterton Lakes National Park (WLNP; originally called Kootenay Lakes Forest Park) and Glacier National Park (GNP) were established in 1895 and 1910, respectively. The WLNP has undergone several boundary changes since it was established as Canada's fourth national park and today it encompasses approximately 525 km^2. The GNP is significantly larger, spanning more than 4,000 km^2. At the time the parks were established, conserving biodiversity was secondary to more exploitative and commercial interests. For example, Alberta's first oil well and, not surprisingly, the first major oil spill in the province, both occurred within the WLNP in 1902. Today, maintaining ecological integrity is the prime mandate of both parks, and the two nations collaborate on management of their shared resources (see Chapter 18 for more details about collaboration).

Both parks received Biosphere Reserve status (the GNP in 1976 and the WLNP in 1979) and the combined Waterton Glacier International Peace Park was designated a World Heritage Site in 1995. Ongoing discussions center on expanding the WLNP's area west across the Continental Divide into the upper Flathead River Basin. The expansion would add approximately 400 km^2 of national park lands in the upper Flathead Valley, an area recognized as one of the most important sources of large carnivores in central North America. This expansion of the WLNP would boost the size of the Peace Park to 4,925 km^2.

Weiglus (2002) estimates, however, that a core reserve of 3,906 km^2 and a no-hunting buffer of 4,650 km^2 (8,556 km^2 total) would be required to protect a minimum viable population of grizzly bears in the upper Flathead Valley.

The CCE has an array of other protected areas that contribute to the conservation of regional biodiversity. The most extensive category of protected area in the CCE, designated wilderness areas, encompasses approximately 7,089 km^2 in Montana (Bob Marshall, Great Bear, Scapegoat, Cabinet Mountains, Mission Mountains, and Rattlesnake). Ten British Columbia provincial parks contribute another 474 km^2, with Elk Lakes (179 km^2), Height of the Rockies (151 km^2), and Akamina-Kishanena (108 km^2) parks accounting for the largest areas. In Alberta, a suite of protected area designations—with varying levels of priority for biodiversity conservation—adds another 32 units and 538 km^2 of legislated protected areas. Several of these units, though, are primarily recreation designations, and resource extraction occurs in several others. The total federal and provincial protected area is approximately 13,100 km^2 or roughly 32% of the CCE (refer to Table 5-1). Although the area is considerable, many of the protected areas consist of rock and ice, which are not necessarily the most productive regions for biodiversity.

Overall, protected areas in the CCE are simply not large enough nor in the right places to fully encompass the variety of life that makes the ecosystem so significant. Other categories of public and private lands clearly require special management consideration to fully conserve the biodiversity of the CCE.

Biodiversity Conservation Policy and Law in the CCE

The terms "policy" and "law" are often confused in discussions of biodiversity conservation. Law refers to the standards set by legislation or administrative rules and the body of case law interpreting those standards. Policy refers to a more mutable collection of positions or guidelines that are largely outside the reach of the courts and fashion the day-to-day work of biodiversity conservation.

The premier species-conservation law in the United States is the Endangered Species Act (ESA) of 1973 (16 USC §§ 1531–43 and subsequent amendments). Ambitious in scope (applying to federal, state, local, and private lands where listed species are found) and firm in standards, the act is controversial. Nonetheless, it is an international standard for biodiversity conservation that seeks to ensure the survival and recovery of species that are either in danger of becoming extinct in the foreseeable future (*endangered* status) or likely to become endangered in the foreseeable future (*threatened* status).

In 2002, Canada enacted a Species at Risk Act (SARA; 2006) that is far less comprehensive than the ESA. SARA has direct jurisdiction only over species on federal public lands, a very small percentage of the Canadian land base. Canadian federal, provincial, and territorial governments, however, agreed on a National Accord for the Protection of Species at Risk (Environment Canada 2006) that commits all jurisdictions to implement laws and programs that are consistent with SARA and complement efforts of neighboring jurisdictions. Furthermore, SARA directs the process for classifying the status of species at risk (Committee on the Status of Endangered Species in Canada [COSEWIC]) and for developing national recovery plans for endangered and threatened species. Six Canadian provinces have enacted endangered species legislation, excluding Alberta and British Columbia, which manage species at risk through existing wildlife legislation and policy.

In addition, Canada and the United States signed a framework agreement in 1997 to better integrate the identification, classification, and subsequent recovery activities of species at risk in both countries (i.e., the Framework for Cooperation between the U.S. Department of the Interior and Environment Canada in the Protection and Recovery of Wild Species at Risk [United States and Canada 2006]). The framework does not include fish, marine mammals, or sea turtles. CCE species listed as being at risk by the agreement include the grizzly bear and swift fox (*Vulpes velox*).

Biodiversity conservation as a whole is broader than single-species conservation. It includes conservation of ecological communities, ecosystems, and landscapes as well as species. Conservation efforts need to provide for species that are not in the "emergency room," but that require some adjustment or management of human activity to avoid reaching such a dire status. In addition, we need to scale up our research and management efforts to focus on key communities and ecological functions.

In the rest of this section, we examine the management of a selected group of listed and nonlisted species, focusing on the similarities and differences in approach for the U.S. and Canadian portions of the CCE. As an example of a terrestrial species, we use grizzly bear, which is listed as threatened in the United States but not in Canada. Aquatic species are represented by bull trout, which is listed under the ESA and also as a species of concern in Canada. Finally, we discuss the northern goshawk, a bird species that is unlisted in either country.

Grizzly Bear

Listed under the ESA in 1975, the grizzly bear has been an icon of ESA protection and the focus of much conservation work in the CCE. In Canada as a whole, the grizzly bear is considered to be a species of special concern. In British Columbia, the grizzly bear is on the "blue list," meaning that it is not immediately threatened but is sensitive to human disturbance. Alberta wildlife managers believe that the grizzly bear "may be at risk" (Kansas 2002). Grizzly bears are legally hunted in Alberta and British Columbia, although mortality concerns have resulted in a moratorium on licenses being issued for the Alberta portion of the CCE. There is a limited-entry hunt for grizzly bears in the Flathead Valley within British Columbia.

Grizzly bear is touted as a "key" or "umbrella" species (Simberloff 1998), which means that it has diverse habitat needs over large areas and that protecting its habitat is a surrogate for managing numerous other species dependent on those habitats. Management issues for grizzly bear are unique because of the species' potential for inflicting human injury. In the CCE, grizzlies are highly dependent on productive streamside habitat, seasonal berry crops, and large ungulate carcasses for food (Servheen 1983; Zager 1983; Mace 1986). Most causes of grizzly bear mortality in the Rocky Mountains are anthropogenic, and include self-defense, property defense, hunting misidentification during black bear season, and mishaps with vehicles (e.g., trains on the south side of the GNP). Bears also die of old age and from contact with other bears (Nielson et al. 2004).

In the U.S. portion of the CCE, each type of landowner is required to contribute to the conservation of the grizzly bear under the ESA and the administrative rule that lists the bear under the act. Federal agencies are required to conserve the species and to avoid actions that lead to the extinction of the species in the wild (16 USC §1536). State and local governments and private landowners may not kill, wound, harass, harm, or capture the animal (16 USC §1538). A recovery plan developed in 1993 sets formal targets for population size and female mortality.

To meet specific conservation obligations for grizzlies, various bodies, standards, and policies have evolved under the ESA. First, an Interagency Grizzly Bear Committee was formed to coordinate the management of this wide-ranging carnivore, with a subcommittee responsible for grizzly habitat in the Northern Continental Divide Ecosystem. This unique multijurisdictional and international body is made up of representatives from federal agencies (U.S. Department of Agriculture [USDA] Forest Service, U.S. Fish and Wildlife Service [FWS], U.S. National Park Service [NPS], and Parks Canada); tribal groups (Salish-Kootenai and Blackfeet tribes); and state/provincial agencies (Montana Department of Fish, Wildlife & Parks [Montana FWP], and British Columbia Ministry of Water, Land, and Air Protection).

Second, bear management units were created on the landscape. Each unit is an estimate of the size of a single female's home range based on three seasons of foraging requirements (Weaver 2004). Bear management units are monitored for their security value for grizzlies, primarily in the form of road density (Flathead National Forest 1995). Following a court order, the USDA Forest Service created a road-density standard that established a roadless secure core of 60–68% of a bear management unit, approximately the size of a female home range, in addition to road densities for grizzly habitat outside the secure core (Flathead National Forest 1995). These densities have served as a goal of landscape management for USDA Forest Service land. Many areas have not attained the goal.

Third, federal agencies are managing recreational access on nonmotorized trails. GNP managers monitor bear activity, post notices of grizzly activity, and close trails to prevent human encounters with bears when warranted. The most aggressive access management occurs in the Mission Mountains Tribal Wilderness, where the Salish-Kootenai Tribe closes a large area of about 4,047 ha to human access from July 15 to October 1, giving grizzlies undisturbed access to food sources. Fourth, the Montana FWP and the Blackfeet Tribal Wildlife Office work with the public to eliminate bear attractants, such as garbage, bird feeders, pet food, birdseed, and livestock carcasses/boneyards on private land within grizzly bear habitat. Grizzlies are managed under a three-strikes policy, which means that any individual grizzly bear involved in three conflicts with humans can be euthanized (subject to approval by the FWS). Finally, in an effort to reduce grizzly mortality through misidentification, the Montana FWP requires hunters to go through a short, online educational program, where they take a test to distinguish black bears from grizzly bears. Managers work especially hard to protect females of breeding age because of their importance for grizzly survival and recovery in the U.S. portion of the CCE.

With the exception of the WLNP, which is under federal jurisdiction, grizzly bear management on the Canadian side of the CCE falls under the jurisdiction of the provincial governments. WLNP managers have protective measures in place for grizzly bears in the region, even though the park is smaller in size than the home range of one adult male grizzly bear. The contribution of private lands and working public lands is essential for the long-term survival of grizzly bears in southern Alberta and British Columbia. The situation for grizzly bears differs significantly east and west of the Continental Divide. In British Columbia, the productive Flathead Valley boasts the highest density and reproductive rates for any grizzly bear populations in the North American interior (McLellan and Hovey 2001). This area is the fountain of grizzly bear production in the CCE and probably serves as the source for surrounding regions. Drier landscapes on the Alberta side of the divide furnish important but less-productive grizzly bear habitat. Connectivity to the foothills and prairies is a unique feature of the eastern-slope habitats. During the early 2000s, bears have been moving back into parts of their historic prairie range via riparian corridors. These moves to the prairies lead to greater interaction with agricultural and recreational property interests and result in significant mortality to grizzly bears. Bear mortality along the eastern edge of their distribution in the CCE is excessive and would not be sustainable without considerable immigration from the west.

Since 1997, efforts have been made to decrease the negative interactions between grizzly bears and humans along the eastern slopes of southern Alberta. The Southwest Alberta Grizzly Strategy (Alberta Grizzly Bear Recovery Team 2005) includes innovative management and educational programs aimed at lowering the high mortality rate. A particularly successful element of the program involves a unique partnership between the WLNP, Alberta Fish and Wildlife, and Volker Stevin, the highway maintenance contractor. Road-killed ungulates are collected through the winter months and then distributed by helicopter to strategic locations within the region in the spring. The purpose of this "intercept feeding program" is to supply a source of high protein needed by bears emerging from hibernation, keep the bears at higher elevations where there are fewer people, and avoid predation on livestock at lower elevations. Initial indications are that the program is achieving some success.

Other facets of grizzly conservation include working with private landowners and having conservation organizations acquire private lands. The flagship example of acquisition on the U.S. side is The Nature Conservancy's (TNC) Pine Butte Swamp Preserve along the Montana Rocky Mountain Front. Prompted by the fact that the area was the last known place that grizzlies roamed the prairies, TNC began acquiring land for the preserve in 1988. TNC actively manages the 7,284-ha preserve, protecting grizzly bear habitat, 40 distinct plant communities, 43 species of mammals, and 150 species of birds. Wildlife management agencies at the state, tribal, and federal levels work to educate ranchers and other large private landholders about the value of their riparian and upland areas for grizzly bears and about ways to avoid livestock depredations by grizzlies. Eliminating boneyards, sanitizing and removing carcasses, and eliminating access to livestock feed have all been key stewardship elements that have allowed grizzlies to range onto private lands in their traditional habitat.

A conservation easement is a voluntary agreement between a landholder and a qualified charitable organization (e.g., the Montana Land Reliance or the Southern Alberta Land Trust). Such easements, which have improved grizzly bear habitat, particularly along the Rocky Mountain Front and in the North Fork of the Flathead River, extinguish development rights on private property while allowing the landowner to retain all other rights (e.g., livestock grazing). Subdivision growth is a particularly troubling trend, especially in the Flathead Valley, where growth rates have approached or exceeded 40% in the last decade. Subdivisions convert formerly undeveloped seasonal bear habitat to residential uses. Although conservation easements are not likely to be sufficient to protect all grizzly habitats, they are an important conservation tool, especially in the low-elevation river bottoms and fescue prairie areas of the CCE that are not federally owned.

The Canadian side of the CCE is enjoying a wide array of private land conservation initiatives as well. The most significant examples of success are the direct acquisition of land and the purchase of conservation easements on lands surrounding the WLNP in Alberta, along with a large property acquisition on Mount Broadwood in British Columbia. Approximately 100 km^2 of private lands have been afforded some level of protection through cooperation with landowners on lands outside the WLNP. These lands contribute substantially to the protection of highly productive lower elevation sites on the eastern side of the CCE and provide critical habitat for grizzly bears, trumpeter swans, northern leopard frogs, and a suite of montane plant and animal communities.

Bull Trout

The bull trout was listed as a threatened species in the United States under the ESA in 1998. It is a species of special concern in Alberta, and considered sensitive (a designation meaning that the species requires special attention to avoid becoming a species at risk) in British Columbia. Currently, bull trout do not have federal status as a species at risk in Canada. The species requires cold, clean water; complex channel structures; and connected river systems to survive and reproduce (Reiman and McIntyre 1993). Because these conditions are most often found in roadless areas of the CCE, bull trout strongholds are associated with these areas.

Unlike grizzly bear, relatively little was known or written about bull trout before the 1980s. Their status and profile have increased significantly as researchers have learned and disseminated more information about the species. Bull trout is truly a transboundary species in the CCE because many of the primary rivers that provide their habitat cross the international border (e.g., St. Mary, Belly, and Flathead rivers). This complicates recovery planning and requires transboundary communication and cooperation.

The goal of bull trout management is to maintain the features of stream systems that are essential for species survival. Because bull trout are sensitive to changes in the hydrologic regime and to sedimentation (Fraley and Shepard 1989), they can be negatively affected by silviculture, mining, residential and commercial development, industrial water discharges, and livestock grazing (Fausch and Young 2004). Moreover, water diversions and dams can disconnect segments of streams that provide habitat for the species. Because of their low reproductive rate and their relatively high "catchability," bull trout are highly susceptible to overexploitation through recreational angling. In addition, direct competition and hybridization with introduced species has proven deleterious to bull trout (Gunckel et al. 2001).

As we write this chapter, the FWS is designating critical habitat for bull trout to achieve its survival and recovery. Because it is subject to the same protection standards that grizzly bears enjoy under the ESA, protection of bull trout can require modifications to land-use practices. "Critical habitat" is essential for the survival and recovery of the species. The service's current plan does not designate any critical habitat for bull trout in the U.S. portion of the CCE. Instead, it relies on the Bull Trout Recovery Plan put in place by the Montana FWP (FWS 1998). That plan forbids fishing for bull trout, requires release of any bull trout incidentally caught when fishing for other trout species, establishes priorities for bull trout recovery watersheds, and designates other watersheds where bull trout populations are to be recovered (Montana Bull Trout Recovery Team 2000). The recovery plan sets a target population of 500 in recovered populations.

Some essential bull trout habitat is protected through wild, recreational, or scenic river designations under the U.S. Wild and Scenic Rivers Act (WSRA; 16 USC §§1271 et seq.). The WSRA places a 1/4-mile-wide buffer on each side of designated river segments. Within the buffer, management must not degrade the wild, recreational, or scenic values. The North and Middle forks of the Flathead River in the CCE, along with part of the South Fork are designated under the WSRA and support healthy populations of bull trout. With their shores at least partially within either GNP or the Bob Marshall Wilderness Complex, these river segments already receive substantial protections along much of their length. The Wilderness Act of 1964 (16 USC §§ 1131 et seq.) designates and governs the management of the Bob Marshall Wilderness, and forbids road building, silviculture, and placement of permanent structures in wilderness areas.

Similarly, the eastern shore of the North Fork of the Flathead River receives protection because it is in the GNP. Because extractive resource use is not permitted in the GNP, sedimentation and channel constraints detrimental to bull trout occur only as a result of maintenance of visitor infrastructure, such as roads and the North Fork entrance to the GNP. This infrastructure existed when the park was established. WSRA restrictions have their greatest impact in those places where the Flathead National Forest enters the 1/4-mile zone to the west of the North Fork. Much of the riverbank on the western shore, however, is privately owned and largely unaffected by WSRA protections. Bull trout conservation on private lands can be accomplished through a variety of mechanisms, such as the habitat conservation plans developed by Plum Creek Timber Company and approved by the FWS.

In Alberta, concern about the plight of the bull trout resulted in the formation of a Bull Trout Task Force (Alberta) in 1993. Subsequently, in 1997, the province developed and implemented a management and recovery plan for the species. Along with the enactment of a province-wide zero-harvest limit, management efforts included a public education campaign to increase awareness and understanding of bull trout and other native salmonids. The bull trout was designated as Alberta's provincial fish in 1997. The campaign to better educate the public about the bull trout, its habitat requirements, and its sensitivity to exploitation has been very successful. Efforts to protect the bull trout and other native salmonids in the CCE illustrate the importance of the human dimension in conserving biodiversity.

Northern Goshawk

Many national forests in the CCE designated the northern goshawk (*Accipiter gentilis*) as a management indicator species for old-growth forests. The goshawk prefers large-diameter trees with high canopy cover for its nesting sites, both of which are key elements of old-growth forests. Under the National Environmental Policy Act (NEPA) of 1969 (42 USC §§ 4321 et seq. and subsequent amendments) forest managers are required to examine impacts on goshawks in developing and analyzing management proposals or permits. Typically, this results in establishing a nest site buffer that maintains large trees and canopy cover for protecting one or more nesting sites. In addition, the goshawk is listed on the sensitive species list for Region 1 of the USDA Forest Service. This requires that managers explicitly consider impacts to the species and mitigation of those impacts in their NEPA environmental analyses.

The northern goshawk is not considered to be at risk in Canada with the exception of the threatened Queen Charlotte subspecies, which does not occur in the CCE. The northern goshawk is recognized as "regionally important" in British Columbia because it is associated with habitats that are becoming rare, and Alberta classifies the species as sensitive. Forest planning in both provinces requires that old-growth habitats be retained.

Emerging Conservation Issues in the CCE

Conservation is as dynamic an enterprise as the very species and communities it seeks to maintain in the region. New information continually informs potential conservation needs and approaches. Often, however, the legislative and policy processes that guide public land management are slow to respond. In this section, we identify some emerging issues that have yet to gain a legal and policy framework as fully developed as the ESA or the Fisheries and Oceans Act (R.S., 1985, c. F-15).

Connectivity

Although mandates exist to protect species and communities, efforts to do so often focus on where the species are found. Even though this is appropriate, studies indicate that natural connectivity between populations or habitats of species is necessary for their genetic viability and their ability to recolonize following disturbance, as well as to maintain natural metapopulation dynamics (Beier and Noss 1998). Knowledge about what kind of habitat is necessary for movement between home ranges of a species is limited, primarily because species are more transient in these locations and therefore harder to study. Nonetheless, accounting for connectivity of species that maintain a natural metapopulation structure is prudent. The policy and legal framework for enhancing ecological connectivity is minimal.

The primary example of agency connectivity planning is the linkage-zone analysis done by the FWS (Servheen et al. 2001). Motivated by the need to eventually reconnect the Northern Continental Divide Ecosystem grizzly bear population, the proposed Bitterroot population, and the Greater Yellowstone Ecosystem population (refer to Figure 15-2), the FWS examined the need for linkage zones across the larger landscape. Using the available landscape and remaining areas of native habitat, the agency mapped areas to consider for linkage-compatible management. Management standards for these areas have not been developed, nor has the linkage concept been formally endorsed by policy or law.

Nonprofit organizations such as American Wildlands and the Yellowstone to Yukon (Y2Y) Conservation Initiative have begun to fill the need for analyzing connectivity through their *Corridors of Life* and conservation planning programs, respectively. American Wildlands has used a least-cost path model to derive potential connectivity areas and is in the process of refining the evaluation of the areas for connectivity value. The Y2Y Conservation Initiative has sponsored a wide array of ecological research on the impacts of roads and the opportunities for mitigation and restoration (see Chapter 15 for details).

Restoration

Plant and animal communities and the landscapes that support them have become degraded over time and need to be restored. Yet agencies have limited funds for restoration. Except for restoration of natural fire cycles to the landscape, policies for watershed, habitat, and community restoration are lacking. Road density within the CCE greatly exceeds that needed for resource utilization, fire management, or recreational access, and has likely exceeded acceptable levels for grizzly bear, elk, and bull trout. Roads increase sedimentation, block fish passage, establish vectors of invasive species invasion, and reduce wildlife security (Trombulak and Frissell 2000). Decommissioning roads is greatly needed to restore connectivity, improve watershed health and function, and reduce the spread of invasive species.

In addition, the need to restore ecological communities is only beginning to be addressed. Whitebark pine, for example, has been devastated by the introduced white pine blister rust disease. Projects to replant blister-rust-resistant varieties of the species have been initiated, but funding is very limited. The Whitepine Bark Ecosystem Foundation has been formed to address the need for research and restoration of this important tree species, which is a key food for grizzlies, Clark's nutcracker, and numerous small mammals in the CCE.

Invasive Species

The bioinvasion of non-native species may be the single greatest threat to biodiversity in the CCE and beyond. The CCE is rife with invasive plants (e.g., leafy spurge [*Euphorbia esula*], knapweed [*Centaurea* spp.], and dalmation toadflax

[*Linaria genistifolia* ssp. *dalmatica*]), animals (e.g., brook trout [*Salvelinus fontinalis*] and brown trout [*Salmo* spp.]), and pathogens (e.g., white pine blister rust). The adverse effects of invasive species on native biodiversity are often poorly understood, but exotic invaders can act as competitors, predators, pathogens, or disrupters of key ecological processes. Protection of native biodiversity from exotic species includes a broad array of programs in the CCE. The Crown Managers Partnership (CMP), a coalition of public land managers in the CCE (refer to Chapter 14 for details), made invasive species the theme of its 2005 annual forum. Integration, collaboration, and resource sharing are all essential for managing the prevention and control of invasive species in the CCE. The CMP has initiated a working group to address issues of exotic plants.

Exotic plant invasion is a significant management issue throughout the CCE. More than 100 exotic plant species are found in the two core protected areas, the GNP and the WLNP (Crown Managers Forum 2005). Primary species posing risks to native biodiversity in the GNP are leafy spurge, spotted knapweed (*Centauria maculosa*), St. Johnswort (*Hypericum* perforatum), timothy (*Phleum pretense*), and oxeye daisy (*Chrysanthemum leucanthemum*) (NPS 1988, 1999). These same species pose threats to livestock and biodiversity on federal, provincial, state, and private lands surrounding the two parks. Human travel is often linked to their spread, with roads and motorized and nonmotorized trails serving as pathways for the dispersal of exotic species. Integrated pest management is generally perceived as the best approach to containing or eradicating these species. This type of management involves using a suite of tools including biocontrol with insects and grazing; cultural techniques such as revegetation, soil disturbance management, and burning; mechanical removal by pulling; and targeted use of herbicides. Managers throughout the region recognize the significance of the issue and are increasingly focused on developing strategies to prevent the spread of these non-native invaders.

Conclusions

The CCE is a spectacular region with a relatively intact ecological system. With respect to biodiversity, the region is of continental and global significance and of great importance to society in terms of its natural resource, environmental, and amenity values. As pointed out in Chapter 4, increasing human populations and associated landscape changes have the potential to erode the ecological integrity of the ecosystem, including biodiversity. Conventional and unconventional petroleum development, coal mining, forestry, recreation, rural residential expansion, and the associated linear disturbances are significant drivers of land-use change, as are bioinvasive species and climate change. To conserve ecological integrity and biodiversity, we must consider human values and thoroughly understand critical ecosystem functions. We need interdisciplinary approaches to science, policy, and management (as discussed in Chapter 4) to develop new methodologies for landscape-scale management of the CCE. New methodologies will require that we integrate competing land-use interests better, gain a greater understanding of cumulative effects, and create more effective frameworks for transboundary management. The challenges are daunting and the window of opportunity is limited. The resource management choices and decisions that managers make in the coming decade will determine the fate of biodiversity in the CCE. Canada and the United States have an opportunity to make this region an international model of effective transboundary management for biodiversity conservation.

References

Alberta Grizzly Bear Recovery Team. 2005. *Draft Alberta Grizzly Bear Recovery Plan 2005–2010.* http://www.srd.gov.ab.ca/fw/bear_management/pdf/GrizzlyRecovery PlanFeb05_Servheen_comments1.pdf (accessed November 16, 2006).
Beier, P., and R.F. Noss. 1998. Do Habitat Corridors Provide Connectivity? *Conservation Biology* 12: 1241–52.
CBD (Convention on Biological Diversity). 2001–2005. Article 2. Use of Terms. http://www.biodiv.org/convention/articles.shtml?a=cbd-02 (accessed November 13, 2006).
Crown Managers Forum. 2005. *Crown of the Continent Managers Partnership 2005 Forum Summary.* Calgary, AB: Crown Managers Forum.
DeLong, D.C., Jr. 1996. Defining Biodiversity. *Wildlife Society Bulletin* 24: 738–49.
Environment Canada. 2006. National Accord for the Protection of Species at Risk. http://www.ec.gc.ca/press/wild_b_e.htm (accessed November 16, 2006).
Fausch, K.D., and M.K. Young. 2004. Interactions between Forests and Fish in the Rocky Mountains of the USA. In *Fishes and Forestry: Worldwide Watershed Interactions and Management.* Edited by T.G. Northcote and G.F. Hartman. Oxford, UK: Blackwell Publishing, 463–84.
Flathead National Forest. 1995. *Forest Plan Amendment #19: Objectives and Standards for Grizzly Bear Management and Timber Production.* Flathead National Forest, Kalispell, MT: U.S. Department of Agriculture (USDA) Forest Service.
Fraley, J.J., and B.B. Shepard. 1989. Life History, Ecology, and Population Status of Migratory Bull Trout (*Salvelinus confluentus*) in the Flathead Lake and River System, Montana. *Northwest Science* 63: 133–43.
FWS (U.S. Fish and Wildlife Service). 1998. Endangered and Threatened Wildlife and Plants: Designation of Critical Habitat for the Klamath River and Columbia River Populations of Bull Trout. *Federal Register* 69: 59996, June 10.
Gunckel, S., J. Li, and A. Hemmingsden. 2001. Effects of Non-Native Brook Trout (*Salvelinus fontinalis*) on the Feeding Behavior of Bull Trout. In *Friends of the Bull Trout Conference Proceedings.* Edited by W.C. Mackay, M.K. Brewin, and M. Monita. Calgary, AB: Bull Trout Task Force (Alberta), Trout Unlimited Canada, 57–68.
Kansas, J. 2002. *Status of the Grizzly Bear* (Ursus arctos) *in Alberta.* Wildlife Status

Report No. 37. Edmonton, AB: Alberta Sustainable Resource Development, Fish and Wildlife Division and Alberta Conservation Association.

Mace, R.D. 1986. Analysis of Grizzly Bear Habitat in the Bob Marshall Wilderness, Montana. In *Proceedings, Grizzly Bear Habitat Symposium.* General Technical Report INT-207. Edited by G. P. Contreras and K. E. Evans. Missoula, MT: USDA Forest Service Intermountain Research Station, 136–49.

McLellan, B.N., and F.W. Hovey. 2001. Habitats Selected by Grizzly Bears in a Multiple Use Landscape. *Journal of Wildlife Management* 65: 92–99.

Montana Bull Trout Recovery Team. 2000. *Restoration Plan for Bull Trout in the Clark Fork River Basin and Kootenai River Basin, Montana.* Helena, MT: Montana Department of Fish, Wildlife & Parks.

Nielson, S.E., S. Herrero, M.S. Boyce, R.D. Mace, B. Benn, M.L. Gibeau, and S. Jevons. 2004. Modelling the Spatial Distribution of Human-Caused Grizzly Bear Mortalities in the Central Rockies Ecosystem of Canada. *Biological Conservation* 120: 101–13.

NPS (National Park Service). 1988. *Management Policies.* Washington, DC: NPS.

———. 1999. *Final General Management Plan and Environmental Impact Statement, Volume 1.* Glacier National Park, MT: NPS.

Reiman, B.E., and J.D. McIntyre. 1993. *Demographic and Habitat Requirements for Conservation of Bull Trout.* General Technical Report INT-302. Ogden, UT: USDA Forest Service, Intermountain Research Station.

SARA (Species at Risk Act). 2006. Canada's Species at Risk Act (SARA). Fisheries and Oceans Canada. http://www.dfo-mpo.gc.ca/species-especes/home_e.asp (accessed November 16, 2006).

Servheen, C. 1983. Grizzly Bear Foods, Movements and Habitat Selection in the Mission Mountains, Montana. *Journal of Wildlife Management* 30: 349–63.

Servheen, C., J. Waller, and P. Sandstrom. 2001. *Identification and Management of Linkage Zones for Grizzly Bears between the Large Blocks of Public Land in the Northern Rocky Mountains.* Missoula, MT: FWS.

Simberloff, D. 1998. Flagships, Umbrellas, and Keystones: Is Single-Species Management Passed in the Landscape Era? *Biological Conservation* 83: 247–57.

Takacs, D. 1996. *The Idea of Biodiversity: Philosophies of Paradise.* Baltimore, MD: The Johns Hopkins University Press.

Trombulak, S.C., and C. Frissell. 2000. A Review of the Ecological Effects of Roads on Terrestrial and Aquatic Ecosystems. *Conservation Biology* 14: 18–30.

United States and Canada. 2006. Framework for Cooperation between the U.S. Department of the Interior and Environment Canada in the Protection and Recovery of Wild Species at Risk. http://www.fws.gov/Endangered/canada/framewrk.htm (accessed November 16, 2006).

Weaver, J.L. 2004. Personal communication with the authors, April 24.

Weiglus, R.B. 2002. Minimum Viable Population and Reserve Sizes for Naturally Regulated Grizzly Bears in British Columbia. *Biological Conservation* 106: 381–88.

Wilson, E.O. (ed.). 1988. *Biodiversity.* Washington, DC: National Academy Press.

Zager, P. 1983. Managing Grizzly Bear Habitat in the Northern Rocky Mountains. *Journal of Forestry* 81: 524–52.

8

Aquatic Ecosystem Health

*F. Richard Hauer, Jack A. Stanford,
Mark S. Lorang, Bonnie K. Ellis,
and James A. Craft*

Ecosystem health is difficult to define. Although we understand the concept of health thoroughly as it applies to animals or humans, it becomes ambiguous when applied to ecosystems. Typically, the term *health* is integrated by specific, well-documented individual physiology (e.g., regulation of temperature and blood pressure). Unlike the homeostasis that defines healthy organisms—such as good nutrition, absence of disease caused by parasites, or freedom from infection—ecosystems contain organisms and processes (e.g., herbivorous insects, parasitic plants, and decomposers of dead material) that we consider symptomatic of disease when they are present at the individual level. Nonetheless, *ecosystem health* is a useful term precisely because we can relate it to the well-understood concept of being either healthy or unhealthy. For practical purposes, an ecosystem is "healthy" when its structure (i.e., components) and function (i.e., processes) are similar to those of the past and are consistent and self-sustainable over time. In addition, in a healthy ecosystem, abiotic and biotic elements are frequent and abundant, and processes occur at rates and in magnitudes that demonstrate long-term quasiequilibrium.

In this chapter, we describe the major factors that influence the health of aquatic ecosystems in the CCE. We also examine the effects of multiple stressors, particularly water and land-use practices, on aquatic ecosystem health in Flathead and Whitefish lakes and their tributaries, which are part of the CCE. We focus on these larger water bodies not only because they are important to the CCE, but also because more is known about these waters than about others. We do not intend this chapter to be an exhaustive treatise on all the factors affecting aquatic ecosystems of the CCE; instead, it highlights the variety and complexity of factors that impinge on the environmental health of these systems.

Pressure on Freshwater Systems

Throughout most of the world's populated areas, freshwater has come under increasing pressure as the demand for clean and abundant water outpaces the ability of the watersheds to supply water. In the United States and Canada, meeting societal needs for power, transportation, and flood control has had a significant impact on many river basins. Dams for hydroelectric generation, water storage, and irrigation have dramatically affected hydrographic regimes and interactions between rivers, riparian floodplains, and downriver lakes. Reservoirs also fundamentally change river systems by converting running water into standing water. Organic and inorganic pollutants are often toxic to aquatic organisms, resulting in disease, reduced reproduction, and stunted growth rates (see also Chapter 9). Nutrient pollution from poor land management or direct-point sources often increases the productivity of noxious or aesthetically objectionable species (e.g., blue-green algae). Widespread introduction of exotic species is fundamentally changing many of North America's ecosystems and degrading native species by dramatically altering food webs or through competitive interactions. Many native populations of plants and animals, once plentiful, are becoming rare.

Federal Clean Water Act

Alarmed by loss of water quality in the nation's wetlands, streams, and lakes, the U.S. Congress passed the Federal Water Pollution Control Act of 1972 (33 USC §1344), which later became known as the Clean Water Act (CWA). This landmark legislation included requirements to improve water quality. Foundational language in the CWA stated that the purpose of the act was to "…restore and maintain the chemical, physical, and biological integrity of the waters of the United States." The overall purpose of the CWA was to achieve clean and healthy water that is "swimmable and fishable." Since 1972, however, it has become clear that restoring and maintaining the chemical, physical, and biological integrity of aquatic ecosystems requires more than removing harmful pollutants. Historically occurring organisms and the physical and biological processes that support ecosystem integrity must also be present and preserved. The ecologically valuable attributes of clean water, then, go far beyond drinkable, swimmable, and fishable. The distribution and abundance of various plants and animals—such as algae, aquatic insects, aquatic macrophytes, fish, birds, and wildlife—are structural attributes of the ecosystem that can be measured and assessed. Primary production, food-web complexity, and biogeochemical cycling are examples of aquatic ecosystem functions that we can characterize by measuring rates of processes (such as productivity) and nutrient cycling and retention. We can also characterize these ecosystem functions by making inferences from biomass and nutrient levels.

At the landscape scale (e.g., >1,000 km^2), aquatic ecosystems are affected by the geology of the catchment(s) they drain and by interactions with adjoining terrestrial ecosystems. Landscape-scale linkages include the terrestrially derived groundwaters that maintain flow in surface waters, the hydrogeomorphic character of wetlands, the longitudinal connections to small streams and large rivers, the lateral exchange of water and materials between river channels and riparian floodplain systems, and the flux of water and materials into and from lake systems. The flux and exchange of water, dissolved nutrients, organic matter, inorganic materials, and organisms characterize the freshwater network that innervates the landscape. These interactions also establish a region's aquatic ecosystems, described as wetlands, streams, rivers, and lakes. Because these systems are complex, we must view them holistically and within the context of hydrologic and geomorphic drivers that establish the character of wetlands, streams, river channels, floodplains, subsurface flow pathways, spring brooks, and lake features. As a result, we must consider aquatic systems and their health not as their separate parts, but rather as an integrated functioning system of groundwaters, wetlands, streams, rivers, riparian and hyporheic zones, ponds, and lakes.

Aquatic Ecosystem Health in the CCE

The CCE is affected by the same major stressors of aquatic ecosystem health that affect other ecosystems (e.g., deteriorating water quality, manipulated hydrographs [graphs showing changes in water flow over time], and expansion of non-native species). Although these stressors are less pervasive in the CCE than in other ecosystems, the decline in aquatic ecosystem health in the region is nonetheless measurable and increasingly acute.

Disturbance is a natural part of the CCE's terrestrial and aquatic systems. For example, fire is a natural, recurring, large-scale disturbance with return intervals of less than 100 years in some CCE forested areas. Similarly, although snow melts and runs off into CCE streams and lakes every spring, periodic heavy floods also inundate the rivers and their floodplains with water, sediment, and stream power. These forces move material and reestablish new, open surfaces where riparian vegetation can regenerate.

The quantity and timing of sediment entering lakes directly affects annual variation in nutrient load and food-web dynamics. Disturbances within historic ranges of spatial and temporal variation contribute to ecosystem heterogeneity and play a fundamental role in maintaining the high level of biological diversity and ecological complexity that characterizes the CCE. In contrast, human disturbances are often sustained or act as long-term factors that place permanent stressors on the ecosystem. Prime examples of human disturbance are logging, building and maintaining transportation corridors, developing residential and commercial areas, and disposing of sewage.

Flathead Lake Morphology and Response to Lake Regulation

Most of the CCE is contained within the Flathead Basin. The basin sits at the southern terminus of the Rocky Mountain Trench, a north–south-oriented fault-block valley that extends from central British Columbia into Montana. The Flathead Valley has been sculpted by a history of geological uplift and multiple glacial periods. Flathead Lake, one of the world's 300 "great lakes," is a dominant feature in the valley. The lake has a surface area of more than 496 km^2, a maximum depth of 113 m, and an average depth of >50 m. The last advance of the Flathead glacial lobe in the late Pleistocene achieved the last major scour of this fault-block basin and deposited terminal moraines along the south and west shores of today's lake. As the glacier retreated from the basin, large sediment loads entered the proglacial lake and were deposited as a thick sequence of deltaic sediments. This sequence spans what is now the north shore and the valley flat from Somers to Bigfork, extending as far north as Kalispell. Before humans controlled the level of the lake, the major landscape feature of the north shore was a large, accreting sandy delta at the mouth of the Flathead River. This delta was created simultaneously with a large island and sand spits that formed at different elevations corresponding to various lake base levels (Lorang et al. 1993a). Wave action caused a succession of sand spits and bars to form at the river mouth. A diverse wetland complex developed within the embayments created by the bars of each spit and within the flood channels of the delta island. This deltaic landscape was flooded annually as snowmelt in the upper basin discharged water into the lake. As a result, water storage increased and the lake level rose. As river discharge subsided, the lake level declined. The magnitude of lake level rise was determined by the peak flow and duration of the runoff.

In the mid-1930s, Kerr Dam was constructed on a bedrock sill about 8 km downstream of the Flathead Lake outlet near Polson. Although the dam plays a major role in controlling lake levels, its control is not absolute. During high-runoff years, water is often discharged from the dam to prevent overfilling of the lake. Lake level is also affected by the rate and magnitude of runoff and discharges from Hungry Horse Dam, located on the South Fork of the Flathead River. Alteration in lake levels has greatly affected shoreline erosion, particularly on the north shore. The silt and sand deposits that comprise the north shore delta, spits, and wetlands have been subjected to dramatic wave erosion and shoreline retreat. Consequently, nearly 1,000 ha of deltaic habitat have eroded into the lake (Lorang et al. 1993b). Shoreline retreat continues along much of the north shore east of the river mouth, although the rate varies from several meters per year in some locations to areas that have stabilized.

Anthropogenic regulation of lake levels has also resulted in the inland expansion of wetlands along the north shore, in eastern portions of Polson Bay, in western portions of Big Arm and Dayton bays, and in other small bays and inlets along the west shore of the lake. These areas were seasonally flooded grassy meadows before they were regulated by the Kerr Dam. Maintaining the lake level for an extended period of time changed these areas into aquatic littoral habitat during the summer and exposed mudflats during winter drawdown of the lake.

The remaining shoreline areas of Flathead Lake fall into two basic categories—exposed bedrock and gravel beaches. Deposits of glacial till are found along most of the east shore and many areas of the west shore, including the island shorelines. Since the glaciers retreated, continuous wave action on the shoreline cobbles of the glacial till has produced a narrow nearshore environment that consists of a cobble-gravel shelf, a steep sandy portion near the shore, and a gravel beach face. Anthropogenic lake-level regulation refocused wave energy at the upper reaches of the pre-dam foreshore, producing an armored zone where small-diameter gravel is in short supply. In addition, the wave energy caused new gravel beaches to form at the full-pool lake level wherever a sufficient supply of gravel material was available. The width and height of the gravel beaches formed around the lake at the "new" full-pool lake level depend on the available wave power and the supply and size of gravel. Many exposed points and similar areas with limited amounts of gravel have never equilibrated to the new lake-level environment and the extended energy of waves that occur during the summer when the lake is at the full-pool level. Indeed, the backshore environments of these areas are still exposed to direct wave action, which causes them to erode, during storms.

Shoreline structures such as seawalls, bulkheads, and riprap built by humans to protect the beach and backshore properties from erosion often interact with the waves by resistance rather than by energy dissipation. This causes end-scouring processes that accelerate erosion of unprotected neighboring properties. When the orientation of the shoreline is oblique to the incoming waves, the energy can move gravel along the shore. Some beaches exposed to wave action from opposite directions can experience no change in sediment deposition over the years. In contrast, some beaches have orientations, or exposures, that cause waves to approach from predominantly one direction. These shoreline areas experience a net transport of gravel in one direction, which can result in a buildup of gravel in one section of the downdrift beach location and a loss of gravel in the updrift beach location.

The restoration of a healthy shoreline largely depends on the extent to which adjacent landowners modify the lakeshore and build docks or cement bulkheads, or both. Freshwater river deltas are unique landscapes that act as biological focal spots. Seventy years ago, the north-shore delta of Flathead Lake was one of the largest freshwater-lake, river-delta, and wetland complexes in the United States. Although the river-delta/wetland complexes have been reduced in size by shoreline erosion in the ensuing years, they remain an important player in Flathead Lake's chemical, physical, and biological integrity. The complexes also serve as critical habitat for fish and wildlife. Pressure to fill wetland areas along the north shore of the lake to accommodate development would have a detrimental effect not only on the integrity of the associated landscape, but also on Flathead Lake's water quality. Economic gain for a few individuals does not justify the loss of irreplaceable and sensitive lakeshore habitat and the associated decline in water quality. Both lakeshore habitat and water quality are major environmental amenities for the whole valley.

Flathead Lake Water Quality

Perhaps there is no better indicator of aquatic health and general environmental health of the CCE than the water quality of Flathead Lake. For most of the year, the waters of Flathead Lake are clear and blue. Numerous variables indicate that water quality is declining, however, even though almost half (42%) of the Flathead Basin is in public land, which includes Glacier National Park (GNP) and the Bob Marshall Wilderness Complex. Water quality is declining because of the rapid development of private land.

More than 80,000 people now live in the Flathead Basin, however, and the basin has roads almost everywhere outside the designated wilderness areas and parks. Residential and commercial developments continue to expand, particularly in the Kalispell Valley between Bigfork and Somers and in the triangle formed by Kalispell, Whitefish, and Columbia Falls. Dominant extractive land uses include timber harvesting and agricultural (crop and livestock) production. Development and extractive industries contribute to the decline in water quality.

Primary productivity in Flathead Lake, which is a robust indicator of water quality, is strongly influenced by external nutrient loadings. Results from more than 25 years of water quality monitoring of the lake indicate that annual primary productivity has varied between 68 and 138 g of carbon (C) per cubic meter of water from 1978 to 2003 with a gradual, but statistically significant increasing trend in concentrations (Stanford and Ellis 2002; Figure 8-1). Primary productivity experiments designed to measure Flathead Lake's ability to grow algae show an increase in algal production over time, indicating a decrease in water quality. These experiments strongly support the conclusion that growth of algae in Flathead Lake is controlled by both nitrogen (N) and phosphorus (P) loadings to the lake (Spencer and Ellis 1990). Food-web changes,

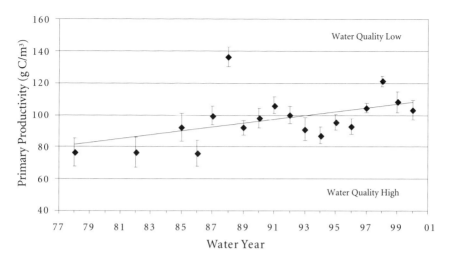

FIGURE 8-1. Annual Primary Productivity in Flathead Lake, 1978–2003

which alter the density of organisms that cycle nutrients within the lake, however, also have a major effect on primary productivity.

Mysis relicta, the opossum shrimp, was introduced into Whitefish, Swan, and Ashley lakes, upstream of Flathead Lake, in the late 1960s with the objective of stimulating the production of kokanee salmon (*Oncorhynchus nerka*). At one time, 90% of the game fish caught in the Flathead system were kokanee, and expenditures related to kokanee fishing amounted to $10 million per year (Stanford and Ellis 2002). Opossum shrimp gradually found their way downstream into Flathead Lake, where they were first captured in 1982. The population experienced a classic exponential growth rate and reached levels of more than 125 individuals per square meter by 1986. The shrimp are a voracious predator of the water flea (Cladocera). Between 1986 and 1990, Cladocera populations declined substantially as the Mysidacea predators became abundant (Spencer et al. 1991). The Cladocera decline led to the collapse of the kokanee fishery and an expansion in populations of lake trout and lake whitefish (see Chapter 2 for scientific names not given in this chapter). The subsequent shift in zooplankton could be responsible for a peak in primary productivity that was observed in 1988.

Nitrogen and phosphorus loadings to Flathead Lake occur via the Flathead River (with a mean discharge of 273 m³/s). In most years, the Flathead River accounts for more than 80% of the annual N and P loadings to Flathead Lake. As a result, during years of very high discharge in the Flathead River, total N and P loadings to Flathead Lake are very high. The Swan River (with a mean discharge of 30 m³/s) contributes, on average, less than 7% of the N and P loadings to the lake. These loadings are attenuated by Swan Lake, which is located in the lower portion of the Swan River drainage. Atmospheric deposition contributes approximately 10% of the annual N and P loadings to the lake, although this percentage is highly variable. During the last two decades, as much as 45% of the annual P load and 24% of the annual N load came to the lake as fallout from the atmosphere, mainly from fugitive dust from rural roads and smoke particulates from forest fires and crop burning. These sources of atmospheric fallout came from both inside and outside the Flathead Basin.

Nitrogen and phosphorus loadings to the Flathead River from the Stillwater River, the Whitefish River, and Ashley Creek, which flow through the major human settlements in the basin, are substantial given the small drainage area of these tributaries compared to the entire Flathead Basin. Measurements over a 15-year period indicate that the Stillwater and Whitefish rivers contributed as much as 24% and Ashley Creek as much as 15% of the total P load carried by the Flathead River to the lake. Nitrogen loadings have been smaller, at a maximum of 11% for the Stillwater and Whitefish rivers and 6% for Ashley Creek. In 1995 and 1996, Stanford and colleagues (1997) conducted a synoptic study of several tributaries in the Flathead River Basin. The investigators carried out the study to improve understanding of the impact of various land-use and other human activities in the lower portion of the basin on water quality in Flathead Lake. Nutrient concentrations and river discharge were determined at sites in headwater reaches above most of the human settlements and at several sites along each river course as it flowed through changing land uses. The researchers derived land-cover data from Landsat™ imagery and calculated proportions of land cover within each watershed (Stanford et al. 1997). Water quality in tributary streams was found to be lower in drainage areas that had land covers indicative of intense human use. In the urbanized Kalispell Valley, nitrate plus nitrite nitrogen loadings were up to 91 times greater than background loadings from agricultural and rural reaches of the Whitefish and Stillwater rivers and Ashley Creek (one synoptic on Ashley Creek was 528 times greater). Increases in total N and P loadings were generally much lower, with maximum increases over background loadings of 5.5 times and 5.1 times, respectively. Background loadings refer to nutrient loadings at the uppermost sites on each river course in regions little affected by human activities. Nitrogen loading from human sources upstream of Flathead Lake has steadily increased over the last three decades, and daily N loadings are weakly and positively correlated with primary productivity. Clearly, managers must consider N loadings from human activities in the basin when developing strategies for improving water quality in Flathead Lake.

Whitefish Lake Water Quality

Whitefish Lake is located at the western base of the Whitefish mountain range at the northern end of the Kalispell Valley. The lake was formed by glacial scour and the impoundment of the basin by Pleistocene morainal deposits. Whitefish Lake is 9.3 km long and 2.2 km wide, with a surface area of 13.2 km². The lake comprises about 4% of its total watershed area.

In 1982 and 1983, Whitefish Lake was in a transitional stage, moving toward eutrophication (in which oxygen content is depleted by organic nutrients; Golnar and Stanford 1984). Most parameters indicating that a lake is becoming more eutrophic (e.g., primary productivity, phytoplankton density and composition, total organic C, total N, and light extinction coefficients) were within the ranges typical of an oligotrophic lake (i.e., one that contains relatively little plant life and nutrients in its waters and is rich in dissolved oxygen from top to bottom throughout the year). Golnar and Stanford observed oxygen depletion, however, in the hypolimnion (the lower and colder layer of water) during late summer stratification. In addition, the researchers saw total P concentrations of greater than 5 μm/L in the epilimnion (the uppermost circulating layer of warm water). These characteristics typically show that a lake is beginning a trend toward enrichment and eutrophication.

A follow-up study in 2001 and 2002 showed that water clarity, hypolimnetic oxygen depletion, and epilimnetic nutrient concentrations were similar to those measured in 1983. Annual primary productivity and phytoplankton biomass, though, had both increased substantially. Since 1986, the annual maximum chlorophyll a concentration has nearly doubled. Green algae (Chlorophyta) have also become more prevalent in the algal community, with significant increases observed in 2002 over the levels seen in 1983. The increases in these metrics strongly indicate that water quality in Whitefish Lake has declined and that the lake is progressing toward eutrophication at a much faster rate than would naturally occur (Craft et al. 2003).

Increased nutrient loading—particularly of P, N, and C—is the major cause of accelerated eutrophication in most lake systems. Total P and total N concentrations in the epilimnion of Whitefish Lake have remained constant over the past 20 years. Soluble forms of N, primarily nitrite and nitrate, are consistently very low (i.e., at or near the detection limit of 0.6 μg of N per liter). Soluble reactive P is also low, but may be increasing slightly. Particulate organic C in Whitefish Lake was consistently low in all the samples collected in 1983, 1986, 1987, 1993, 2001, and 2002. Dissolved organic C concentrations have been fairly constant over the years. Total organic C within the water column is within the ranges for oligotrophic lakes (less than 1–3 mg/L), indicating that hypolimnetic oxygen depletion may indeed result from external nutrient loadings that stimulate in-lake algae production, rather than from C loading from wetlands in the catchment area above the lake.

About 40–70% of the annual bioavailable P loading and 60–70% of the annual total N loading to Whitefish Lake come from Swift Creek, which is the largest tributary (fourth-order stream) to the lake. Swift Creek drains 63% of the total watershed (208 km²). Bioavailable P concentrations in the creek appear to have increased slightly over the past 20 years. Total N concentrations also appear to be rising slightly, with more concentrations recorded above 100 µg/L in 2002 than in any other year.

The other significant tributary flowing into Whitefish Lake is Lazy Creek, a meandering lowland (900–1,200 m) second-order stream that drains 42 km^2 or 13% of the total watershed. Each year, Lazy Creek contributes approximately 10% of the total P and N loadings to Whitefish Lake. Although Lazy Creek's smaller size reduces the impacts of total loadings, it has the highest mean annual concentrations of P, N, and C of all the tributaries, which is a function of the creek's low-gradient wetland characteristic. Over the past 20 years, N concentrations have increased slightly and total P concentrations have remained fairly stable in Lazy Creek. Mean annual concentrations of dissolved organic C were very similar between water years 1983 and 2002, but peak concentrations in 2002 were double those measured in 1983 (Golnar and Stanford 1984). On the other hand, the mean annual concentration of particulate organic C was nearly three times lower in 2002 than in 1983.

Total P and N loadings from bulk precipitation (atmospheric fallout) in 2002 accounted for 15% of the total nutrient load to Whitefish Lake. Mean annual concentrations of total P were 4–9 times higher and mean annual N concentrations were 6–13 times higher in bulk precipitation samples than in stream samples.

Nutrient loading from shoreline groundwater sources makes a negligible contribution to the total annual nutrient load to Whitefish Lake. Jourdonnais and colleagues (1986) found that only 0.3% of the total P load was attributed to groundwater sources even though elevated nutrient inputs were documented in localized areas of land disturbance. It is possible that groundwater is now a more significant contributor to nutrient loading because of increased conversions of land from forested to urban or exurban uses since 1986. Of greatest concern, however, is the localized impact of elevated nutrient loadings to shoreline growth of periphyton (aquatic plants and animals that live attached to rocks and other submerged objects). In the 1983 study (Golnar and Stanford 1984), higher accumulations of benthic biofilms (periphyton) were observed in recently developed areas of the shoreline compared to forested shorelines. These accumulations are apparently related to nutrient-laden groundwater seeping into the shoreline.

Stream Aquatic Health in Pristine Watersheds

Nutrient dynamics in watersheds have been used for more than three decades as an overall indicator of how well ecosystems function or respond to disturbance (e.g., Bormann and Likens 1970). Typically, investigators evaluate nutrient dynamics to determine ecosystem responses to logging practices or other land-use changes. Nutrient dynamics have also been related to changing physical and biological characteristics. The concentrations and dynamics of N, P, and particulate organic C in McDonald Creek watershed, a large pristine watershed in the GNP, provide a good water quality baseline for the CCE.

McDonald Creek is a fourth-order stream that drains about 443 km^2 of the total McDonald watershed. Watershed and stream-channel morphology are greatly influenced by the sedimentary bedrock and glacial history of the area. Maximum elevation in the drainage is 2,912 m along the Continental Divide. Above Lake McDonald, McDonald Creek drains an area of 279.4 km^2 and has a mean annual discharge of 16.4 m^3/s, an annual maximum discharge of about 75–130 m^3/s, and a minimum discharge of approximately 1–3 m^3/s. Lake McDonald, a glacial lake on the valley floor near the terminus of the McDonald watershed, has an elevation of 961 m, a surface area of 2.78 km^2, and a maximum depth of 486 m.

Accumulation of deep snow at high elevations in the upper McDonald watershed is characteristic of many of the headwater streams of the Flathead Basin. Perhaps no other stream exemplifies this more than McDonald Creek. Snow water equivalent (SWE) accumulation within the watershed generally begins in mid-October, with maximum SWE occurring in late April. Average discharge in the stream typically follows a pattern established by the melting snowpack, with peak average discharge occurring in late May. Average maximum discharge from the creek is 131.2 m^3/s. Although the hydrograph of the stream above the lake tends to fluctuate more rapidly than that of the stream below the lake, the general pattern of discharge is nearly identical. The several small sub-basins that drain directly into Lake McDonald cause the stream below the lake to have a slightly higher discharge.

Nutrient loading throughout the drainage is exceptionally low (Hauer et al. 2002). Lake McDonald is extremely oligotrophic, with very high water clarity for most of the year. Phosphorus dynamics in the stream are typically different above and below the lake. Above the lake, total phosphorus (TP) concentrations in the stream increase approximately 10 times between the low-flow discharge that precedes spring snowmelt and the high discharge (Figure 8-2) because small amounts of P-containing sediment are mobilized by the higher discharge. In contrast, below the lake, TP concentrations in the stream remain remarkably constant regardless of discharge volume from the lake or the season because P entering the lake is rapidly metabolized. Nitrogen dynamics in the stream are distinctly dissimilar above and below Lake McDonald (Figure 8-3). Nitrate N comprised more than 95% of the total nitrogen (TN) concentration transported into or from the lake during all seasons. Comparatively higher TN concentrations are seen, though, during base flow from the upper basin than are observed leaving the lake. This pattern appears to result from relatively high nitrate concentrations in groundwaters that sustain stream discharge during base flow, along with the processing of nitrogen in Lake McDonald.

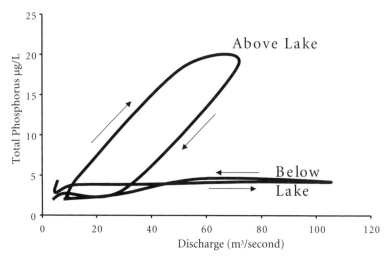

FIGURE 8-2. Total Phosphorous Concentration Entering and Departing Lake McDonald

TN concentration is generally 50 to 100 times higher than that of TP, and soluble reactive phosphorus (SRP) is generally one-half to one-tenth that of TP in the McDonald watershed, depending on the season and the rate of discharge. Thus, the ratio of nitrate to SRP is generally between 100:1 and 1,000:1.

Why does the upper McDonald watershed produce such a proportionately high concentration of nitrate N, and why is the ratio of N to P so high? Answers to these questions depend on N-fixation dynamics in the extensive alpine meadows that are so abundant in the McDonald watershed and the robust N-fixing ability of the alder trees that profusely cover avalanche chutes in the mountains. The interactions between stream-channel and shallow-aquifer waters associated with upstream floodplains may play a role as well. At any rate, the total concentrations of both P and N in the McDonald tributaries, the main creek, and Lake McDonald are extremely low. The significance of this is twofold. First, the McDonald watershed is typical of many of the pristine sub-basins of the Flathead Basin, which have very low levels of nutrients that are carried downstream to Flathead Lake. Second, because the nutrient levels are so low, even moderate nutrient loading from human sources can decrease water quality and aquatic health in ways that are scientifically measurable and directly observable by people who have lived in the basin for a few decades.

Stream Aquatic Health in Logged Watersheds

CCE resource managers have major concerns about land-use practices on the region's federal, state, and private forestlands. A controversy surrounding forest-management practices and the potential impact of those practices on streams

Aquatic Ecosystem Health

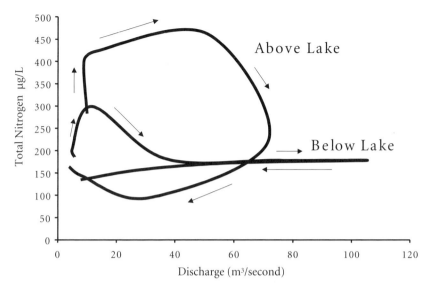

FIGURE 8-3. Total Nitrogen Concentration Entering and Departing Lake McDonald

and lakes has been ongoing for several decades. The controversy is particularly acute in the CCE because more than 80% of the region is in public or private forestlands. Studies in the mountainous regions of the United States indicate that loadings of fine inorganic sediment (a nonpoint-source pollutant) to streams located in watersheds where logging occurs have reached levels that are likely to impair aquatic resources (Hauer and Blum 1991). But sedimentation of streams and the potential direct consequences on fish populations and habitat are not the only concerns. Timber-harvesting practices also influence other determinants of stream ecosystem structure and function, including nutrient loading (Hauer and Blum 1991) and recruitment of woody debris (Hauer et al. 1999). These practices also affect riparian vegetation (Gregory et al. 1991) and macroinvertebrates (Van Sickle et al. 2005).

Examination of stream water chemistry for undisturbed versus logged/roaded watersheds shows distinctive responses in P and N loadings. Disparities in water chemistry values between streams often result from the heterogeneity between watershed geology and stream size. Differences in nutrient concentrations could, however, be directly attributable to differences in land-management practices. Even though nitrate and bioavailable P do not vary between comparable pristine and logged streams, TP and TN are significantly higher in streams that drain logged watersheds. This occurs because the biologically available sources of P and N are rapidly absorbed by algae growing in the streams, but the P and N then appear in particulate forms as TP and TN in the algae that either attach to the bottom of the stream or float down the stream (after sloughing from the stream bottom and banks). Figure 8-4 shows the

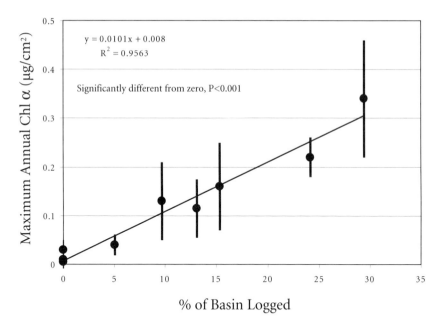

FIGURE 8-4. Effects of Logging on Annual Algal Biomass in Watersheds of the CCE

Note: Expressed as Chlorophyll a density (µg/cm²).

strong logarithmic relationship between the percentage of a watershed that is logged and the quantity of algae that grows on the surface of rocks in the pristine and logged watershed streams. We can see that logging and other disturbances cause more nutrients to be mobilized than are absorbed and used by algae. Increased algal growth is a distinct and clear expression of a decrease in water quality and aquatic ecosystem health.

River Floodplains and Aquatic Ecosystem Health

River ecosystem integrity and the aquatic health of a river are maintained across a variety of landscape scales. Variations in natural processes and human impacts on river ecosystems influence river health. To gain an understanding of the CCE's large rivers and floodplains, we must first take into account a landscape concept—the *shifting habitat mosaic* (SHM). The SHM concept gives us a framework for understanding how the dynamic nature of the river maintains the physical and ecological diversity of its habitats, its biotic communities, and its ecosystem integrity (Stanford et al., forthcoming). Although each river system has unique physical and ecological characteristics, many human activities and ecological effects can be expressed within the SHM concept. For example, societal needs for power generation, transportation, water management, and urban and agricultural development often alter natural processes of hydrologic regimes and material transport and deposition. These factors affect interactions between the river channel and the surrounding river–riparian corridor.

At the landscape scale, rivers change through interaction along three spatial dimensions associated with landscape character. The flow of water, materials, and organisms is primarily upstream–downstream among confined river reaches. Among unconfined floodplain reaches, however, there is also strong connectivity of flow both laterally, across the surface of the floodplain (especially during floods), and vertically, in which water from the channel penetrates into the layers of gravel that have filled the floodplain basin. Water and materials move dynamically and change with variation in landscape features, such as geomorphology, fluvial characteristics, and surface- and groundwater exchange. These landscape-scale characteristics directly affect the CCE's hydrographic patterns, thermal regimes and nutrient dynamics, along with the quantity and quality of organic matter sources and the distribution and abundance of plants and animals.

The degree of interactions among the river channel, the surrounding riparian areas, and subsurface hyporheic zones affect rates of change and the spatial extent of the SHM. The latter is maintained by associations among hydrographic regimes, distribution of stream power, sediment supplies, cut-and-fill alluviation, and the temporal interactions with the characteristic riparian vegetation (Hauer and Lorang 2004). Human-induced changes in river ecosystem structure and function at the landscape scale are varied and pervasive. Dams for hydroelectric generation, water storage, and irrigation have dramatically affected the longitudinal continuity of large river systems. Dam tailwaters are fundamentally different from unregulated river waters. Dams directly affect river hydrology, thermal regimes, biogeochemical cycling, and community structure of the riverine and riparian organisms. Hydrologic regimes that have been modified by regulation affect rates of cut-and-fill alluviation required to maintain the SHM, as well as water and material flux. Geomorphic alteration of river floodplains caused by agriculture, transportation corridors, urbanization, and other human activities fundamentally change river–riparian corridors. Our uses of land in a watershed can significantly degrade river ecosystems as sediment and nutrients are rapidly routed into the transport system.

Conclusions

National aquatic ecosystem restoration projects, whether undertaken in the United States or Canada, generally attempt to reduce nutrient loadings in lake systems and reintroduce heterogeneity to stream and river systems. Many of these efforts focus on engineering solutions, but because evaluations are rarely performed at the ecosystem level, the net effects of these restoration efforts on water quality and stream biota are unclear. Recovery of fish populations (e.g., trout abundance) is often used as the barometer of success. Gore (1985) warns, however, that a fish-centered view of recovery is myopic because lower trophic levels (e.g., macroinvertebrates or periphyton) must first recover to provide the food base for a stable fish community.

As a direct result of federal environmental protection, we have been able to reduce point-source chemical stressors on many lake and river systems. Yet numerous factors operating at the landscape scale continue to constrain the structure and function of aquatic systems. In restoring aquatic systems, we should focus on returning hydrographic and geomorphic dynamics—including the timing of high-water events—to magnitudes and durations that are within the natural range of variation. At the reach scale, we must restructure river floodplain confinement of otherwise unconfined reaches to permit lateral and vertical exchange of water and materials, as well as movement of organisms. We must manage land usage to reduce the movement of nonpoint sources of sediment, nutrients, and toxic pollutants into river systems. To restore lakes and rivers, we must integrate management strategies across appropriate spatial and temporal scales and carefully evaluate restoration approaches that are often landscape-level in their context.

Fortunately, because the CCE is positioned strategically at the headwaters of major river systems, the region's aquatic systems serve as a geographical focal point for the distribution of many plants and animals. The ecological integrity of these aquatic systems is threatened primarily by nutrient loading from human sources of pollution; introductions of exotic species that permanently change food webs; and river and lake regulations that manipulate lake levels, create reservoirs out of rivers, or alter hydrographic regimes in the downstream river systems. Although land uses have changed primarily in the valleys, timber harvesting and associated road construction threaten trout-spawning streams in national and state forests. Coal extraction and coal-bed methane development in the Canadian North Fork of the Flathead River (see Chapter 18) loom on the near horizon. Although the environmental impacts of widespread energy development in the Canadian North Fork are uncertain, we have seen significant environmental impacts of energy development in other locations. High-density well drilling, for example, requires construction of well sites, roads, and pipelines, all of which cause major surface disturbances.

The current status of aquatic health in the CCE is mixed. On the one hand, streams, rivers, and lakes in the CCE are healthy compared to other regions of the country. On the other hand, aquatic health has been significantly degraded from its original condition. The dominant species of fish in Flathead Lake are non-native and the introduction of Mysis has permanently altered the base of the food web.

The streams and rivers flowing out of the GNP and the Bob Marshall Wilderness Complex are among the most pristine anywhere in the world's central latitudes. But intensive human development ranging from private housing and commercial developments to large-scale resource extraction will continue to threaten the aquatic health of Flathead and Whitefish lakes, as well as the other lakes and streams in the Flathead Basin, in the years to come.

References

Bormann, F.H., and G.E. Likens. 1970. The Nutrient Cycles of an Ecosystem. *Scientific American* 223: 92–101.
Craft, J.A., J.A. Stanford, and B.E. Jackson. 2003. *Draft Whitefish Lake Water Quality 2003.* Open file report 177-03. Prepared for Whitefish County Water and Sewer District, Whitefish, Montana. Polson, MT: Flathead Lake Biological Station (FLBS) of The University of Montana.
Golnar, T.F., and J.A. Stanford. 1984. *Limnology of Whitefish Lake, Montana.* Open file report 080-84. Polson, MT: FLBS of The University of Montana.
Gore, J.A. (ed.). 1985. *The Restoration of Rivers and Streams: Theories and Experience.* Stoneham, MA: Butterworth Publishers.
Gregory, S.V., F.J. Swanson, W.A. McKee, and K.W. Cummins. 1991. An Ecosystem Perspective of Riparian Zones. *BioScience* 41: 540–51.
Hauer, F.R., and C.O. Blum. 1991. *The Effect of Timber Management on Stream Water Quality.* Open file report 121-91. Polson, MT: FLBS of The University of Montana.
Hauer, F.R., G.C. Poole, J.T. Gangemi, and C.V. Baxter. 1999. Large Woody Debris in Bull Trout Spawning Streams of Logged and Wilderness Watersheds in Northwest Montana. *Canadian Journal of Fisheries and Aquatic Sciences* 56: 915–24.
Hauer, F.R., D.B. Fagre, and J.A. Stanford. 2002. Hydrologic Processes and Nutrient Dynamics in a Pristine Mountain Catchment. *Verh. Internat. Verein. Limnol.* 28: 1490–93.
Hauer, F.R., and M.S. Lorang. 2004. River Regulation, Decline of Ecological Resources, and Potential for Restoration in an Arid Lands River in the Western USA. *Aquatic Sciences* 66: 1–14.
Jourdonnais, J.H., J.A. Stanford, F.R. Hauer, and R.A. Noble. 1986. *Investigation of Septic Contaminated Groundwater Seepage as a Nutrient Source to Whitefish Lake, Montana.* Open file report 090-86. Polson and Kalispell, MT: FLBS of The University of Montana and Montana Bureau of Mines and Geology, Montana College of Mineral Science and Technology.
Lorang, M.S., P.D. Komar, and J.A. Stanford. 1993a. Lake Level Regulation and Shoreline Erosion on Flathead Lake, Montana: A Response to the Redistribution of Annual Wave Energy. *Journal of Coastal Research* 9: 494–508.
Lorang, M.S., J.A. Stanford, F.R. Hauer, and J.H. Jourdonnais. 1993b. Dissipative and Reflective Beaches in a Large Lake and the Physical Effects of Lake Level Regulation. *Ocean and Coastal Management* 19: 263–87.
Spencer, C.N., and B.K. Ellis. 1990. Co-Limitation by Phosphorus and Nitrogen, and Effects of Zooplankton Mortality, on Phytoplankton in Flathead Lake, Montana, U.S.A. *Verh. Internat. Verein. Limnol.* 24: 206–09.
Spencer, C.N., B.R. McClelland, and J.A. Stanford. 1991. Shrimp Stocking, Salmon

Collapse and Eagle Displacement: Cascading Interactions in the Food Web of a Large Aquatic Ecosystem. *BioScience* 41: 14–21.

Stanford, J.A., and B.K. Ellis. 2002. Natural and Cultural Influences on Ecosystem Processes in the Flathead River Basin (Montana, British Columbia). In *Rocky Mountain Futures: An Ecological Perspective.* Edited by J.S. Baron. Washington, DC: Island Press, 269–84.

Stanford, J.A., B.K. Ellis, J.A. Craft, and G.C. Poole. 1997. *Water Quality Data and Analyses to Aid in the Development of Revised Water Quality Targets for Flathead Lake, Montana.* Open file report 142-97. Prepared for the Flathead Basin Commission, Kalispell, Montana. Polson, MT: FLBS of The University of Montana.

Stanford, J.A., M.S. Lorang, and F.R. Hauer. Forthcoming. The Shifting Habitat Mosaic of River Ecosystems. *Verh. Internat. Verein. Limnol.*

Van Sickle, J., C.P. Hawkins, D.P. Larsen, and A.T. Herlihy. 2005. A Null Model for the Expected Macroinvertebrate Assemblage in Streams. *Journal of the North American Benthological Society* 24: 178–91.

9

Conserving Water Resources in the CCE

James M. Byrne and Stefan Kienzle

Most people in North America perceive the CCE as an ultimate pristine place—a sanctuary of nature where abundant cold, fresh, and pure water, captured from Pacific winds by lofty peaks, nourishes a plethora of flora and fauna. Millions of visitors to the CCE's parks, wilderness, and recreational areas will testify to the quality of the environment and to the restorative powers that such dynamic wilderness brings to the body and soul. People travel great distances in all seasons to visit the CCE and enjoy its environmental amenities. In late spring, summer, and early fall, huge numbers of tourists descend on the major parks and wilderness areas in and near the CCE to hike, camp, and tour the region's mountains, lush forests, alpine meadows, lakes, rivers, streams, and waterfalls. In the winter, deep snows make backcountry and downhill skiing and other types of winter recreation popular activities.

The CCE has plenty of water because precipitation is abundant and because the water requirements of a cool alpine ecosystem are much lower than those of the surrounding warmer plains and valleys. For these reasons, the CCE has a positive water balance, which means it receives more water than it needs to sustain the alpine forests, meadows, and wetlands. In essence, water is the lifeblood of any ecosystem, and the CCE's lifeblood is robust.

We must, however, take care to preserve and protect these valuable water resources. In this chapter, we evaluate (1) the importance of water in the CCE; (2) the scientific basis for the challenges, problems, and threats to its water resources; and (3) the relationships between water and ecosystems in the region.

Many Rivers Run Through the CCE

The CCE contains a unique three-way continental divide west of Browning, Montana. Three major rivers flow from this point into three basins that cover much of the center of the continent and drain west to the Pacific Ocean, south and east to the Gulf of Mexico and the Atlantic Ocean, and northeast to Hudson Bay. The Saskatchewan River is a critical water source to the three prairie provinces: Alberta, Saskatchewan, and Manitoba. The Columbia River supplies water to British Columbia in Canada and to Idaho, Washington, and Oregon in the United States. The Missouri River flows through Montana, the Dakotas, Nebraska, Iowa, and Missouri before joining the Mississippi River near St. Louis.

Within the CCE, tributary watersheds are managed for water supply and other ecosystem goods and services. West of the Continental Divide, the Blackfoot River drains the southern regions of the CCE. The Flathead River drains most of the CCE west of the divide, and in the far north, the Elk River carries snowmelt and rainfall off the northern Rocky Mountains. All three rivers are tributaries of the mighty Columbia River, which flows west to the Pacific Ocean and forms the boundary between Washington and Oregon. The Milk and Two Medicine rivers and Cut Bank Creek flow east off the divide. Cut Bank Creek and Two Medicine River join to form the Marias River, which flows, along with the Milk River, into the Missouri River. The latter eventually joins the Mississippi River that drains to the Gulf of Mexico. The St. Mary River flows out of Glacier National Park (GNP) through St. Mary Lakes and north into Canada to join the Waterton, Belly, Castle, Crowsnest, and Oldman rivers, all tributaries of the Saskatchewan River system. The Highwood River drains the far northern reaches of the CCE, and joins the Bow River below Calgary. The Bow River traverses east and south across Alberta to join the Oldman River, where it becomes the South Saskatchewan River, which is part of the Saskatchewan-Nelson River System that runs to Hudson Bay.

Climate Interactions Result in Environmental Amenities

The environmental amenities that characterize the CCE result from interactions between the Rocky Mountains and the regional climate. Warm, moist Pacific air masses flow eastward over the CCE for most of the year. The soaring alpine peaks deflect these winds high into the atmosphere where rapid cooling causes vast quantities of water vapor to condense and fall as snow and rain. Surplus water runs off into streams and rivers or becomes groundwater. Precipitation in the high elevations of the CCE is much greater than in the adjacent plains and valleys. These interactions are the source of the water that makes the CCE such a verdant place.

The Importance of Water Resources

The CCE acts as a massive water collection and storage system. In late winter (March and April), vast amounts of water are stored in the CCE in the form of snow. With a meter of fresh snow being equivalent to about 10 cm of water,

3 m of fresh snow amount to approximately 30 cm of water equivalent. With the onset of regular above-freezing temperatures, this snowpack begins to melt. In most years, almost the entire snowpack melts within a few weeks, generating the equivalent of 30 cm of rainfall. Although this amount of precipitation as rainfall would be rare, these spring snowmelt events are common, and they cause most of the river runoff in the CCE. The most damaging floods happen in rare years when a substantial snowmelt coincides with major spring rainstorms across the region's watersheds.

On the prairies, *sublimation* is a critical process that depletes the snowpack. In this process, snow or ice passes directly from a solid phase to a vapor phase without becoming a liquid. The prairies immediately east of the CCE are famous for sublimation losses of snow to regular Chinook winds during the winter. The airflow warms by as much as 20°C during the 2,000-m drop-off from the Rockies to the plains below. Warm, dry Chinook winds often sublimate the entire snowpack for several hundred kilometers eastward. Chinooks have sublimated snowpacks for centuries, even millennia, as evidenced by the origin of the term. *Chinook* is a Blackfeet word meaning "snow-eater," a direct reference to the impact of Chinook winds on the plains of Montana and Alberta, the ancestral home of the Blackfeet Nation.

The Chinook effect has a major impact on water availability in the plains east of the CCE. Winter snows in these regions are minimal, and subject to decline and even elimination because of routine Chinook winds. In the mountains above the plains, cooler temperatures prevent sublimation of the high alpine snowpacks. Studies in the Upper Oldman watershed in southern Alberta indicate that sublimation has little impact at high elevation (see, for example, Lapp et al. 2005). As a result, snowmelt runoff from the CCE is the major source of water for the Alberta and Montana plains.

Snowpack in the CCE functions as a water storehouse that generates runoff in the spring and early summer. This runoff flows to lands as far away as Idaho, Washington, Montana, North and South Dakota, and states in the lower Missouri River Basin, as well as to the Canadian provinces of British Columbia, Alberta, Saskatchewan, and Manitoba. Snow accumulations in the CCE are often more than ten times those in the surrounding lowlands. The vast snow and ice fields of the CCE are a primary source of water for the environment and economy of much of the continent. Polluting the waters of the CCE can result in contamination that spreads to the ecosystems and populations of a number of states and provinces. Conserving the water resources of the CCE, then, is critical at both local and continental scales.

Global Environmental Change and the CCE

In geologic and biologic systems, natural change is the only constant, ordinarily taking place in geologic time frames of thousands to millions of years. The

CCE was heavily glaciated twelve thousand years ago, and the glaciers subsequently retreated slowly. More recently, the rate of change has accelerated. The GNP of 150 years ago contained 150 glaciers. Only 27 of those glaciers remain, and their ice mass has declined to about 10% of their ice volume in 1850 (see Chapter 12). The scientific literature indicates that this extremely rapid loss was accelerated by human-induced climate warming. Climate change is potentially the most significant environmental challenge facing humanity today (IPCC 2001), and the CCE is sensitive to even modest climate changes.

Climate change is not, however, the only environmental challenge facing the CCE. Human activities pose a range of threats to the region's aquatic ecosystems and water resources (Environment Canada 2001), including

- Biological and chemical contamination
- Physical disruptions and associated problems
- Persistent organic pollutants and mercury
- Endocrine-disrupting substances
- Nutrients such as nitrogen and phosphorus
- Urban runoff and municipal wastewater effluents
- Aquatic acidification
- Waterborne pathogens and enteric infections
- Agricultural and forestry impacts
- Effects of dams and diversions
- Climate change.

In the discussion that follows, we elaborate on many of these threats, particularly those occurring at or near alpine environments in the CCE.

Persistent Organic Pollutants and Mercury

The cold climate of the CCE is a natural trap for many atmospheric pollutants. Chemicals emitted by agriculture, industry, transportation, cities, power plants, and other human activities and constructs often remain in the atmosphere as gases for substantial periods of time. When these gases encounter cold climates, such as the high alpine areas of the CCE, they condense or are scavenged from the atmosphere by precipitation. As a result, deposition rates increase with elevation, a phenomenon known as *elevation deposition*.

Persistent organic pollutants (POPs) comprise a group of chemicals that degrade slowly in the environment, bioaccumulate, and have toxic properties (Environment Canada 2001). POPs have been identified as a critical contaminant trapped by orographic precipitation in Rocky Mountain snow and ice along the northern limits of the CCE. POPs in meltwater have concentrated in small crustaceans in alpine lakes (Blais et al. 2003), resulting in biomagnification in fishes near the top of the food web (Blais et al. 1998, 2001). Other studies (Davidson et al. 2003) report increased concentration of POPs in spruce tree needles at higher elevations in Alberta and British Columbia, adding further support to the elevation-deposition hypothesis. These researchers have clearly demonstrated that high alpine regions are traps for a number of volatile organic pollutants, including polychlorinated biphenyls (PCBs), hexachlorocyclohexanes (HCHs such as lindane), and endosulfans (a highly toxic crystalline insecticide).

Current evidence suggests that some of these POPs have traveled great distances in the atmosphere. Long-range transport (LRT) of pollutants is a well-known process that has been identified as a key driver of acid deposition. LRT is problematic in many areas of the world, including eastern North America, Europe, and Asia. One of the key pollutants in the Rocky Mountains is DDT (dichloro-diphenyl-trichloroethane), a pesticide banned in the United States in 1972. Air-mass tracking suggests that Asia and possibly Africa and South America, where DDT continues to be used for controlling agricultural pests and protecting against disease, are the current sources of DDT (Davidson et al. 2003).

POPs are also being transported within North America. Lindane, for example, is an insecticide that was used primarily to protect canola crops on the Canadian prairies until 2003. Ten to thirty percent of the chemical typically vaporizes and is transported elsewhere by the atmosphere. LRT modeling indicates that 25% of the time, the most likely transport route for this chemical is the front ranges of the Rocky Mountains. Extensive agriculture on the prairies was likely a key source of lindane, and possibly other agricultural chemicals, in the CCE (Waite et al. 2005). Donald and colleagues (1999) strengthen the evidence that substantial quantities of POPs deposited in Rocky Mountain snow and ice come from continental sources. During the growing season in North America (May through July), air-mass transport over a Rocky Mountain site just north of the CCE originated within the continent 50% of the time. This time period coincides with the period of greatest use of agricultural chemicals across the continent.

Endocrine-Disrupting Substances

Endocrine-disrupting substances (EDS) are pollutants that affect growth, development, and reproduction in biota, including humans, even at extremely low concentrations. PCBs are an acknowledged EDS that are found in alpine aquatic environments near the CCE (Blais et al. 2001, 2003). Because the same deposition mechanisms are active in the CCE, we can assume similar deposition rates. The disruptive capacity of other POPs/EDS deposited in the Rocky Mountains is unclear, but of substantial concern. Many metals, such as cadmium (Cd), zinc (Zn), copper (Cu), lead (Pb), and nickel (Ni), which are reportedly EDS (Levesque et al. 2003), are released into the environment from smokestacks at mining and smelting operations. Organochlorines such as lindane are transported as vapors from agricultural lands where they are applied by airborne or surface sprayers. EDS have an impact on a range of wildlife. Investigators have examined effects on populations of trout and perch (Levesque et al. 2003; Lacroix and Hontela 2004), white suckers (*Catostomus commersonii*; Dorval et al., 2005), and amphibians (Goulet and Hontela 2002). Mammals and birds that consume aquatic vertebrates and invertebrates are also at substantial risk (Damstra et al. 2002).

Our discussion here has identified likely sources for many of these contaminants in relative proximity to the CCE. New studies to investigate possible impacts of EDS/POPs on the CCE are under way or planned.

Nutrients such as Nitrogen and Phosphorus

Nutrients flushed into aquatic ecosystems, mainly nitrogen and phosphorus, increase the risk of eutrophication, a process where excessive nutrient loads promote aquatic vegetation and algal blooms. Eutrophication reduces dissolved oxygen content through the decomposition of vegetation, which causes aquatic fauna to die off and adversely affects habitat requirements of fish and their food resources (such as macroinvertebrate populations). Potential sources of nutrient contamination are forestry clear-cuts, agriculture, recreational areas (especially golf courses), and residential developments.

In the Canadian Rockies north of the CCE, effluents from municipal wastewater treatment plants result in elevated total phosphorus (TP) availability that enhances benthic algae abundance by 4- to 30-fold over natural occurrence (Bowman et al. 2005). Such TP loading could have a serious impact on aquatic species.

Urban Runoff and Municipal Wastewater Effluents

There are a number of towns and numerous smaller communities within the CCE, such as Elkford, Fernie, Crowsnest, and Waterton in Canada, and Whitefish, Kalispell, Columbia Falls, and Evergreen in Montana. Large proportions of these urban areas have sealed surfaces (e.g., streets, parking lots, driveways, and rooftops) that prevent water infiltration into the soil. This causes two major problems. First, because more precipitation becomes surface runoff, less water infiltrates the soil or recharges the groundwater, which often results in increased flooding downstream of an urban area. Second, any substance that is present on the sealed surface can be flushed untreated into the nearest stream through the urban storm-water system. Potential urban pollutant sources include lawns and parks that receive pesticides, fertilizers, pet feces, or all three. Other substances, such as oil, gasoline, tire rubber, heavy metals, and road salt, can run off from streets or driveways. Educating the public about these effects is essential in minimizing water pollution from urban runoff.

Aquatic Acidification

Deposition of acid pollutants was recognized as a critical environmental issue in the 1970s. At that time, acid emissions from coal-fired power plants and fossil-fuel-powered vehicles created acid precipitation that seriously harmed terrestrial and aquatic ecosystems in eastern North America and much of Eurasia. Acid precipitation had the greatest impact on areas with naturally acidic environments typical of the mixed-wood forests of those regions. Scientists thought that western North America was less affected by acid pollution because the soils and geology are basic rather than acidic, giving rise to a substantial buffering capacity. In 1985, however, the journal *Science* reported that the American West, including the Rocky Mountains, was experiencing serious acid pollution problems. More recent work reports substantial acid pollution in the CCE (Nanus et al. 2003), as well as deposition increases with elevation that result from orographic precipitation (Beniston 2000). These researchers found that sulfur (SO_4) compounds are deposited in amounts of up to 8 kg/ha and nitrogen (NO_3) levels reach up to 2 kg/ha. Higher deposition is taking place to the south of the CCE, and upper elevations in Colorado, Utah, and Wyoming receive up to 12 kg/ha of SO_4 and 2.5 kg/ha of NO_3.

Evaluating the impact of acid pollutants in the CCE is difficult. The U.S. Environmental Protection Agency classifies the northern Rocky Mountains, including the CCE, as highly sensitive to acid precipitation (Keller 1999). Water resources are likely to be most sensitive at a local scale where acidification of small alpine lakes and streams cannot always be neutralized because of thin soils. Acid pulses, typical of areas where acidic snowpacks melt rapidly, cause problems in aquatic ecosystems, particularly for biota in the rapid-development phase of life.

Little information is available in the CCE on the impacts of ozone pollution in the lower atmosphere or ozone depletion in the stratosphere, both of which lead to increased exposure to UV-B (ultraviolet radiation at 280–320 nm). Diamond and colleagues (2005) estimated UV-B exposure by amphibians in wetlands of the GNP but these exposure levels did not produce substantive negative effects for amphibian populations (Adams et al. 2005). Work in Europe indicates that spruce and pine forests can be damaged by elevated exposure to UV-B, increased ozone concentrations in the lower atmosphere, or both (Schnitzler et al. 1999). Emissions from vehicles and coal-fired power plants are likely to contribute to lower-atmospheric ozone in the CCE.

Waterborne Pathogens and Enteric Infections

Intestinal or enteric infections in humans and animals are typically caused by food- or waterborne pathogens. Waterborne pathogens, which include a series of microbes, are clearly an issue given the rapid increase in infection rates that we have seen in the last several decades. A number of urban centers in North America have experienced major infectious outbreaks (Medema et al. 2002). Some of the most prominent outbreaks have resulted from bacteria (*Escherichia coli*; *Salmonella*), protozoa (*Giardia, Cryptosporidium*), and viruses (*Norwalk, hepatitis, rotavirus*).

Little work has been done on the occurrence and/or persistence of waterborne pathogens in the CCE. Research in Alberta (Johnson et al. 2003) established that livestock operations are a primary source of waterborne pathogens, and most of the contamination found was proximal to higher livestock densities away from the alpine areas. Wherever possible, steps must be taken to limit livestock access to watercourses to prevent accumulations of manure that can contaminate streams and lakes during and after significant runoff events.

In some cases, pathogen-related infections can be traced to wildlife. For example, a major urban infectious outbreak occurred in British Columbia when a raw water source in an upland area became contaminated with *Toxoplasma gondii* oocysts. Infected cougars were most likely the source of the infectious agents (Isaac-Renton et al. 1998). Drinking raw (untreated) water anywhere in the CCE (or anywhere else) is not advised except under emergency circumstances.

Agricultural and Forestry Land-Use Impacts

Agriculture is limited in many areas of the CCE by climate, steep terrain, and public land ownership in protected areas. Converting native grassland and forestland to cropland creates several water problems, including increased contamination by sediment and nutrients (derived from both the soil and fertilizers) and chemical pesticides and herbicides. Livestock grazing on steeper slopes and alpine meadows causes soil erosion and compaction, which change water absorption and runoff characteristics. In addition, livestock often carry pathogens that may reach watercourses as we discussed earlier.

Clear-cutting of forests has a profound impact on surface runoff. For example, runoff increased by 70 mm from a conifer forest when 20% of the forest cover was removed; clear-cutting (i.e., 100% removal of forest cover) increased runoff up to 350 mm (Binkley and MacDonald 1993). Increased surface runoff is always associated with increased erosion and sediment loading to the streams. Impacts on individual streams vary widely depending on topography, soils, land cover, and precipitation. During recent decades, improved road drainage, use of buffer strips, and other best management practices (BMPs) have substantially reduced the effect of forest-management activities on erosion and sediment yields (Binkley and MacDonald 1993). Sediment yields from inappropriate (non-BMP) forest-management practices, however, can be 2 to 50 times higher, with road erosion being the most important source (Reid 1993).

Clear-cut areas have lower water losses from transpiration and interception, which translates into higher soil moisture and runoff. Snow-catch efficiency can be increased when the width of the gap between trees is equivalent to only two to three tree heights. But clear-cuts usually result in much larger gaps that reduce snow-catch efficiency. Snow typically melts earlier in openings than under the forest canopy because of rapid warming from solar radiation, turbulent transfer of heat to the site, and water vapor transfer away from the clear-cut areas.

Effect of Dams and Diversions

Dams and water diversions in the Rocky Mountains disrupt and segment ecosystems. In the Montana Rockies alone, there are more than 2,700 dams. Impacts of dams and diversions on water include

> ...thermal stratification within the reservoir and modification of downstream water temperatures; eutrophication; promotion of anoxic conditions in hypolimnetic water and related changes in metal concentrations in outflow; increased methylation of mercury; sediment retention; associated changes in TDS, turbidity and nutrients in the reservoir and discharged water; increased erosion/deposition of downstream sediments and associated contaminants. Flow diversions can also produce major changes in water quality. The most dramatic shifts result from mixing of waters from disparate hydro-ecological systems (e.g., across major hydrologic divides or from freshwater to estuarine environments), resulting in changes in chemistry, temperature and sediment. In addition, the transfer of fish, parasites, and pathogens can accompany such mixing. (Environment Canada 2001, 69)

Increasing demand and declining supply of water often results in calls for more storage of water. There are demands for more dams and reservoirs, primarily to support expansion of irrigated agriculture. We address the uncertainty of future water supply in the next section, offering scientific evidence as to why it may be inappropriate to build more dams and water storage in the CCE.

Climate Change

Throughout this chapter, our discussion has demonstrated that water is vital to the CCE. It defines the nature of the ecosystem and makes the Rocky Mountains a source of water for regions both near and far. Because climate and water are intimately linked, continuing global atmospheric change will bring changes to the CCE that have not been observed for thousands of years.

Developing scenarios to describe how climate change might affect alpine regions is challenging. Because rapid changes in elevation, slope, and orientation alter climate inputs to the landscape, climate varies over very short distances in alpine areas. In addition, local variations in soils and other land characteristics make land cover and plant communities spatially diverse. Generally, rainfall supplements soil water and increases biomass productivity, and snowfall accumulates and creates runoff in the spring and summer. Climate warming in the CCE is expected to make winters shorter and warmer and to reduce snowpack, resulting in less runoff. Leung and Ghan (1999) used the linked GCM-RCM (Global Circulation Model-Regional Climate Model) to simulate regional climate in the Pacific Northwest (PNW) over terrain that has substantial relief. Their results suggest that precipitation will increase in the PNW and a warmer climate will cause snowpack to decline by up to 50%. Other researchers (Lapp et al. 2005) forecast a decline of up to 40% in average winter snow accumulations in the Upper Oldman Basin, Alberta, near the northern extent of the CCE. Hayhoe et al. (2004) report widely variable but substantial declines in alpine snow accumulations and resulting decreases in water supply from the Sierra Nevada Mountains in California. In south–central British Columbia, Leith and Whitfield (1998) predict that climate change will result in more rainfall in late fall and early winter, causing a modest increase in rainfall runoff during this period, but substantially reducing early spring runoff and streamflow in the subsequent summer months. Forecasts of less snow accumulation are in keeping with observations that ice accumulations are shrinking rapidly. The GNP, which is at the core of the CCE, has lost most of the glaciers that existed in 1850, and those remaining have lost up to 90% of their ice mass.

Climate change will have a dramatic impact on, for example, the ski industry. Shortened ski seasons and mid-winter snowpack melting on lower slopes will shorten the ski season and reduce the industry's contribution to economic output and employment. We saw this phenomenon during the winter of 2004–2005, when some ski hills had poor or nonexistent snowpacks. Lapp and colleagues (2005) modeled alpine snowpack variation with elevation under a climate-warming forecast. They found that a complete melting of snow at lower elevations during the mid-winter period is likely even in years with substantial winter precipitation. In the winter of 2004–2005, at least one ski hill was rained out of operation. The most important limiting variable is not precipitation, however, but temperature.

Conclusions and Recommendations

Climate, location, and elevation interact to make the CCE a critical water capture zone in western North America. A clean, abundant water supply is an integral component of any ecosystem. Snowmelt and rainfall from the CCE feeds into three major continental river systems that flow to Hudson Bay and the Pacific and Atlantic oceans. More importantly, the CCE furnishes water for riparian ecosystems, agriculture, forestry, industry, recreation, and urban uses.

Human activities, including agriculture, forestry, tourism, and residential development, are having serious impacts on water quantity and quality in the CCE. Global climate change has reduced snow accumulations and forced the melting of glaciers throughout the region. Cold mountain climates capture volatile chemicals from the atmosphere, resulting in contamination of snow, ice, and aquatic ecosystems in the remote areas we perceive as pristine. Acid pollutants from coal-fired power plants and the transportation industry, along with increased energy consumption in western North America, are changing water chemistry and having deleterious effects on aquatic life in the CCE. The region's natural and economic health and wealth are intimately linked to the quality and quantity of water resources.

Dams, reservoirs, and diversions fragment riparian ecosystems, change flow regimes, degrade water quality, and alter the thermal properties of rivers. Consequently, changes in water temperature are now altering entire aquatic systems that evolved with more normal water temperatures. Increasing pressure to achieve more secure water supplies is expected to increase the demand for water storage capacity. Building additional capacity would result in additional environmental impacts on the CCE.

If our negligence continues, the safety and sustainability of water supplies and the health of the CCE ecosystem will be impaired for current and future generations. Because current mitigation strategies alone will not significantly alter the process of climate change, we must develop and implement adaptation strategies. The time has come to adopt a new paradigm that addresses water resource issues in the CCE and beyond.

In the list that follows, we outline several recommendations for scientists, managers, and indeed, society at large, to consider:

- Sources of POPs and EDS must develop new control techniques to reduce these pollutant loadings. The EDS research we cited in this chapter shows that these toxins adversely affect the biosphere (including humans) at concentrations that were undetectable only a few decades ago.
- Wastewater treatment plants must not dispose of nutrients from sewage effluent in rivers and lakes. Agricultural and forestry operations and urban/industrial complexes must develop means, such as the BMPs that have proven successful, to reduce or efficiently treat nutrient loads. James M. Byrne has toured a number of urban wastewater treatment operations in which the plant discharges equivalent or better quality water back to the environment, or better yet, uses the output stream as a source of industrial or irrigation water (or both). Further treatment options include using reverse osmosis technology that creates high-quality water (Comerton et al.

2005). Better practices and improved treatment of waste streams also minimize threats from waterborne pathogens. Finally, treatment plants should incorporate ozonation and/or UV disinfection for pathogens, which are highly effective, into treatment processes.
- Power plants and the transportation industry must find ways to reduce acid and metal pollutants. High alpine ecosystems are extremely sensitive to deposition or biomagnification of these compounds.

Climate change is possibly the single greatest threat to alpine ecosystems because it can dramatically reduce water supply in some years. In addition, it increases evapotranspiration because of extended growing season length and intensity, and is likely to reduce the dilution capacity of pollutant concentrations. These potential climate change impacts increase the urgency of addressing the issues presented in this chapter.

References

Adams, M.J., B.R. Hossack, R.A. Knapp, P.S. Corn, S.A. Diamond, P.C. Trenham, and D.B. Fagre. 2005. Distribution Patterns of Lentic-Breeding Amphibians in Relation To Ultraviolet Radiation Exposure in Western North America. *Ecosystems* 8: 488–500.

Beniston, M. 2000. *Environmental Change in Mountains and Uplands*. New York: Oxford University Press, 88–9.

Binkley, D., and L. MacDonald. 1993. *Forests as Non-Point Sources of Pollution, and Effectiveness of Best Management Practices*. Technical bulletin. New York: National Council for Air and Stream Improvement.

Blais, J.M., F. Wilhelm, K.A. Kidd, D.C.G. Muir, D.B. Donald, and D.W. Schindler. 2003. Concentrations of Organochlorine Pesticides and PCBs in Amphipods (*Gammarus lacustris*) along an Elevation Gradient in Mountain Lakes of Western Canada. *Environmental Toxicology and Chemistry* 22: 2605–13.

Blais, J.M., D.W. Schindler, D.C.G. Muir, M. Sharp, D.B. Donald, M. Lafreniere, E. Braekevelt, M. Comba, and S. Backus. 2001. Melting Glaciers are a Dominant Source of Persistent Organochlorines to Subalpine Bow Lake in Banff National Park, Canada. *Ambio—A Journal of the Human Environment* 30: 410–15.

Blais, J.M., D.W. Schindler, D.C.G. Muir, D.B. Donald, and B. Rosenberg. 1998. Accumulation of Persistent Organochlorine Compounds in Mountains of Western Canada. *Nature* 395: 585–88.

Bowman, M.F., P.A. Chambers, and D.W. Schindler. 2005. Epilithic Algal Abundance in Relation to Anthropogenic Changes in Phosphorus Bioavailability and Limitation in Mountain Rivers. *Canadian Journal of Fisheries and Aquatic Sciences* 62: 174–84.

Comerton, A.M., R.C. Andrews, and D.M. Bagley. 2005. Evaluation of an MBR–RO System to Produce High-Quality Reuse Water: Microbial Control, DBP Formation and Nitrate. *Water Research* 39: 3982–90.

Damstra, T., S. Barlow, A. Bergman, R. Kavlock, and G. Van Der Kraak. 2002. *Global

Assessment of the State of the Science of Endocrine Disruptors. WHO/PCS/EDC/02.2. Geneva, Switzerland: World Health Organization.

Davidson, D.A., A.C. Wilkinson, J.M. Blais, L.E. Kimpe, K.M. McDonald, and D.W. Schindler. 2003. Orographic Cold-Trapping of Persistent Organic Pollutants by Vegetation in Mountains of Western Canada. *Environmental Science and Technology* 37: 209–15.

Diamond, S.A., et al. 2005. Estimated Ultraviolet Radiation Doses in Wetlands in Six National Parks. *Ecosystems* 8: 462–77.

Donald, D.B., J. Syrgiannis, R.W. Crosley, G. Holdsworth, D.C.G. Muir, B. Rosenberg, A. Sole, and D.W. Schindler. 1999. Delayed Deposition of Organochlorine Pesticides at a Temperate Glacier. *Environmental Science and Technology* 33: 1794–98.

Dorval, J., V. LeBlond, C. DeBlois, and A. Hontela. 2005. Oxidative Stress and Endocrine Endpoints in White Suckers (*Catostomus commersoni*) from a River Impacted by Agricultural Chemicals. *Environmental Toxicology and Chemistry* 24:5.

Environment Canada. 2001. *Threats to Sources of Drinking Water and Aquatic Ecosystem Health in Canada.* NWRI Scientific Assessment Report Series No. 1. Burlington, ON: National Water Research Institute.

Goulet, B., and A. Hontela. 2002. Toxicity of Cadmium, Endosulfans and Atrazine in Adrenal Steroidogenic Cells of Two Amphibian Species, *Xenopus laevis* and *Rana catesbeiana. Environmental Toxicology and Chemistry* 22: 2106–13.

Hayhoe, K., D. Cayan, C.B. Field, P.C. Frumhoff, E.P. Maurer, N.L. Miller, S.C. Moser, et al. 2004. Emissions Pathways, Climate Change, and Impacts on California. *Proceedings of the National Academy of Sciences* 101: 12422–27.

IPCC (Intergovernmental Panel on Climate Change). 2001. *Climate Change 2001: Impacts, Adaptation, and Vulnerability Technical Summary.* A Report of Working Group II of the Intergovernmental Panel on Climate Change 2001, WMO and UNEP. Cambridge, UK: Cambridge University Press.

Isaac-Renton, J.,W.R. Bowie, A. King, G.S. Irwin, C.S. Ong, P. Fong, M.O. Shokeir, and J.P. Dubey. 1998. Detection of *Toxoplasma gondii* Oocysts in Drinking Water. *Applied and Environmental Microbiology* 64: 2278–80.

Johnson, J.Y.M., J.E. Thomas, T.A. Graham, I. Townshend, J. Byrne, B. Selinger, and V.P.J. Gannon. 2003. Prevalence of *Escherichia coli* O157:H7 and *Salmonella* spp. in Surface Waters of Southern Alberta and its Relation to Manure Sources. *Canadian Journal of Microbiology* 49: 326–35.

Keller, E.A. 1999. *Introduction to Environmental Geology.* Upper Saddle River, NJ: Prentice Hall.

Lacroix, A., and A. Hontela. 2004. A Comparative Assessment of the Adrenotoxic Effects of Cadmium in Two Teleost Species, Rainbow Trout, *Oncorhynchus mykiss*, and Yellow Perch, *Perca flavescens. Aquatic Toxicology* 67: 13–21.

Lapp, S., J. Byrne, I. Townshend, and S. Kienzle. 2005. Climate Warming Impacts on Snowpack Accumulation in an Alpine Watershed: A GIS-based Modeling Approach. *International Journal of Climatology* 13: 521–36.

Leith, R.M., and P. Whitfield. 1998. Evidence of Climate Change Effects on the Hydrology of Streams in South-Central B.C. *Canadian Water Resources Journal* 23: 219–30.

Leung, L.R., and S.J. Ghan. 1999. Pacific Northwest Climate Sensitivity Simulated by a

Regional Climate Model Driven by a GCM. Part II: $2 \times CO_2$ Simulations. *Journal of Climate* 12: 2031–53.

Levesque, H., J. Dorvaland, and A. Hontela. 2003. Hormonal, Morphological and Physiological Responses of Yellow Perch (*Perca flavescens*) to Chronic Environmental Metal Exposures. *Journal of Toxicology and Environmental Health*, Part A 66: 657–76.

Medema, G.J., P. Payment, A. Dufour, W. Robertson, M. Waite, P. Hunter, R. Kirby, and Y. Andersson. 2002. Chapter 1: Safe Drinking Water: An Ongoing Challenge. In *Resolving the Global Burden of Gastrointestinal Illness: A Call to Action*. Edited by P. Payment and M.S. Riley. Washington, DC: American Academy of Microbiology.

Nanus, L., D.H. Campbell, G.P. Ingersoll, D.W. Clow, and A.M. Mast. 2003. Atmospheric Deposition Maps for the Rocky Mountains. *Atmospheric Environment* 3: 4881–92.

Reid, L.M. 1993. *Research and Cumulative Watershed Effects*. General Technical Report PSW-GTR-141. Albany, CA: U.S. Department of Agriculture (USDA) Forest Service, Pacific Southwest Research Station.

Schnitzler, J.-P., C. Langebartels, W. Heller, J. Liu, M. Lippert, T. Döhring, B. Thorsten, S. Günther, and H. Sandermann. 1999. Ameliorating Effect of UV-B Radiation on the Response of Norway Spruce and Scots Pine to Ambient Ozone Concentrations. *Global Change Biology* 5: 83–95.

Waite, D., T. Hunter, G. Fraser, and B.J. Wiens. 2005. Atmospheric Transport of Lindane (τ-hexachlorocyclohexane) from the Canadian Prairies—A Possible Source for the Canadian Great Lakes, Arctic and Rocky Mountains. *Atmospheric Environment* 39: 275–83.

PART IV
Ecosystem Dynamics

10
Paleoperspectives on Climate and Ecosystem Change

*Greg Pederson, Cathy Whitlock,
Emma Watson, Brian Luckman,
and Lisa Graumlich*

The rugged beauty of the glacial landscapes, the assemblage of plants and animals that currently inhabits the CCE, and the resulting array of ecosystem services are the result of the intricate interplay of past geological, climatic, and ecological processes across a complex landscape. To assess the sustainability of these resources, we must first understand the effects of past, present, and future climate variability and change on the ecosystem. For instance, forests have long lags in responding to climatic shifts, which means that what we see as "normal" is a combination of past and present conditions. If land managers are to sustain current resources, they need paleoperspectives that give them a realistic view of what those resource conditions should be. Paleoenvironmental research seeks to document and understand past responses to environmental change and serves as a benchmark for evaluating the current state of the region. In this chapter, we examine conditions that have shaped the region in the 20,000 years since the last glacial maximum during which the modern climate, landscape, and vegetation patterns have developed. These patterns are also susceptible to projected future climate changes.

Mountain landscapes such as the CCE are particularly suited to paleoenvironmental research because strong climate and altitudinal gradients juxtapose vastly different ecological communities. The steep environmental gradients result in sharply defined ecotone systems (e.g., upper and lower treeline) that respond relatively rapidly to climate changes as species shift their distributions. The CCE also contains watersheds in various stages of deglaciation, which range from the major valleys that were vacated by glaciers 15,000 years ago to high-elevation valleys in which glaciers are retreating today. These landscapes offer a "natural" laboratory in which we can study ecological impacts associated with past and present climate change and the disappearance of glaciers and perennial snowfields. Because they have experienced dramatic changes in the past, these climatic, topographic, and ecological systems afford a unique opportunity to investigate the causes of these changes. Such studies are of great societal interest because we can expect the systems to undergo similar changes in the future.

In paleoenvironmental research, the greatest challenges are identifying the causes of climate change, the regional climate signal, and the spatial and temporal scales of the environmental response. Paleoenvironmental research in the CCE has identified climate variations that range in length from seasonal to multimillennial over the last 20,000 years. The shorter climate variations are superimposed on the longer ones, and each scale of variation has left an imprint on the CCE's present-day environment and ecosystems (Whitlock et al. 2002). Climate variations on time scales from 10,000 to 100,000 years are attributed to slowly varying changes in the earth's orbit (Kutzbach et al. 1998). Superimposed on these long-term variations are centennial-scale (i.e., 100-year) climate changes. Although the causes of these centennial-scale variations are not well known, researchers have suggested changes in solar activity, greenhouse gases, and volcanic activity (see, for example, Overpeck et al. 2005). The last millennium has given us two examples: (1) a period of warm and dry conditions in the western United States from approximately 900 to 1300 ad, known as the Medieval Warm Period (Bradley et al. 2003); and (2) the Little Ice Age (LIA), which lasted from approximately 1300 to 1850 ad (Carrara 1989). The LIA is defined as a period when glaciers in the northern hemisphere reached their greatest extent during the Holocene (the past 10,000 years), and it includes some of the coldest periods experienced in the Holocene. The magnitude of climate changes during these centennial-scale events is less than those on longer time scales, but environmental responses still include changes in glacial extent, disturbance frequency, and species distributions.

On shorter time scales, twentieth-century instrumental records and proxy climate data for several past centuries indicate that changes at decadal to multidecadal scales are a well-defined component of the climate history of the CCE and western North America. Decadal shifts in climate are particularly well expressed in precipitation reconstructions, and many records indicate historical decadal-scale droughts or wet periods that surpassed twentieth-century events in magnitude, intensity, and duration. These events, which were often regional to subcontinental in scale, began and ended rapidly (within a few years). The timing of rapid climate changes (or regime shifts), and their spatial expression suggests that such events are related to changes in sea-surface temperature and pressure anomalies in both the Atlantic and Pacific oceans (see, for example, Gray et al. 2003). In other words, events that begin in both oceans are likely to affect the climate of the CCE.

These decadal shifts in moisture and temperature cause regional ecological responses. For example, a recurring multidecadal shift in Pacific sea-surface temperatures (e.g., the Pacific Decadal Oscillation [PDO]; Mantua et al. 1997) has driven changes in distribution and abundance of fish populations along North America's northeastern Pacific Coast. Decadal-scale droughts in the semiarid woodlands of the American Southwest have resulted in widespread bark beetle outbreaks, forest fires, and tree mortality (e.g., Allen and Breshears 1998). Likewise, decadal shifts in moisture and temperature have led to rapid changes in glacier mass balance in both coastal and continental glacier systems (e.g., Bitz and Battisti 1999; Pederson et al. 2004; Watson and Luckman 2004a). The paleoenvironmental record of rapid changes in climate and their consequences holds important messages for natural resource managers interested in protecting essential ecosystem goods and services.

Interannual climate variability is a shorter, well-known mode of climate change that is largely attributed to the El Niño Southern Oscillation (ENSO). ENSO is driven by ocean–atmosphere teleconnections related to patterns of warming (El Niño) and cooling (La Niña) in the central and eastern equatorial Pacific. These patterns come into play about every 2 to 7 years (Chang and Battisti 1998). The strength of these teleconnections appears to vary through time and they typically result in differing regional moisture conditions. For example, El Niño events have resulted in warm, dry conditions in the CCE and the U.S. Pacific Northwest and in higher levels of precipitation in the southwestern United States. La Niña, on the other hand, has generally coincided with cool, moist conditions throughout the CCE and U.S. Pacific Northwest while triggering droughts in the Southwest. This mode of climate variability appears strongest in the Southwest (e.g., Swetnam and Betancourt 1998) and seems to be relatively weak and spatially variable in the Pacific Northwest (e.g., Dettinger et al. 1998). Consequently, the paleorecord of ENSO impacts in the CCE is relatively weak. In the discussion that follows, we focus mainly on longer-term climate changes that span decades to millennia.

Reconstructing Environmental History

Many natural archives have paleoenvironmental data to offer. The data we describe here were recovered from sites within and around the CCE. Information on past changes in vegetation assemblages and abundance comes primarily from the sediments of lakes and wetlands, which incorporate pollen, charcoal, and other plant remains as the sediment accumulates. Because most lakes in the northern Rocky Mountain region were formed following late-Pleistocene glaciation, they provide a sedimentary record spanning the last 15,000 years or longer, depending on the time of ice retreat. Sediment cores are retrieved from modern lakes and wetlands using anchored platforms in the summer or from the ice surface in the winter (Figure 10-1, top right).

The sedimentary record of most lakes, which is usually 5–10 m in thickness, is recovered in a series of core segments. Laboratory analyses of the extracted

FIGURE 10-1. Reconstructing Environmental History

Notes: (Left) Melissa Hornbein and Karen Holzer of the U.S. Geological Survey sample an ancient drought-sensitive Douglas fir in Glacier National Park, Montana (photo by Greg Pederson). (Top right) Cathy Whitlock, Montana State University, assisted by several research scientists, prepares an anchored coring platform for sampling. (Bottom right) Brian Luckman, University of Western Ontario, and several research assistants cut a cross-section from ancient deadwood that recently emerged from the snout of the Columbia Glacier in Alberta.

cores typically involve detailed examination of the lithology (i.e., rock composition and texture) and microfossils. Samples for pollen and charcoal analyses are removed from the cores at regular intervals that depend on the detail and temporal resolution required. In the laboratory, the pollen extracted from the sediment is chemically treated, identified under the microscope, and tallied for each sediment level sampled. Pollen counts are converted to percentages of terrestrial pollen, which are then plotted as pollen-percentage diagrams. The reconstruction of past vegetation (and climate) from pollen percentages rests on the relationship between modern pollen rain and present-day vegetation and climate. After modern pollen samples were collected at lakes throughout North America, this information was calibrated to the modern vegetation and climate.

Past fire activity is inferred from the analysis of particulate charcoal, which is extracted and tallied from the sediment cores (Whitlock and Bartlein 2004). High-resolution charcoal analysis involves extracting continuous samples from the sediment core such that each sample represents a decade or less of sediment accumulation. These samples are washed through sieves and the charcoal residue is tallied under a microscope. Examining these relatively large particles enables local fire regimes to be reconstructed, because large particles do not travel far from a fire. Charcoal counts are converted to charcoal concentrations, which are then divided by the deposition time of each sample (year per centimeter) to yield charcoal accumulation rates (particle per square centimeter per year). Detection of fire events involves identification of charcoal accumulation rates above background levels. The fire-history record inferred from lake sediments extends thousands of years into history and supplements shorter records reconstructed from tree-ring data and historical fire atlases.

The chronology for these long records is established from a series of radiocarbon dates obtained on organic matter. These calibrated radiocarbon years are presented as "cal yr BP," which stands for "calendar years before present." By convention, the "present" began in 1950 ad, the standard baseline for radiocarbon dating. Calibration programs are used to convert radiocarbon years to calendar years, allowing lake-sediment chronologies to be compared with tree-ring data and other records.

Additional age determinations are derived from the identification of tephra deposits (i.e., volcanic ash layers) within these sediments. In the CCE region, tephra from Mount Mazama, which erupted approximately 7,627 cal yr BP in southwestern Oregon, and Mount St. Helens and Glacier Peak (dated between about 13,200 and 13,300 cal yr BP) are common markers.

Because tree growth is highly sensitive to environmental changes, tree-ring records are powerful tools for investigating annual to centennial climate variations. Tree-ring chronologies are used to reconstruct past climates such as growing season temperature and precipitation. The most sensitive trees are those growing in extreme environments where subtle variations in moisture or temperature can have a significant impact on growth. For example, precipitation or drought reconstructions, or both, are often derived from extremely dry sites or from sites at forest–grassland boundaries, where moisture is the strongest growth-limiting factor (Figure 10-1, left). Similarly, sites at altitudinal and latitudinal treelines are often targeted for temperature-sensitive chronologies. The year-to-year variability in individual tree-ring-width series (or other tree-ring parameters such as density) from long-lived stands of trees are combined to produce site histories (or chronologies) that span centuries or millennia. These chronologies contain considerable replication (e.g., two cores per tree, minimally 10–15 trees per site) and dating accuracy is rigorously verified by comparing ring-width patterns among trees, known as "crossdating." Crossdating also allows tree-ring series from ancient dead wood (found in old dwellings, in lakes and sediments, and on the surface in cold and dry environments) to be combined with overlapping records from living trees, thereby extending records further back through time. Statistical relationships established between annual tree ring-width chronologies and instrumental climate records are used to hindcast estimates of precipitation and/or temperature back through time.

Historically, studies of the nature and timing of glacier fluctuations have been a primary source of much of the information we have on preinstrumental climate changes. Changes in glacier length and area are excellent—and obvious—physical indicators of climate change at decadal to multimillennial time scales. Glacier changes are primarily driven by changes in mass balance, which is the difference between mass input (accumulation) and mass loss (ablation) over time. For temperate alpine glaciers, the primary climate controls of these processes are winter snowfall and summer temperatures (McCabe and Fountain 1995). Changes in glacier mass balance reflect regional changes in climate because the temporal and spatial variability of temperature and precipitation are functions of ocean–atmosphere synoptic circulation patterns (Bitz and Battisti 1999), particularly during the winter months. In the CCE the small size of the high-elevation cirque glaciers results in rapid response of glacier mass balance to relatively small changes in climate, making them excellent indicators of climate variability on decadal and longer time scales.

The timing of historical glacier advances and retreats can be reconstructed from periods of moraine formation and estimates of changes in glacier size (i.e., volume, length, and coverage). Formation dates for more recent moraines in front of present-day glaciers can be estimated from old photographs, paintings, or documentary records, or by determining the age of trees or lichens growing on the moraine surfaces. Ages of older events may be dated using trees tilted, killed, or overridden during glacier advances (Figure 10-1, bottom right). The glacier record of past climates is by its nature, however, incomplete. For example, throughout the Cordillera the most extensive Holocene advances were <150–200 years ago, and these advances destroyed evidence of earlier events. Recently, inferences on past glacier change have been developed from estimates of paleo-massbalance based on tree-ring series that are sensitive to winter precipitation and summer temperatures (Pederson et al. 2004; Watson and Luckman 2004b).

The CCE's Environmental History in the Last 20,000 Years

At the time of the last glacial maximum, approximately 20,000 years ago, summer and winter insolation in the northern hemisphere were close to present-day values. The large Laurentide and Cordilleran ice sheets of North America affected regional climates of the northwestern United States and Canada in three ways (Bartlein et al. 1998). First, the presence of the ice sheets depressed temperatures (by approximately 10°C for areas south of the ice sheets) and steepened the latitudinal temperature gradient. Second, the presence of the large ice sheets displaced the jet stream south of its current position, greatly reducing winter precipitation in the northwestern United States and Canada. The third element of the full-glacial climate was the stronger-than-present surface easterlies (related to a strong anticyclone that formed over the ice sheets), which resulted in decreased precipitation and airflow from the west. All these factors probably caused cold, dry, and windy climates in the unglaciated areas south of the ice sheets, including the northwestern United States and the CCE (Bartlein et al. 1998).

Large-scale controls of climate changed during the glacial/Holocene transition, causing significant environmental and biotic adjustment during the period from 16,000 to 11,000 cal yr BP. Summer insolation in the northern hemisphere and at the latitude of the CCE increased during this period. Around 11,000 cal yr BP, summer insolation reached a maximum that was 8.5% higher than today's levels. At the same time, winter insolation was 10% lower than today's levels.

The regional climate influence imposed by the glacial maximum ice sheets also waned. For example, the position of winter storm tracks shifted to the north, bringing increased winter moisture but warmer conditions than before the period from 16,000 to 11,000 cal yr BP (Bartlein et al. 1998). During this transitional period, alpine glaciers and ice sheets retreated rapidly, and meltwater lakes formed in glacial outwash plains and behind end moraines (e.g., Flathead Lake, Lake McDonald, Waterton Lake, and Lake St. Mary).

New areas became available for plants, and low- and middle-elevation areas were gradually colonized by tundra, then forest parkland, and finally by forest and steppe. During this general ice recession, a brief period of alpine glacier advance is registered between about 12,200 and 12,900 cal yr BP in regions surrounding the CCE. This "Crowfoot Advance" is named for a series of moraines at Crowfoot Lake. Similar late-glacial advances have been described in the Canadian and Colorado Rockies (e.g., Menounos and Reasoner 1997) but not in glacial or regional vegetation records of the Glacier National Park (GNP) area (Osborn and Davis 1985). Where expressed, evidence indicates the Crowfoot advances were nearly as extensive as those of the LIA. Significant climate changes were likely occurring.

Greater-than-present levels of summer insolation in the early Holocene (11,000 to 6,000 cal yr BP) caused most areas to experience warmer- and drier-than-present conditions. During the middle and late Holocene, summer insolation decreased and winter insolation increased gradually to current levels. The cool, wet conditions that characterize the late Holocene were established about 5,000 years ago, and the onset of modern plant communities and renewed glacial advances (during a period known as the Neoglacial, discussed later) coincided with this change in climate.

An Example: The History of Johns Lake

An analysis of a sediment core from Johns Lake (located west of the Continental Divide in the GNP) illustrates the long-term shifts in climate and vegetation

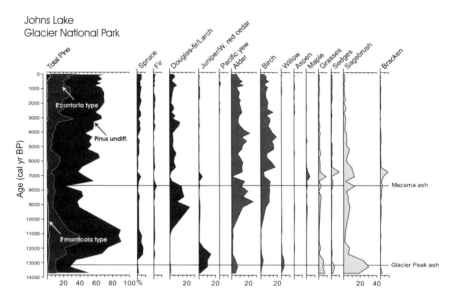

FIGURE 10-2. Pollen-percentage Diagram from Johns Lake, Glacier National Park

Notes: Selected conifers (dark gray); deciduous trees and shrubs (gray); and herbaceous and fern taxa (light gray). Note the change from high herbaceous taxa as well as alder, birch, and willow before 13,000 cal yr BP, followed by juniper, then by pine and fir, and finally by Douglas fir and larch.

that can be inferred from fossil pollen data (Whitlock 1995; Figure 10-2). The sediment core extracted from the center of Johns Lake contained ashes from the eruptions of Mount Mazama, Glacier Peak, and Mount St. Helens. Based on series of radiocarbon dates and the known ages of these eruptions, the record was determined to have covered about the last 14,000 cal yr.

Today, Johns Lake is located in a mixed forest that includes western red cedar (*Thuja plicata*) and western hemlock (*Tsuga heterophylla*). In addition, grand fir (*Abies grandis*) and Pacific yew (*Taxus brevifolia*) grow on wetter sites, and lodgepole pine (see Chapter 2 for scientific names not given here), Douglas fir, and western larch occur in areas of recent fires. Cottonwood (*Populus balsamifera*), quaking aspen (*Populus tremuloides*), alder (*Alnus incana*), and willow grow near the lake, and paper birch (*Betula papyrifera*) is also present. The presence of western red cedar and western hemlock attests to the mild wet winters at the site, and the climate of Johns Lake is ameliorated by the presence of nearby Lake McDonald.

During the late-glacial period from 14,000 to 10,500 cal yr BP, pollen assemblages show high values of sagebrush (*Artemisia* spp.), grass (Poaceae), and sedge (Cyperaceae). The dominant conifer taxa were juniper (*Juniperus* spp.), spruce (*Picea* spp.), fir, and pine. High percentages of alder, birch, and willow were also present, suggesting shrubby margins along waterways and on boggy ground. These pollen assemblages indicate that a tundra parkland landscape and cold conditions, which supported juniper and Engelmann spruce, existed during this period (Figure 10-2).

The early Holocene (from 10,500 to 7,000 cal yr BP) was characterized by an increase in conifer and riparian taxa. Pine is the dominant pollen taxa over this period at the Johns Lake and other sites, and most of the pollen is attributed to lodgepole pine. Douglas fir or larch, alder, and members of the saltbush family (Chenopodiaceae) attain their highest pollen percentages in this interval. The presence of bracken fern (*Pteridium aquilinum*) spores suggests frequent forest disturbances. Although the pollen of Douglas fir and larch are indistinguishable, needle remains from Johns Lake suggest that Douglas fir was abundant and western larch was less prevalent at this time. Charcoal records from the Bitterroot Range in western Montana, from Yellowstone National Park (YNP), and from southwestern British Columbia show that there were more fires at this time than there are today. The pollen and macrofossil assemblages imply an increase in montane taxa at the expense of subalpine plants and a general closing of the forest. Douglas fir and lodgepole pine increased across the region, suggesting an intensification of summer drought conditions and probably of fires as well. The occurrence of saltbush (a diverse plant family) and the persistence of sagebrush also offer evidence of dry habitats. The great increase in riparian taxa such as alder and birch may indicate a period of shallower water conditions, which created habitat for these trees along the lake margin. The evidence shows that, between 7,500 and 4,000 cal yr BP, percentages of spruce and fir pollen increased; those of Douglas fir, larch, and sagebrush decreased. Birch and alder pollen percentages remained high, and needle fragments from Douglas fir, western larch, and western red cedar indicate that all these trees grew locally. The vegetation composition suggests that summer temperatures were higher in the middle Holocene than they are today, and further indicates that precipitation increases were approaching modern levels. The last 4,000 years marks the establishment of modern vegetation, and the pollen data indicate an increase in spruce, fir, western red cedar, and western white pine. This shift implies cooler, wetter conditions, consistent with the decrease in summer insolation and the onset of renewed glacier activity during the Neoglacial.

Vegetation history records similar to those from Johns Lake have been studied throughout the northern Rocky Mountains. Together, they paint a picture of long-term changes in vegetation that occurred at different elevations. For example, in the region of Banff and Jasper national parks, pollen records have shown that the subalpine forest was historically dominated by Engelmann spruce or white spruce (*Picea glauca*), subalpine fir, lodgepole pine, and whitebark pine (e.g., Reasoner and Huber 1999). Sites from the foothill regions to the east describe the history of the forest–grassland ecotone (e.g., White and Mathewes 1986). The Banff region deglaciated prior to 16,000 cal yr BP and tundra communities were the first to colonize the region. Warming and drying in the early Holocene allowed for the spread of lodgepole and whitebark pine and confined spruce and fir to higher elevations. Herbaceous and shrub taxa are abundant in most pollen records, furnishing evidence that the forest was more open in the early Holocene than it is today.

Pollen studies and radiocarbon ages on fossil wood found beyond today's treeline suggest that upper and lower treeline were higher than their current elevation in the early and middle Holocene. The shifts in treeline suggest a longer growing season and decreased effective moisture in the past. Detrital wood washed out of Athabasca and Dome glaciers also indicates that valley floors were covered by forest in the middle Holocene (ca. 8,300–9,144 and 6,907–7,020 cal yr BP; Luckman et al. 1993) and therefore glaciers were less extensive than they are now. In the last 5,000 cal yr, subalpine forests have become more closed and spruce has become an important constituent at all elevations. Moreover, both upper and lower treeline have shifted downslope in response to cooler and more humid conditions in the Neoglacial. Upper treeline first shifted downslope below today's levels during the period from 4,904 to 5,170 cal yr BP (Luckman and Kearney 1986) and has subsequently fluctuated between 0 and 100 m below its current position, coinciding with Neoglacial glacial advances that began after 4,780 cal yr BP.

The Neoglacial

The Neoglacial (covering the last 5,000 years) was the time of renewed glacial activity within the northern Rocky Mountains. The period culminated with the LIA, when many glaciers reached their maximum Holocene extent. Although evidence of the latest phases of the LIA is preserved in distinct moraines, information about earlier, less-extensive glacier advances is based on a few fragmented records that are revealed as glaciers recede. Overridden (in situ) trees, buried forests, and terminal and lateral moraines indicate a complex history of glacier activity during the late Holocene. The earliest of the Neoglacial glacier advances are evident at the Robson Glacier, where recent glacier recession has exposed large trees and a paleosol overridden shortly after the period from 5,470 to 5,589 cal yr BP (Luckman et al. 2005). Evidence from glacier forefields in the southern Canadian Rockies suggests that advances took place between 3,500 and 2,800 cal yr BP and shortly after 1,500 cal yr BP. These early glacial advances were apparently short-lived and separated by long periods of ice recession. Glacier advances became progressively more extensive, culminating in the events of the LIA. The onset of the LIA is generally accepted to have begun after the somewhat controversially defined Medieval Warm Period (Hughes and Diaz 1994).

The framework for the CCE's environmental history was initially defined from studies of glacier fluctuations and pollen records. In recent years our knowledge of conditions during the Medieval Warm Period and the LIA has increased considerably with the development of annually resolved proxy temperature and precipitation reconstructions derived from tree rings. These precisely dated records extend several centuries beyond the start of the instrumental climate record and expand our knowledge of climate variability on decadal to centennial time scales. In the discussion that follows, we present tree-ring-based proxy records of summer temperature and precipitation for the CCE and adjacent areas. We conclude with an integrated discussion of tree-ring-based reconstructions of glacier-mass-balance history.

Temperature Variability

The CCE currently lacks a local millennial-length reconstruction of temperature. Twentieth-century seasonal temperature variability has been shown, however, to be strongly coherent in instrumental records from climate stations situated along the spine of the Rockies from Jasper to the GNP (Watson et al., forthcoming). Given the regional similarity of current temperature records, a long summer temperature reconstruction developed for the Canadian Rockies (the "Athabasca" reconstruction, 950–1994 ad; Luckman and Wilson 2005) can be considered representative of temperature variability in the CCE. The Athabasca summer (May–August) maximum temperature reconstruction was developed from annual ring-width and density measurements of living and dead trees (primarily Englemann spruce) sampled near upper treeline at a number of sites near the Peyto, Athabasca, and Robson glaciers, which are located north of the CCE. The Athabasca reconstruction, calibrated using a regional instrumental temperature record spanning 1895 to 1994 ad, shares many common features with several other millennial-length northern-hemisphere temperature reconstructions (Luckman and Wilson 2005).

The most striking feature of this temperature reconstruction is the anomalously warm temperatures of the twentieth century (Figure 10-3, top). The only two intervals when temperatures approached twentieth-century values occurred during the periods from 1000 to 1050 and 1380 to 1425 ad. The earlier interval coincides with the Medieval Warm Period, but these warm intervals are separated by two prolonged cool periods that extended from about 1050 to 1350 ad and from approximately 1450 to 1850 ad. The first cooling period coincides with evidence that glaciers advanced into forest at Robson, Peyto, and Stutfield glaciers in the Canadian Rockies.

Following a sharp temperature decrease in the 1450s, an extended cool period lasted until the middle of the nineteenth century (ca. 1850 ad). Many proxy temperature records show this cooling, which coincided with LIA glacier advances locally and throughout the Americas. In most areas, glaciers reached their greatest extent over the last 10,000 years of this period (Luckman and Villalba 2000). In the reconstructed Athabasca temperature record, the coldest

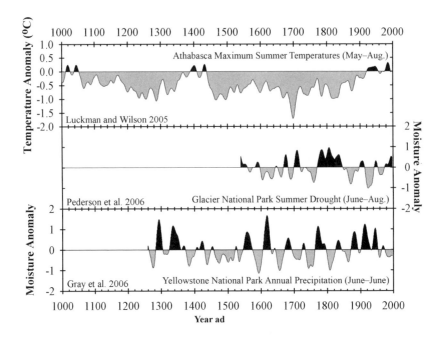

FIGURE 10-3. Select Tree-ring-based Temperature and Moisture Reconstructions Available in the CCE

Notes: The summer (May–August) maximum temperature reconstruction is expressed as anomalies from the 1901–1980 mean. The Yellowstone National Park annual precipitation reconstruction and the Glacier National Park drought reconstructions are expressed in standard deviation units from the mean. All series have been smoothed with a 25-year spline to highlight low-frequency variability.

conditions occur between 1685 and 1704 ad, when temperatures were about 1.5°C below the twentieth-century mean (Luckman and Wilson 2005). Several glaciers in the Jasper area experienced their maximum LIA advance at this time (just slightly beyond the later mid-nineteenth-century advance) with a concurrent major dieback at upper treeline near the Columbia Icefield. The glacier advances of the early 1700s and the mid-nineteenth century were associated with periods of decreased insolation resulting from sunspot activity (Luckman and Wilson 2005). Tree-ring evidence indicates that winter snowpack (a driver of winter mass balance) was exceptionally high for several decades at this time (Pederson et al. 2004; Watson et al., forthcoming).

The Athabasca temperature reconstruction compares well with an independent reconstruction of maximum temperatures developed for an adjacent area of interior British Columbia and several reconstructions of northern hemisphere and global temperatures (Luckman and Wilson 2005). This suggests that temperature variability in the CCE has mirrored that seen over much larger spatial scales. The 1690s were particularly severe at the Columbia Icefield, though, possibly reflecting short-term volcanic forcing, low solar activity, or both. The recent warming trend began in the mid-nineteenth century and continued through the twentieth century, as is also seen in the northern hemisphere temperature reconstructions.

Precipitation Variability

Several precipitation variables—encompassing the past three to seven centuries—have been reconstructed for the CCE and the surrounding region. Figure 10-3 (middle and bottom) presents a summer (June–August) drought reconstruction for the GNP region and a 13-month June–June precipitation reconstruction for the YNP based on ring widths from Douglas fir and limber pine. Although the reconstructions are for different seasons (summer versus annual) and use different measures of precipitation or drought, they are nevertheless comparable because tree growth in Douglas fir primarily reflects growing-season conditions (Watson and Luckman 2002).

The two hydroclimatic records share several common long-duration wet and dry periods (e.g., wet—the 1680s, 1770–1800, the 1840s, and 1900–1915; and dry—the 1650s, 1750, and the 1980s). The records often exhibit opposing moisture regimes, though, because the Yellowstone reconstruction generally experiences moisture anomalies linked to those of the American Southwest, whereas the GNP record mirrors that of the northern Rockies and the Pacific Northwest. These moisture reconstructions, however, lack the level of centennial-scale variability evident in the Athabasca temperature reconstruction. Instead, the prominent mode of variability is concentrated in decadal (20- to 30-year) to interdecadal (> 50-year) wavelengths (Pederson et al. 2006). This mode of variability is consistent among regional records and reflects the strong interdecadal shifts in precipitation (i.e., decadal moisture regimes) that are evident in both the instrumental and paleoclimatic precipitation records throughout western North America.

The reconstructed precipitation records within the general vicinity of the CCE and throughout the larger region of the Pacific Northwest reveal many interesting spatial differences and similarities in historical moisture anomalies. Reconstructed spatial moisture anomalies by decade (from 1700 to 1989 ad) were developed using the network of tree-ring-based precipitation reconstructions for Canada (see Figure 7 in Watson and Luckman 2004a) and drought reconstructions from U.S. Palmer Drought Severity Index (PDSI) values (Cook et al. 2004). Two major intervals of above-average precipitation during the twentieth century show different patterns. The longest event lasted from the 1940s through the 1960s, and the wettest event of the twentieth century (and the past ~300 years) spanned the interval from 1898 to 1916 (see decade 1900–1909). Other notably wet decades for the CCE include the 1770s, the

1780s, and the period from 1800 to 1840, which is consistent with conditions exhibited in the GNP record.

Spatially extensive and sustained drought events carry major economic, social, and ecological impacts. For the CCE and the Pacific Northwest, records demonstrate that such events commonly extend over large parts of the region and last from 10 to more than 20 years (see Figure 7 in Watson and Luckman 2004a). The most prolonged and spatially extensive dry period extended over 25 years (from ca. 1917 to 1941). The early period of this drought (the 1920s) resulted in extremely dry conditions in, and to the northwest, of the CCE. In the 1930s, the drought expanded east of the GNP along the U.S.–Canadian border. This result, which is consistent with other large-scale moisture reconstructions for the United States, caused the widespread crop failures and farm abandonments of the "dust bowl," and increased fire size and frequency in the CCE (see Chapter 13). Only the shorter reconstructed drought of the 1790s appears to have been as spatially extensive.

Three other decadal-scale droughts of note occurred in the mid- to late-nineteenth century (the 1850s, from 1868 to 1875, and from 1889 to 1897). The drought of the 1850s coincided with the severe drought on the western Great Plains, which, in conjunction with other factors, led to the decline of the regional mountain bison populations (Woodhouse et al. 2002). The drought in the 1860s and 1870s contained the single worst reconstructed drought year (1868–1869), when there were grasshopper plagues on the Canadian prairies (Phillips 1990) and extensive fires in the Caribou region of southern British Columbia (Iverson et al. 2002). During the eighteenth century, conditions generally appeared to be average to dry, with the exception of the extremely wet period in the 1780s. These spatial reconstructions of drought and the associated social and ecosystem impacts show that severe multidecadal droughts that span large areas are common. This result must be considered in plans for managing natural resources and developing sustainable western communities.

In the future, a drought of the magnitude of the 1917–1941 event might have serious impacts on tourism and the rapidly growing population centers within the CCE. Because the CCE region is semiarid, it relies heavily on winter snowpack for water resources throughout the dry summer season. Documented and projected warmer annual temperatures in the CCE, coupled with a shorter winter season, reduced winter snowpack, and consequent earlier runoff will increasingly limit future water resources (see Chapter 12). Multidecadal droughts are associated with large, stand-replacing fires in the montane forests of the CCE (Pederson et al. 2006). Independent of any anthropogenic triggers, severe summer drought combined with low winter snowpack results in dry forests where large crown fires are more probable. The main drivers of fire behavior are climate and weather conditions (e.g., Westerling et al. 2003). This observation is strengthened by findings that twentieth-century droughts in western North America are trending toward the longer and more spatially extensive droughts characteristic of the Medieval Warm Period (Cook et al. 2004). In addition, climate projections signifying a return to conditions similar to those of the early Holocene (Whitlock et al. 2002) suggest the possibility of dramatic ecosystem-wide changes and large, more frequent fires in the future. The record of past environmental changes demonstrates the potential future impacts of climate change on resources (and consequently people) within the CCE, on the structure and function of western forests, and on the terrestrial and aquatic wildlife that dwell in those forests.

Glacier Variability

The mass balance of glaciers is linked to temperature and precipitation variables that can be reconstructed using the biological proxies we described previously. Tree-ring-based reconstructions of climate can therefore be used to reconstruct both annual and seasonal glacier mass balances (Figure 10-4). Such reconstructions yield continuous records of glacier fluctuations that can be compared with the independently derived, discontinuous, more traditional records we discussed in earlier sections of this chapter (Pederson et al. 2004; Watson and Luckman 2004b). These reconstructed mass-balance records also allow investigation of the relative importance of the seasonal climate variables that drive observed fluctuations in glacier size. Finally, they can be used to evaluate the linkages among glacier fluctuations, long-term insolation, and ocean-basin-driven climate changes.

Mass-balance records available for the CCE show that the LIA glacial maximum in the region resulted from a combination of climatic factors operating over different temporal and spatial scales. The glaciers appear to have been highly sensitive to decadal- and multidecadal-scale variability in both summer and winter climate. Summer temperature seems to have been the primary driver of glacial fluctuations, with winter snowpack moderating or exaggerating glacier fluctuations (Pederson et al. 2004; Watson and Luckman 2004b). Glaciers are therefore sensitive to climate fluctuations that are driven by conditions in the northern Pacific Ocean (e.g., the PDO). These conditions exert a strong control on regional winter snowpack and consequently on winter mass balance (e.g., Bitz and Battisti 1999). For example, in the early twentieth century (from the 1920s to the 1940s) and since the 1980s, extremely hot summers and extended periods of low-snowpack winters resulted in severe negative net mass balance and exceptional frontal recession for the Peyto Glacier and glaciers in the GNP (Figure 10-4). On the other hand, cooler summer temperatures coupled with high winter snowpack from about 1800 to 1850 resulted in an extended period of positive net mass balance, which was responsible for most glaciers reaching their Holocene maximum extent during that period.

The glaciers of the CCE region are also highly sensitive to centennial-scale variability and long-term trends in summer temperatures. Comparison of the

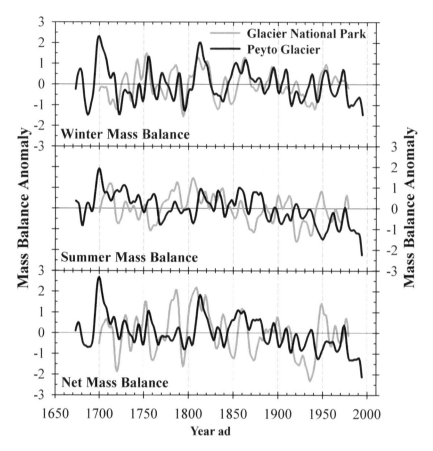

FIGURE 10-4. Tree-ring-based Reconstructions of Seasonal and Net Mass Balance for Peyto Glacier and the Glaciers of Glacier National Park

Source: Watson and Luckman 2004a; Pederson et al. 2004.

Athabasca temperature reconstruction with the record of glacier advances shows that periods of glacier advance and moraine-building were associated with known intervals of lower sunspot activity and cooler summer temperatures (Luckman and Wilson 2005). The extremely cold period during the 1700s, for example, probably triggered a major glacial readvance that caused many glaciers (from more northerly regions) to reach their Holocene maxima at that time (Luckman and Wilson 2005). Lower-than-present summer temperatures were sustained through 1825, coinciding with the period of greatest glacial advance for the majority of glaciers in the CCE and northern Rockies. Since the mid-nineteenth century, however, regional temperatures have increased by 1.5°C (and continue to rise). Glaciers have responded to this rise with exceptional retreat rates and evidence suggests that the current retreat rates are perhaps the greatest and most severe of the Holocene.

Conclusions

Climate changes in the CCE over the last 16,000 years have driven extensive biological changes and reorganizations. First, tundra parkland vegetation became established, followed by forest cover. Warmer summer temperatures from 11,000 to 5,000 cal yr BP favored drought-adapted species and resulted in higher-than-present treeline and frequent fire activity. The ameliorated climate conditions allowed in-migration and ultimately human settlement, which began around 11,000 cal yr BP (see Chapter 3). Long-term solar and orbital variations caused cooler and wetter conditions between 4,000 and 5,000 cal yr BP. These conditions resulted in more extensive glacier advances (during the Neoglacial), a downslope shift of treelines, more closed forests, increases in subalpine taxa, and less-severe fire regimes.

The regional temperature history is known in greatest detail for the last 1,000 years. Evidence from tree-ring data suggests that the warmest temperatures of the last millennium occurred in the twentieth century and that this pattern is similar to that of the northern hemisphere. Regional reconstructions of precipitation extend back several centuries and show widespread shifts between periods of wet and dry conditions that persisted across multiple decades. Increased fire activity has been associated with many of the decadal droughts of the past three centuries in southern British Columbia and the CCE (Pederson et al. 2006). Sizes of area burned and frequency of past fires were low within the CCE in recent centuries, but the twentieth century was marked by several large stand-replacing fires (e.g., fires in 2003 burned approximately 74,000 ha, or approximately 13% of the GNP). Identifying a single cause for this increased fire activity is confounded by the co-occurrence of decadal-scale shifts in precipitation and changes in forest-management practices (e.g., fire suppression and logging) during this period. Many forests were established during the cooler and wetter conditions of the LIA when fires were less frequent. Although forests at middle and high elevations now support species that are especially sensitive to fire, it is difficult to demonstrate that these forests have been greatly altered by fire-suppression management because historical fires tended to be infrequent and stand-replacing. The impacts from human activities are probably secondary to those caused by climate variations. At low elevations where fires have been frequent and low in severity, fire-suppression policies may have led to alterations of forest composition and structure. Here, the human factor is an important variable that needs to be considered in assessing ecosystem health.

The complex geographic response within CCE is evident in the records of past vegetation, tree growth, and glacial fluctuations. The environmental history at one elevation is not necessarily that of another elevation. Our information from east and west of the Continental Divide as well as north and south within the CCE suggests that past climate changes gave rise to ecological responses that were spatially heterogeneous and temporally variable. Given the complexity of this history, we can clearly see that a single approach to ecosystem management would not be appropriate everywhere. Instead, the paleoenvironmental record suggests that we need local histories that are informed by multiple data sets to help untangle the relative influence of human activity and climate change in the CCE, a key step in preparing for the future.

Information derived from studies of past climate and ecosystem conditions gives us estimates of the natural range of variability that can be expected for physical and biological processes within the CCE and the northern Rockies. These are important benchmarks against which managers and policymakers can gauge the current status of resources and the likelihood of achieving desired outcomes. Climate varies at all time scales in response to an interacting hierarchy of diverse physical controls. These complex drivers, even in the absence of human-induced greenhouse gases, result in a present-day climate that is unique in history. Anthropogenic factors, both local and global, have now been added to the equation and may exert an overriding influence on the climate of the next few centuries (e.g., Overpeck et al. 2005). The paleoenvironmental records discussed here demonstrate the interrelationships between past climate and ecosystem processes, giving us some understanding of how climate change may influence these ecosystems in the future. We have summarized the highlights of what we know about the natural history of the CCE. But much remains to be done to allow us to better understand the development of today's landscape and to anticipate and adjust to future changes in this ecoregion.

References

Allen, C.D., and D.D. Breshears. 1998. Drought-Induced Shift of a Forest-Woodland Ecotone: Rapid Landscape Response to Climate Variation. *Proceedings of the National Academy of Sciences* 95: 14839–42.

Bartlein, P. J., K.H. Anderson, P.M. Anderson, M.E. Edwards, C.J. Mock, R.S. Thompson, R.S. Webb, T. Webb III, and C. Whitlock. 1998. Paleoclimate Simulations for North America over the Past 21,000 Years: Features of the Simulated Climate and Comparisons with Paleoenvironmental Data. *Quaternary Science Reviews* 17: 549–86.

Bitz, C.M., and D.S. Battisti. 1999. Interannual to Decadal Variability in Climate and Glacier Mass Balance in Washington, Western Canada, and Alaska. *American Meteorological Society* 12: 3181–96.

Bradley, R.S., M.K. Hughes, and H.F. Diaz. 2003. Climate in Medieval Times. *Science* 302: 404–5.

Carrara, P.E. 1989. Late Quaternary Glacial and Vegetative History of the Glacier National Park Region, Montana. *U.S. Geological Survey Bulletin* 1902.

Chang, P., and D.S. Battisti. 1998. The Physics of El Niño. *Physics World* 8: 41–7.

Cook, E.R., C. Woodhouse, C.M. Eakin, D.M. Meko, and D.W. Stahle. 2004. Long-Term Aridity Changes in the Western United States. *Science* 306: 1015–18.

Dettinger, M.D., D.R. Cayan, H.F. Diaz, and D.M. Meko. 1998. North–South Precipitation Patterns in Western North America on Interannual-to-Decadal Timescales. *Journal of Climate* 11: 3095–111.

Gray, S.T., J.L. Betancourt, C.L. Fastie, and S.T. Jackson. 2003. Patterns and Sources of Multidecadal Oscillations in Drought-Sensitive Tree-Ring Records from the Central and Southern Rocky Mountains. *Geophysical Research Letters* 30: 1316.

Hughes, M.K., and H.F. Diaz. 1994. Was There a "Medieval Warm Period" and, if so, Where and When? In *The Medieval Warm Period*. Edited by M.K. Hughes, and H.F. Diaz. Dordrecht/Boston/London: Kluwer Academic Publishers, 109–42.

Iverson, K.E., R.W. Gray, B.A. Blackwell, C. Wong, and K.L. MacKenzie. 2002. *Past Fire Regimes in the Interior Douglas-Fir Dry Cool Subzone Fraser Variant*. IDFdk3. Report to Lignum, Ltd., Williams Lake, British Columbia.

Kutzbach, J. E., R. Gallimore, S. Harrison, P. Behling, R. Selin, and F. Laarif. 1998. Climate and Biome Simulations for the Past 21,000 Years. *Quaternary Science Reviews* 17: 473–506.

Luckman, B.H., and M.S. Kearney. 1986. Reconstruction of Holocene Change in Alpine Vegetation and Climate in the Maligne Range, Jasper National Park, Alberta, Canada. *Quaternary Research* 26: 244–61.

Luckman, B.H., and R. Villalba. 2000. Assessing Synchroneity of Glacier Fluctuations in the Western Cordillera of the Americas during the Last Millennium. In *Interhemispheric Climate Linkages*. Edited by V. Markgraf. San Diego: Academic Press, 119–40.

Luckman, B.H., and R.J.S. Wilson. 2005. Summer Temperatures in the Canadian Rockies during the Last Millennium—a Revised Record. *Climate Dynamics* 24: 131–44.

Luckman, B.H., G. Holdsworth, and G.D. Osborn. 1993. Neoglacial Glacier Fluctuations in the Canadian Rockies. *Quaternary Research* 39: 144–53.

Luckman, B.H., M. Masiokas, and A.C. Aruani. 2005. The Neoglacial History of Robson Glacier, B.C. *Programme with Abstracts Annual Meeting, Canadian Association of Geographers*. London, ON: University of Western Ontario, A104.

Mantua, N.J., S.R. Hare, Y. Zhang, J.M. Wallace, and R.C. Francis. 1997. A Pacific Decadal Climate Oscillation with Impacts on Salmon. *Bulletin of the American Meteorological Society* 78: 1069–79.

McCabe, G.J., and A.G. Fountain. 1995. Relations between Atmospheric Circulation and Mass Balance at South Cascade Glacier, Washington, USA. *Arctic and Alpine Research* 27: 226–33.

Menounos, B., and M.A. Reasoner. 1997. Evidence for Cirque Glaciation in the Colorado Front Range during the Younger Dryas Chronozone. *Quaternary Research* 48: 38–47.

Osborn, G.D., and P.T. Davis. 1985. Age of Pre-Altithermal Cirque Moraines in the American and Southern Canadian Cordillera. *Symposium on the Paleoenvironmental Reconstruction of the Late Wisconsin Deglaciation and the Holocene*. Program with Abstracts and Field Guide. Calgary, AB: University of Calgary, 50.

Overpeck, J., J. Cole, and P. Bartlein. 2005. A "Paleoperspective" on Climate Variability and Change. In *Climate Change and Biodiversity*. Edited by T. Lovejoy and L. Hannah. New Haven, CT: Yale University Press, 91–108.

Pederson, G.T., D.B. Fagre, S.T. Gray, and L.J. Graumlich. 2004. Decadal-Scale Climate Drivers for Glacial Dynamics in Glacier National Park, Montana, USA. *Geophysical Research Letters* 31: L12203.

Pederson, G.T., S.T. Gray, D.B. Fagre, and L.J. Graumlich. 2006. Long-Duration Drought Variability and Impacts on Ecosystem Services: A Case Study from Glacier National Park, Montana, USA. *Earth Interactions* 10: 1–28.

Phillips, D. 1990. *The Climates of Canada*. Ottawa, ON: Minister of Supply and Services.

Reasoner, M.A., and U.M. Huber. 1999. Postglacial Paleoenvironments of the Upper Bow Valley, Banff National Park, Alberta, Canada. *Quaternary Science Reviews* 18: 475–92.

Swetnam, T.W., and J.L. Betancourt. 1998. Mesoscale Disturbance and Ecological Response to Decadal Climatic Variability in the American Southwest. *Journal of Climate* 11: 3128–47.

Watson, E., and B.H. Luckman. 2002. The Dendroclimatic Signal in Douglas-fir and Ponderosa Pine Tree-Ring Chronologies from the Southern Canadian Cordillera. *Canadian Journal of Forest Research* 32: 1858–74.

———. 2004a. Tree-Ring Based Reconstructions of Precipitation for the Southern Canadian Cordillera. *Climatic Change* 65: 209–41.

———. 2004b. Tree-Ring-Based Mass-Balance Estimates for the Past 300 Years at Peyto Glacier, Alberta, Canada. *Quaternary Research* 62: 9–18.

Watson, E., G.T. Pederson, B.H. Luckman, and D.B. Fagre. Forthcoming. Glacier Mass Balance in the Northern U.S. and Canadian Rockies: Paleo-perspectives and 20th Century Change. In *Darkening Peaks*. Edited by B. Orlove, B.H. Luckman, and E. Wiegandt. Berkeley: University of California Press.

Westerling, A.L., A. Gershunov, T.J. Brown, D.R. Cayan, and M.D. Dettinger. 2003. Climate and Wildfire in the Western United States. *Bulletin of the American Meteorological Society* 84: 595–604.

White, J.M., and R.W. Mathewes. 1986. Postglacial Vegetation and Climatic Change in the Upper Peace River District, Alberta. *Canadian Journal of Botany* 64: 2305–18.

Whitlock, C. 1995. *The History of Larix occidentalis during the Last 20,000 Years of Environmental Change in Ecology and Management of Larix forests*. Intermountain Research Station General Technical Report GTR-INT-319. Ogden, UT: U.S. Department of Agriculture Forest Service, 83–90.

Whitlock, C., and P.J. Bartlein. 2004. Holocene Fire Activity as a Record of Past Environmental Change. In *Developments in Quaternary Science*. Volume 1. Edited by A. Gillespie and S.C. Porter. Amsterdam: Elsevier, 479–89.

Whitlock, C., M.A. Reasoner, and C.H. Key. 2002. Paleoenviromental History of the Rocky Mountain Region during the Last 20,000 Years. In *Rocky Mountain Futures: An Ecological Perspective*. Edited by J.A. Barron. Washington, DC: Island Press, 41–59.

Woodhouse, C.A., J.J. Lukas, and P.M. Brown. 2002. Drought in the Western Great Plains, 1845–1856. *Bulletin of the American Meteorological Society* 83: 1485–93.

11
Modeling and Monitoring Biophysical Dynamics and Change in the CCE

Dan Fagre

Before we can project an ecosystem's future responses to ongoing climate change, increases in disturbance, or regional landscape fragmentation, we must first figure out how the entire ecosystem works. An appropriate analogy is the physician's need to understand the functions of a healthy human body before predicting a body's response to external stressors. An ecosystem's immense complexity is paralleled by its geographic size, which makes detecting change and understanding the causes of that change extremely difficult. Fortunately, new technologies such as remote sensing and ecological modeling have been rapidly developed over the past few decades. These technologies have given us powerful new capabilities for monitoring change in an ecosystem, tracking its dynamics, and understanding its inner workings. In this chapter I will describe the application and development of the new technologies in the CCE but with a focus on Glacier National Park, Montana, where mountain ecosystems have been intensively studied since 1991.

Remote Sensing through the Years

Now that we actually can use the Internet to "Google" photographs of our own homes taken from space, the idea of obtaining critical monitoring information about an ecosystem from a satellite is not exotic. Although civilian use of remote-sensing technologies has become possible just in the past three decades, it is likely to become all the more important as we learn to better interpret information from spaceborne sensors. Military use of remote sensing dates back to World War I, when pilots of canvas, wire, and wood biplanes used bulky cameras to take photographs while flying over enemy lines. When surplus aircraft became available after World War II, aerial photographs became more widely available. Ecologists began using aerial photography, including infrared, to map and monitor the spatial characteristics of mountain ecosystems. The first aerial photographs of Glacier National Park (GNP) were taken in 1946 during a reconnaissance mission to look for a suitable site to build a dam. Although the aerial photographs from 1946 do not cover the entire park, and they were not intended to map or establish reference conditions for the park, they have been useful for contrasting changes over time (Klasner and Fagre 2002).

In 1936, scientists from the University of California, Berkeley conducted a vegetation-mapping project within the GNP, aided by neither aerial photographs nor accurate maps. The aerial photographs used to create the first U.S. Geological Survey (USGS) topographic map of the GNP were taken in 1966. When we consider how long the park had existed before 1966 and the degree to which the landscape had already been altered by, for example, road- and dam-building, town development, logging, and farming, this first complete aerial snapshot of the GNP seems relatively recent. Because the 1966 photographs have been used over the years to assess glacier-size changes, alpine–treeline movements, forest–grassland boundary shifts, floodplain changes, and historical forest-fire perimeters, they have served as a critically important baseline for assessing changes not only in the GNP but also in the larger CCE. High-resolution aerial photographs continue to play an important role in detecting changes in the CCE. In recent years, resource managers and planners have relied on such photographs to develop a continually evolving, detailed vegetation map of the GNP, as well as to plan the reconstruction of the Going-to-the-Sun Road.

The advent of satellite-based multispectral sensors in the 1970s greatly enhanced our ability to monitor and understand mountain ecosystems. The Advanced Very High Resolution Radiometer (AVHRR), for example, provided a coarse-scale (1-km^2) perspective of vegetation distribution for large landscapes such as the entire CCE. The Landsat Thematic Mapper was particularly useful for establishing ecosystem attributes within specific watersheds at 30-m resolution, which could not be directly measured with aerial photography. The Normalized Difference Vegetation Index (NDVI), a ratio between two bands measured by the Landsat sensors, showed where vegetation growth was most vigorous and biomass accumulation was greatest. Other sensors, such as Synthetic Aperture Radar, can map soil moisture and detect areas of freezing or thawing. These types of information have many other applications. For example, these tools can be used to calculate fuel loads for spatially predicting potential forest-fire intensity or to examine yearly changes in "green up" over large areas.

Newer sensors and techniques can pinpoint very-fine-resolution details (down to 1 m) on more specific ecosystem characteristics. For instance, Walsh et al. (2004) used IKONOS imagery (IKONOS is available commercially from Space Imaging Inc., Thornton, Colorado), to map and analyze individual avalanche paths in the John F. Stevens Canyon that lies between the GNP and the Bob Marshall Wilderness Complex along U.S. Highway 2. This level of specificity, made possible by sensors on satellites or aircraft, gives us an unparalleled ability to measure the extent, dynamics, and long-term changes of complex mountain ecosystems at multiple spatial scales across the entire expanse of the CCE. The newest technologies, such as hyperspectral imagery of the Flathead River corridor, provide 5- to 10-cm resolution in several bands of the light spectrum and open up the possibility of monitoring ever-smaller parts of the ecosystem with greater precision. Remote sensing will likely become even more indispensable to natural resource managers over time.

Remote-Sensing Data Are Validated by On-the-Ground Observations

As useful as remote-sensing technologies are for monitoring ecosystem processes, we cannot always believe our interpretations without collecting some data on the ground. Called "groundtruthing," this critically important monitoring step is too often undervalued or even ignored. Monitoring ecosystem dynamics in place (i.e., on the ground) is very expensive, and even the newer automated devices require much human attention. Ground-based monitoring also suffers from the fact that any point chosen for monitoring is unlikely to completely represent the spatial variation in what is being monitored. Consider snow, for example—canopy interception by trees and redistribution by wind call into question data gathered from monitoring the snow depth at a single point. Spatially extensive monitoring at many points becomes necessary, driving up the costs.

All these issues are especially acute in mountainous terrain where it is very difficult to travel to sites, access is seasonal, weather can be extreme, and spatial variation is so great that choosing sites to monitor is problematic. In addition, mountain landscapes are often the last to be instrumented, not only because of cost and effort, but also because there may be a perception that the absence of humans makes monitoring mountain dynamics less important. Even today, more than 90% of the weather stations in the mountainous western United States are in valley bottoms and do not represent higher elevations. Around the globe, very few high-elevation weather stations are maintained year-round.

The last monitoring problem, shared with remote sensing, is that monitoring data do not reach very far back in time. Systematic monitoring of many landscape-level phenomena has been conducted only for the past 20 to 50 years and has been sporadic in many cases. Because ecosystem properties change slowly over decades to centuries, much of our monitoring-based understanding of mountain ecosystems is limited. This has necessitated the development of "proxy" measures, such as tree-ring-based estimates of precipitation, to extend our knowledge of ecosystem dynamics over longer periods.

On-the-Ground Monitoring in the CCE

In this section, I describe the various on-the-ground monitoring activities that are taking place in the CCE. I intend this discussion to be representative rather than exhaustive. New types of monitoring are continually spurred by new programs, regulations, and research questions, along with the development of new or better technologies.

Climate (in the long term) and weather (in the short term) are the most dynamic factors that affect ecosystem dynamics, human activity, and the interactions between the two. Consequently, climate records go back the farthest and weather is the most widely monitored (and talked about) phenomenon in the CCE, as well as across the globe.

In the Montana part of the CCE, climate data were first systematically collected in 1896 in the town of Kalispell in the Flathead Valley. Since 1896, the U.S. National Weather Service (NWS) has established additional weather stations around the perimeter of the GNP as part of the agency's continued expansion. Although the first climate data were collected within the GNP in 1913, the first reliable data are from West Glacier in 1932 when daily observations were made. Finklin (1986) summarized the existing data for the GNP, Waterton Lakes National Park (WLNP), and surrounding areas, describing basic climatic patterns that influence all the biophysical processes in the mountains. Since 1986, a growing network of remote, automated climate-data stations has provided more complete spatial coverage. Sixteen stations (mostly at high elevations) and additional sensors and instruments have been added to address specific climate-related research questions, to monitor trends, and to furnish data for parameterizing and improving computer models (Fagre et al. 1997). To improve regional analyses, National Oceanic and Atmospheric Administration (NOAA) scientists have been operating a high-precision U.S. Climate Reference Network station in the GNP since 2003 (see http://www.ncdc.noaa.gov/oa/climate/uscrn/). Historical climate data from the region have been compiled and statistically reviewed. Climatic data products (e.g., DAYMET, University of Montana, Missoula; Thornton et al. 1997) geospatially extrapolate climate variables, making important climate information available even for remote areas that have never had direct measurements because they are difficult to access. Several Web-based data clearinghouses make access to climate data easier than ever before, and real-time access capabilities have recently been added to high-elevation stations with radio links to hosted Web sites. For example, from the comfort of your office—wherever it may be—you can learn that the Logan Pass climate station has recorded sustained 160 km/h (99 mph) winds with gusts to 207 km/h (129 mph).

A spectrophotometer that measured ultraviolet (UV) radiation energy in all bands was operated in the GNP near St. Mary, Montana, for seven years (from 1997 to 2004) to assess potential harm to sensitive species such as amphibians

(Diamond et al. 2005). The spectrophotometer also produced baseline data against which we can compare future global changes in UV exposure resulting from ozone-layer depletion. At the latitude of the U.S.–Canada border, for example, we have seen an 8% increase in UV energy since 1979, enough to cause concern about increases in skin cancer rates.

Hydrological Monitoring in the CCE

As the "water towers of the world," mountains bestow significant benefits on humans and downstream ecosystems through stream and river discharge. Hydrological data include the USGS gauging stations on major rivers that are part of a national network of many thousands of stations. Using the World Wide Web, we can access data from these stations in real time. Within the GNP, these are augmented by automated gauging stations that produce data at a finer scale. These stations are located on streams within two major watersheds, Lake McDonald and Saint Mary Lake. Downloading the data from these stations several times per year for the last ten years has created a decade-long database of daily discharge and temperature. Since 1993, this database has been augmented with frequent water chemistry and biological inventories (Hauer et al. 2002). All these data have been used to parameterize and validate ecosystem models and to determine the relative contribution of the GNP to regional rivers. In addition, the data have been related to climatic and snow-pack trends, allowing managers to translate future climate changes into projected changes in regional water supply. Another addition is the Hydrologic Benchmark Network (HBN) site in the GNP's Many Glacier Valley, which monitors streamflow and water quality. Because the Many Glacier HBN site is part of a national network of the U.S. Geological Survey (USGS), observed changes in the GNP can be interpreted within a broader context. Detecting changes in regional water quality has been the goal of several lake surveys (e.g., Western Lakes Survey; Clow et al. 2003). Finally, the intensive studies of pristine streams flowing out of the GNP (described in Chapter 8) also include a floodplain study on the Middle Fork of the Flathead River (see http://www.umt.edu/flbs/Research/Biocomplexity.htm). This study, which sought to define and explain biocomplexity as a function of biophysical complexity, is a model of interdisciplinary cooperation.

Snow Dominates in the CCE's Mountains

In the mountains of the CCE, from 70 to 90% of the region's annual precipitation falls as snow, and many of the high-elevation areas are snow-free for only one or two months a year. Perennial snowfields and glaciers have disappeared through time, but not because of reduced precipitation. In fact, the past century has seen a 10% increase in annual precipitation but overall, during the past 50 years, snowpack water content in the spring has declined (Selkowitz et al. 2002). This is because a greater portion of the annual precipitation now falls as rain rather than snow because of rising temperatures. For the same reason, snow melts into runoff two weeks earlier. Because of its importance to river and water supply forecasting, researchers have used many techniques to monitor snow in the CCE.

The SNOTEL System and More

The U.S. Department of Agriculture's (USDA's) Natural Resources Conservation Service (NRCS) operates and maintains an extensive automated system to collect snowpack and related climatic data in the western United States. Called SNOTEL (for SNOwpack TELemetry), the system evolved from a congressional mandate in the mid-1930s, which called for the NRCS "to measure snowpack in the mountains of the West and forecast the water supply" (USDA NRCS n.d.). SNOTEL uses meteor burst communications technology to collect and communicate data in near-real time. The technology does not rely on satellites but rather on very-high-frequency (VHF) radio signals, which are reflected at a steep angle off the band of ionized meteorites that extends from about 50 to 75 miles above the earth. Eleven western states are home to more than 730 SNOTEL sites, which are generally located in remote high-mountain watersheds where access may be difficult or restricted. The battery-powered sites, recharged with solar cells, are designed to operate unattended and without maintenance for a year. The condition of each site is monitored daily.

The SNOTEL sites automatically measure the weight of snow that has accumulated on top of a liquid-filled pillow. These data are translated into the water equivalent of the snow and transmitted daily to a Web site along with air temperature. In the GNP, additional sensors have been added to the two SNOTELs to report on other relevant ecological information such as wind speed and direction, solar radiation, relative humidity, and soil temperature. The Flattop SNOTEL site has been operating daily since 1969, producing a valuable long-term record of high-elevation snow trends. Monthly snow-depth and snow-water-equivalent measurements have been taken at 110 sites within the Lake McDonald and Saint Mary Lake watersheds since 1993 (Fagre et al. 2002). This spatially distributed data set augments daily data collected at single SNOTEL sites and also evaluates the effects of forest-cover snow interception. Finally, it supplements annual data that have been collected from seven NRCS snow-survey sites since 1922.

From these latter records, which span more than 80 years, maximum snowpack accumulation trends have been correlated with sea-surface-temperature anomalies in the Pacific Ocean that have shown consistent multidecadal pat-

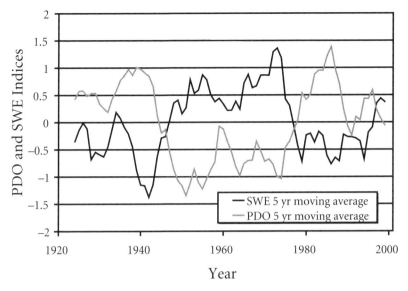

FIGURE 11-1. Pacific Decadal Oscillation (PDO) and Snow Water Equivalent (SWE) in Glacier National Park

Note: Measured on May 1st each year.
Source: Selkowitz et al. 2002.

terns (Selkowitz et al. 2002). The Pacific Decadal Oscillation (PDO) drives 20- to 30-year cycles of above- and below-average snowfall in the CCE (Figure 11-1). These cycles help to explain many ecosystem dynamics, such as forest-fire frequency. Knowledge of this PDO pattern may be useful for planning any snow-related human activities, ranging from operating a ski area to managing aquatic species. Scientists have noted a trend toward later maximum snowpack accumulation and earlier snowmelt, potentially resulting in more-intense spring runoff and flooding. Models of snow distribution in complex topography have been devised and tested (e.g., Geddes et al. 2005) and a study of forest-canopy-cover interception of snow, along with wind transport, has been concluded (Selkowitz et al., in review). Because snow controls so many ecosystem processes in the CCE, these research activities all aim to translate the impact of past and future climate change on the spatial and temporal dynamics of snow.

As part of the National Atmospheric Deposition Program/National Trends Network, which has more than 250 cooperators, researchers have monitored atmospheric deposition at a single low-elevation site in the GNP since 1980. Orographic uplift brings tremendous snow amounts to the higher elevations of the CCE. Up to nine months of atmospheric deposition is contained in the annual snowpack at high elevations and released during spring melt. Because climate change affects the timing and amount of snow, research projects have examined the spatial distribution of major ion snow chemistry since 1998. These projects, which are nested within a larger-scale study of the Rocky Mountains, have assessed elevation and the different climatic regimes on either side of the Continental Divide (Fagre et al. 2000). Two projects have evaluated the presence of persistent organic pesticide residues in snow. These residues may be transported from as far away as Southeast Asia. Because the atmospheric transport system may be greatly affected by climate change, researchers are concerned about this residue transport. Monitoring atmospheric deposition patterns, the chemical content of snow, and lake chemistry in the CCE will provide an integrated assessment for detecting future changes to the region's mountain ecosystems.

Geographic Information Systems Monitoring

Small alpine glaciers are rapidly disappearing in the CCE, primarily in the GNP. Because glaciers are strong indicators of climatic trends that influence ecosystem dynamics and natural resources, geographic information systems (GIS) have been used to capture and analyze historical data on recession. Series of maps that show the sizes of glaciers have been produced and show that over 72% of the largest glaciers have disappeared during the past century (Key et al. 2002). Monitoring of glaciers includes repeating photographs taken of glaciers in the past and comparing the images; analyzing the areal size using aerial photographs (i.e., photogrammetry); measuring the margins with high-precision global positioning systems (GPS); determining ice thickness with ground-penetrating radar; and measuring the loss of snow and ice in the summer (i.e., ablation studies). These monitoring efforts will continue until 2030 when geospatial modeling techniques suggest that all glaciers will be gone (Hall and Fagre 2003).

Recognizing that the CCE is not an entirely discrete ecosystem and that the surrounding regional landscapes provide critical inputs, I have, with many collaborators, expanded the spatial scale of the monitoring several times. We have analyzed regional snow and climatic data sets for trends, comparing the CCE to other relatively unaltered mountain landscapes for approximately 400 km along the spine of the Rocky Mountains in Canada and the United States (Watson et al., forthcoming). Ecosystem monitoring, including three years of intensive field studies to support modeling, has been extended along a transect from the CCE to the Pacific Ocean and includes two other bioregions surrounding national parks, Olympic and North Cascades (Fagre and Peterson 2002). This 800-km transect spans three distinct mountain ranges and a variety of land uses. We have conceptually described the Western Mountain Initiative that proposed to link existing mountain research across all the mountain areas of the western United States (see, for example, http://www.cfr.washington.edu/research.fme/wmi/). The initiative is also designed to find common and differing trends in mountain ecosystems through monitoring. These mountain programs are geographically well distributed across the different mountain ranges and include both arid and temperate rainforest extremes in climate and vegetation communities. Finally, the GNP is a part of the United Nations Education, Scientific and Cultural Organization (UNESCO) Biosphere Reserve network (described in Chapter 2); has joined the Global Observation Research Initiative in Alpine Environments (GLORIA; described in Chapter 6); and will participate in global-scale mountain research through the Mountain Research Initiative based in Switzerland (see, for example, http://mri.scnatweb.ch).

Ecological Modeling Complements Remote-Sensing Monitoring

Although remote sensing is a valuable tool, it is limited to information reflected back to sensors in space. It can detect change and measure spatial relationships, but it does not tell us how the ecosystem works. Biophysical monitoring, on the other hand, gives us high-quality data about parts of the ecosystem but is generally expensive and limited in its relevance to the entire system. Enter ecological modeling, which combines the strengths of remote sensing and monitoring. This type of modeling can describe ecosystem function, identify thresholds for change, examine sensitivity to external forcing such as climate change, extrapolate across scales, and give us clues as to what the future may hold. We must also recognize that modeling is hypothetical and must be relentlessly tested before we can rely on the results.

With that said, however, it is an effective way to organize vast amounts of data in an integrated fashion that addresses very complex systems. Modeling output, particularly maps and other visual tools, also gives concrete dimension to our understanding of ecosystem dynamics and the scale and magnitude of potential future changes. Modeling alerts scientists and land managers to apparently counterintuitive outcomes of ecosystem responses to climate change or management decisions. For instance, in an early modeling exercise for northwest Montana, Running and Nemani (1991) found that streamflows decreased 30% in the Swan Range even when the modeling scenario increased precipitation by 10%. This resulted from increased temperatures, which caused the snow to melt earlier and extended the growing season, thereby enhancing forest growth and increasing evapotranspiration. Because mountains are particularly complex, scientists have been modeling parts of the CCE for more than a decade (Fagre et al. 2005). Finally, fine-resolution modeling of entire mountain ranges (i.e., the CCE scale) is not as common as global- or continental-scale modeling at one end of the spectrum or watershed- and catchment-scale modeling at the other end. Most land managers, though, have to make decisions about natural resource issues along the spectrum between large and small scales, making ecosystem modeling particularly germane to the CCE.

Past and Present Modeling Work

Collaborators Steven Running, of The University of Montana Numerical Terradynamic Simulation Group (NTSG), and Robert Keane, of the U.S. Department of Agriculture (USDA) Forest Service Intermountain Fire Sciences Laboratory (IFSL), have been implementing an integrated ecosystem modeling system at the GNP since 1991. The USGS Global Change Research Program at the GNP has sponsored this work (attributed from this point forward to "the research team" and summarized by, among others, Fagre et al. 2005). The integrated modeling program for GNP was built on previous research efforts and has coalesced into two related modeling approaches, the Regional Hydro-Ecological Simulation System (RHESSys) and FIREBGC (Fire BioGeoChemical), both described below. The research team then applied both approaches at larger scales.

RHESSys is an evolving group of models that the team modified to supply appropriate input and output to each other and to collectively address ecosystem patterns and processes (White et al. 1998). Remotely sensed data from satellite platforms are used to estimate the distribution and density of vegetation on the ground. MT-CLIM (a Mountain Climate simulator; Hungerford et al. 1989) is used to generate daily meteorological variables, which are combined with landscape features to make spatially explicit calculations of daily tree growth. Daily processes such as gross primary productivity are mapped across the mountain landscape and averaged on an annual basis. For a specific watershed, this integrated modeling system can simulate net primary productivity (NPP) as well as soil-carbon pools, carbon-to-nitrogen ratios, and annual hydrologic discharges.

FIREBGC is a closely related integrated model that emphasizes structural components of the mountain forest ecosystem. It includes forest fires as a key disturbance factor (Keane et al. 1996), but does not yet include hydrologic routing and discharge estimates. This model simulates a forest that is similar to what we see rather than estimating photosynthetic activity that we do not directly perceive. This allows FIREBGC to project rates and trajectories of postdisturbance succession and to simulate important compositional and structural attributes such as tree-species dominance, stand age, and coarse woody debris levels. These latter attributes are critical for estimating the potential occurrence of large, stand-replacing wildfires and their intensity using companion models such as FARSITE (Fire Area Simulator; Finney and Ryan 1995). For a mountain ecosystem such as the GNP, FIREBGC can map mosaics of forest stands of different ages and compositions resulting from both climatic variability and forest fire frequency. It can also provide details of tree-stand dynamics such as the depth of duff and litter on the forest floor. In addition, FIREBGC interfaces well with other forest-science research and models, allowing us to examine forest-management scenarios. An example is the amount of smoke and the level of air-quality deterioration that result from different fire-suppression policies (Keane et al. 1997).

To determine whether these models were returning reasonable estimates and portraying an accurate picture of this mountain ecosystem's dynamics, the research team gathered field data on key ecosystem outputs for 10 years and compared these data with the estimates from the models. For watershed-scale simulations, snow estimates and hydrologic discharge simulations tracked daily data closely except for storm events, where some overprediction was noted (White et al. 1998). In the GNP, only 4 of 84 watersheds have as much as 3% of their area covered by glacial ice and 18 have only 1%. Nonetheless, in watersheds with remnant glaciers, higher observed values during the late summer underscored both the contributions of glacial meltwater to streamflow and the need to include this source in future models of the region's mountain hydrology. In addition, modeled daily estimates of stream temperatures throughout the watershed closely matched daily measurements from seven monitored streams (Fagre et al. 1997). Carbon-budget estimates for the watersheds indicated close agreement with observed values for soil carbon dioxide (CO_2) effluxes and productivity for both low- and high-elevation forests that cover 75% of the watersheds. White and coworkers (1998) concluded that RHESSys generated reasonable estimates of ecosystem processes for these watersheds. Some ecosystem process estimates such as NPP were much less sensitive to scale than hydrologic discharge. The team compared results from FIREBGC simulations to field data and found that the simulations were making reasonable estimates of most ecosystem attributes and processes for which the team had recorded field observations. Validating these models by comparing outputs to observed data gave the team confidence that the models were accounting for most major ecosystem processes in the GNP.

Results of Modeling with Future Scenarios

With the models' capacities established to simulate the GNP mountain ecosystem and its responses to current climatic variability, the research team applied various climate scenarios to estimate potential future conditions in the park. This predictive, or forecasting, ability does not actually predict the future, but it does give managers a valuable tool for assessing a range of possibilities for the GNP. The team found that RHESSys and FIREBGC translate possible future climate change into spatially explicit effects on the park landscape with a reasonable degree of confidence.

One scenario of climate change was based on the team's evaluation of four general circulation models and several downscaling approaches to result in a "most likely" climate change scenario for the GNP. This scenario projects a 30% annual precipitation increase and a 0.5°C annual temperature increase by 2050. This resulted in tree-species shifts in distribution and dominance, including a reduction in subalpine fir (see Chapter 2 for scientific names not given in this chapter) concomitant with treeline rise and a significant expansion of Engelmann spruce at the expense of lodgepole pine. Another scenario incorporated an extremely variable climate but no long-term increases in temperature or precipitation. After 120 years, long-term conifer NPP in the GNP decreased 4% on the western side of the Continental Divide and 13% on the eastern side (White et al. 1998). Broad-leaved shrubs and alpine vegetation increased 2–7%, but grass NPP at the forest–grassland ecotone decreased by 4%. In fact, the lower treeline (the forest–grassland ecotone) rises under this scenario, permanently reducing the amount of forest cover in the Saint Mary Lake watershed. This reduces the fuel load for large fires but probably increases the frequency of fires.

Ecosystem models can also suggest changes in limiting factors that drive plant interactions. Water and nitrogen indices, which are calculated as part of the carbon-allocation process, integrate information about water stress, nutrient availability, and the potential shoot-to-root growth in vegetation. Under current conditions, growth was limited by nitrogen availability for conifer forests and shrubs, but grasses were limited by water availability. Under the extremely variable climate scenario, however, some limitations changed. For instance, alpine vegetation is water-limited under the current climate but became nitrogen-limited under the variable climate scenario. The relative nitrogen limitation for conifer forests decreased slightly and grasses at lower treeline became much more water-limited. These shifts in limitations, as simulated by ecosystem models, can give ecologists more specific constraints with which to predict how individual species will fare and how biodiversity patterns will change for different vegetation types. For example, as atmospheric deposition of nitrogen increases in alpine environments, the alpine vegetation that is nitrogen-limited under the variable climate scenario will respond differently than it does under the current climate.

Modeling of Future Fire Regimes

Wildland fire, the primary disturbance process in CCE forests, greatly influences carbon cycles in mountain ecosystems. FIREBGC performed well in estimating the present distribution and abundance of forest tree species using historical climate and fire-frequency data (Keane et al. 1996). The research team applied a variety of climatic and fire management scenarios to the GNP to look at possible future outcomes.

Under all future climate scenarios the team examined, FIREBGC clearly indicates that the resulting, more-productive forest landscapes will generate more frequent and severe fires than the same landscapes experienced historically even with the increase in annual precipitation (Keane et al. 1997). This nearly doubles smoke emissions in the future, jeopardizing the pristine air quality that the GNP area currently enjoys and posing a management challenge for park managers who need to restore historical fire frequencies. Because humans have altered fire frequency throughout the CCE for more than a century, fuel loads have built up to levels that could lead to higher fire intensities. Keane et al. (1997) examined the interplay of different fire-management policies coupled with different climate scenarios. With fire present (not suppressed), after 250 years of simulation, fires burned more than 55% and 67% of the GNP landscape under a current and future climate scenario, respectively. The resulting landscapes were more productive and diverse than the landscapes that developed during the simulation without fire. These latter landscapes became marginally productive and tended to respire more of the CO_2 fixed by photosynthesis than did the communities dominated by fire. Keane et al. (1997) assert that the fire-maintained early successional communities create overall landscapes that release less carbon to the atmosphere than landscapes without fire. This is true under both current and future climate scenarios and even when carbon emissions during fires are considered. Long-term fire suppression has implications for global carbon balance if other landscapes behave similarly to those simulated by the research team.

Keane and coworkers (1999) also used FIREBGC to examine spatial attributes of ecosystem processes instead of the more-common spatial analyses performed on structural components such as cover types. The researchers' rationale was that cover-type maps do not give insights into landscape evolution and trends, whereas the spatial distribution of NPP may better project future ecosystem conditions. Keane and colleagues found that, under both current and future climate scenarios, patch density for vegetation compositional landscape maps decreased or remained stable when fire was present, meaning that the type of forest did not change much. Patch density *increased* for NPP, however, indicating that fire increased the spatial heterogeneity of productivity. Thus, these researchers mapped areas where the forest became more productive to indicate important shifts in dynamic ecosystem outputs. The ability to "see" and spatially analyze processes, in addition to visible structural components of ecosystems, gives scientists and managers a useful tool for gaining new understanding of the drivers of biodiversity in mountains and the potential effects of climatic change on natural resources.

A Look Ahead

The NTSG research team used another model, BIOME-BGC (University of Montana, Missoula) to look at the impact of possible future climate change on water balance across the U.S. portion of the CCE. We were particularly interested in the spatial distribution of drought sensitivity in the future. Comparisons between vegetation attributes (leaf area index) measured in the field and those predicted by BIOME-BGC under current climates was good ($r^2 = 0.94$), giving us confidence to use the model to examine future relationships (Fagre et al. 2005). We applied a climate change scenario that decreased current summer precipitation and increased annual temperature. Not unexpectedly, this resulted in reduced water supplies (outflows) from mountain areas and across the region. The greatest effects, though, were at mid-elevation sites where most of the forests grow and where the increasing human population tends to build new homes. Under this scenario, the potential for increased wildfire hazards to humans will grow and the amount and predictability of the regional water supply will lessen in landscapes undergoing rapid transformation. The low-lying areas, which are already relatively dry, showed no major changes in this simulation. The highest elevations, however, which have miniscule regional spatial representation, actually increased outflow. We can conclude from this work that the highest mountains may be even more critical in providing ecosystem services to people in the future.

Modeling has yielded valuable insights into the inner working of mountain ecosystems. As we continue to develop new remote-sensing technologies, acquire more monitoring information, and implement ever more sophisticated models, we will be able to expand on these insights. Activities such as these should contribute to sustainable strategies for living in the CCE for a long time to come as they become incorporated into decisionmaking and policy formulation.

References

Clow, D.W., J.O. Sickman, R.G. Striegl, D.P. Krabbenhoft, J.G. Elliott, M. Dornblaser, D.A. Roth, and D.H. Campbell. 2003. Changes in the Chemistry of Lakes and Precipitation in High-Elevation National Parks in the Western United States, 1985–1999. *Water Resources Research* 39: 1171, doi:10.1029/2002WR001533.

Diamond, S.A., et al. 2005. Estimated Ultraviolet Radiation Doses in Wetlands in Six National Parks. *Ecosystems* 8: 462–77.

Fagre, D.B., P.L. Comanor, J.D. White, F.R. Hauer, and S.W. Running. 1997. Watershed Responses to Climate Change at Glacier National Park. *Journal of the American Water Resources Association* 33(4): 755–65.

Fagre, D.B., K. Tonnessen, K. Morris, G. Ingersoll, L. McKeon, and K. Holzer. 2000. An Elevational Gradient in Snowpack Chemical Loading at Glacier National Park, Montana: Implications for Ecosystem Processes. In *A Merging of Theory and Practice; Proceedings of the International Snow Science Conference, Big Sky, MT*. Edited by K. Birkeland. Bozeman: Montana State University, 462–67.

Fagre, D.B., and D.L. Peterson. 2002. Modeling and Monitoring Ecosystem Responses to Climate Change in Three North American Mountain Ranges. In *Mountain Biodiversity: A Global Assessment*. Edited by C. Körner and E.M. Spehn. London: Parthenon Publishing Group, 249–59.

Fagre, D.B., D. Selkowitz, B. Reardon, K. Holzer, and L. McKeon. 2002. Modeling and Measuring Snow for Assessing Climate Change Impacts in Glacier National Park, Montana. In *Proceedings of International Snow Science Workshop*, Penticton, B.C., Canada, Sept. 29–Oct. 4, 2002. Victoria, B.C.: International Snow Science Workshop Canada, Inc., 417–24.

Fagre, D.B., S.W. Running, R.E. Keane, and D.L. Peterson. 2005. Assessing Climate Change Effects on Mountain Ecosystems Using Integrated Models: A Case Study. In *Global Change and Mountain Regions: An Overview of Current Knowledge*. Edited by U.M. Huber, H.K. Bugmann, and M.A. Reasoner. Dordrecht, Netherlands: Springer, 489–500.

Finklin, A. 1986. *A Climatic Handbook for Glacier National Park—with data for Waterton Lakes National Park*. U.S. Department of Agriculture Forest Service General Technical Report INT-204. Ogden, UT: Intermountain Research Station.

Finney, M.A., and K.C. Ryan. 1995. Use of the FARSITE Fire Growth Model for Fire Prediction in U.S. National Parks. In *Proceedings of The International Emergency Management and Engineering Conference (TIEMEC'95)*, May 9–12, 1995, Nice, France, 183–89.

Geddes, C., D.G. Brown, and D.B. Fagre. 2005. Topography and Vegetation as Predictors of Snow Water Equivalent (SWE) Across the Alpine Treeline Ecotone at Lee Ridge, Glacier National Park, Montana. *Arctic, Antarctic, and Alpine Research* 37: 197–205.

Gray, S.T., L.J. Graumlich, and J.L. Betancourt. Forthcoming. Annual Precipitation in the Yellowstone National Park Region since ad 1173. *Quaternary Research*.

Hall, M.P., and D.B. Fagre. 2003. Modeled Climate-Induced Glacier Change in Glacier National Park, 1850–2100. *Bioscience* 53(2): 131–40.

Hauer, F.R., D.B. Fagre, and J.A. Stanford. 2002. Hydrologic Processes and Nutrient Dynamics in a Pristine Mountain Catchment. *Verh. Internat. Verein. Limnol.* 28: 1490–93.

Hungerford, R.D., R.R. Nemani, S.W. Running, and J.C. Coughlan. 1989. MTCLIM: A Mountain Microclimate Simulation Model, INT-414. Ogden, UT: Intermountain Research Station.

Keane, R.E., K.C. Ryan, and S.W. Running. 1996. Simulating Effects of Fire on Northern Rocky Mountain Landscapes with the Ecological Process Model FIRE-BGC. *Tree Physiology* 16: 319–31.

Keane, R.E., C.C. Hardy, K.C. Ryan, and M.A. Finney. 1997. Simulating Effects of Fire on Gaseous Emissions and Atmospheric Carbon Fluxes from Coniferous Forest Landscapes. *World Resource Review* 9(2): 177–205.

Keane, R.E., P. Morgan, and J.D. White. 1999. Temporal Patterns of Ecosystem Processes on Simulated Landscapes in Glacier National Park, Montana, USA. *Landscape Ecology* 14(3): 311–29.

Key, C.H., D.B. Fagre, and R.K. Menicke. 2002. Glacier Retreat in Glacier National Park, Montana. In *Satellite Image Atlas of Glaciers of the World, Glaciers of North America—Glaciers of the Western United States*. Edited by R.S. Williams, Jr., and J.G. Ferrigno. Washington, DC: U.S. Government Printing Office, J365–J381.

Klasner, F.L. and D.B. Fagre. 2002. A Half Century of Change in Alpine Treeline Patterns at Glacier National Park, Montana, U.S.A. *Arctic, Antarctic, and Alpine Research* 34(1): 49–56.

Running, S.W., and R.R. Nemani. 1991. Regional Hydrologic and Carbon Balance Responses of Forests Resulting from Potential Climate Change. *Climate Change* 19: 349–68.

Selkowitz, D.J., D.B. Fagre, and B.A. Reardon. 2002. Interannual Variations in Snowpack in the Crown of the Continent Ecosystem. *Hydrological Processes* 16: 3651–65.

Selkowitz, D., A. Nolin, and D.B. Fagre. In review. Variable Snow Cover Accumulation Associated with Vegetation Type and Density in Glacier National Park, Montana, USA. *Hydrological Processes*.

Thornton, P.E., S.W. Running, and M.A. White. 1997. Generating Surfaces of Daily Meteorological Variables over Large Regions of Complex Terrain. *Journal of Hydrology* 190: 214–51.

USDA NRCS (U.S. Department of Agriculture National Resources Conservation Service). No date. SNOTEL Data Collection Network Fact Sheet. http://www.wcc.nrcs.usda.gov/factpub/sntlfct1.html (accessed November 24, 2006).

Walsh, S.J., D.J. Weiss, D.R. Butler, and G.P. Malanson. 2004. An Assessment of Snow Avalanche Paths and Forest Dynamics Using IKONOS Satellite Data. *Geocarto International* 19: 85–93.

Watson, E., G.T. Pederson, B.H. Luckman, and D.B. Fagre. Forthcoming. Glacier Mass Balance in the Northern U.S. and Canadian Rockies: Paleo-Perspectives and 20th Century Change. In *Darkening Peaks*. Edited by B. Orlove, E. Weigandt, and B. Luckman. Berkeley: University of California Press.

White, J.D., S.W. Running, P.E. Thornton, R.E. Keane, K.C. Ryan, D.B. Fagre, and C.H. Key. 1998. Assessing Simulated Ecosystem Processes for Climate Variability Research at Glacier National Park, USA. *Ecological Applications* 8(3): 805–23.

12
Ecosystem Responses to Global Climate Change

Dan Fagre

Global climate change is a topic of urgent and widespread interest as we face an uncertain future and realize that we have played a role in causing fundamental and rapid change. Climate change now dominates many scientific endeavors, influences geopolitical dynamics, and is debated in international venues. We can assess the state of the entire biosphere by detecting trends at the global scale and running global simulation models. At the other extreme, many examples of biological change at the local scale are attributed to climate change. Often, however, we neglect to address these regional scales, where populations interact with their environment in complex ways and where critical decisions are made that drive global systems and constrain local impacts and responses. Mountain systems are an example of globally dominant regions that have an integrated response to climate change at multiple scales. Mountains comprise 20% of the world's terrestrial land surface and are home to at least 10% of the human population. A better understanding of mountain–climate relationships is clearly important for understanding and addressing global-scale environmental change such as the climate shifts seen in the past 50 years.

Mountains provide numerous ecosystem products (such as timber) and services (such as wildlife viewing, tourism and recreational opportunities, and biodiversity conservation) to society. Although mountains have long been recognized as critical entities in regional climate and hydrological dynamics, their importance in storing terrestrial carbon has now been underscored as well (Schimel et al. 2002). Of particular importance to most people is the clean water the mountains supply each year to users who are often far downstream from the mountains themselves. Globally, 50% of the freshwater we use comes from mountains. In the western United States, an estimated 85% of the water people use comes directly from mountains (USGCRP 2000). As a result, changes to mountain ecosystems that are driven by global climate change have the potential to affect water supplies for entire regions. Mountains are also subject to greater climate variability than intermountain lowland environments and have experienced greater rates of climate change in the past 100 years (Diaz et al. 2003). Consequently, mountains are likely to be more sensitive to global-scale climate change. These changes should also become obvious in high-elevation mountain environments earlier than in other environments.

Mountain ecosystems are excellent indicators of climate change because of their topographic complexity and strong environmental gradients. Climatic controls on biota occur over relatively short distances, allowing us to detect subtle long-term changes more easily. With the different ecosystem "parts" in close proximity to each other, abundance and distribution changes (such as shifts in the elevation occupied) can be attributed to climate rather than other factors that operate over large geographic scales (e.g., the different day lengths associated with different latitudes). By studying climate change impacts to relatively intact, protected mountain areas, such as Glacier National Park (GNP), we can gain a better understanding of the mechanisms of ecological response and to monitor rates of change.

Examining the GNP as Representative of the CCE

As an example, the area around the GNP in the larger CCE reflected a 1.6°C (2.88°F) increase in mean annual summer temperature from 1910 to 1980 (Hall and Fagre 2003). This increase is nearly three times the global average of 0.6°C (USGCRP 2000). Researchers have detected similar upward trends in the broader CCE, where the minimum spring and summer temperatures have climbed steadily for the past century (Watson et al., forthcoming). Globally, the upward temperature trend has continued unabated—the five highest recorded global temperature averages have occurred (in order of rank) in 2005, 1998, 2002, 2003, and 2004 (NASA 2006).

Because people usually do not live in the highest elevations of mountainous areas such as the CCE, many of the changes attributable to global temperature increases go unnoticed. In addition, the climate changes accompanying a warming globe are often complex. Some mountain areas near the Front Range in Colorado have actually cooled, even as other western mountain areas have warmed dramatically. Because projections for future change indicate even more warming, it is important for us to understand how the CCE has fared with climate change over, for example, the past 200 years. All the future climate scenarios used by the U.S. Global Change Research Program (USGCRP 2000) and the 2001 Intergovernmental Panel on Climate Change (IPCC) incorporate various degrees of warming across the globe because the evidence for continued warming is so compelling. For the CCE, this warming means a shift in familiar climate patterns around which we have built our expectations of agriculture, sustainable forestry, forest-fire frequency, snow-based recreation, and so forth.

Climate forcing in mountain ecosystems occurs at multiple spatial and temporal scales, and some of these changes are distinctly nonlinear. One type of change is expressed as a threshold effect, in which climate has relatively minor impacts until a critical level is reached, leading to rapid (and continuing) change. Another type of change is cyclical, in which mountain ecosystems react to alternating climate states. Both phenomena have produced measurable changes in the CCE's mountain forest ecosystems. Land managers and others in the CCE must be aware of these nonlinear changes as they make decisions of increasing complexity, often working within shorter time frames. These decisionmakers need current information about climate dynamics and future projections.

The GNP has had an active global change research program since 1990. This program is part of the larger USGCRP, which has incorporated national parks in its research strategy. Under the GNP program, scientists seek to (1) understand the effects of past climate variability on the park's mountain resources, (2) document and comprehend the recent changes that are attributable to climate warming, and (3) project future changes by integrating available knowledge and information. In this chapter, I describe some of the past and recent climate-change-driven responses of the GNP and surrounding areas. Because past changes are analogs for future change, evidence of past climate change and the response of mountain ecosystems is key to making integrated projections for the future.

Glacier Recession

Although forests, meadows, and rivers in the GNP have all responded to past climate variability, the most obvious and characteristic change in mountainous regions has been seen in glaciers. Often the highest part of the ecosystem, glaciers interact with weather systems without the complication of vegetation and other biota and are therefore excellent physical indicators of climate change. As evidenced by the park's name, glaciers are also symbolic icons of the area.

Small alpine glaciers offer a number of advantages as climate change signals. Intuitively, we grasp that warming temperatures melt ice, and the extent of the ice in late summer is easy to see. When we compare what we see today with previous photos of a given glacier, we are left with powerful images of change (see, for example, the photos of Shepard Glacier in Figure 12-1). Glaciers, unlike organisms and biological systems, do not adapt to changing climates by altering their responses to those changes, which makes interpreting glacier change relatively straightforward. Because they represent an average of recent years of snowfall, rather than reflecting only one year as a seasonal snowpack does, glaciers retain a "memory" of recent climate trends. As a result, the consistent recession of glaciers is a strong indicator of sustained climate trends. Finally,

FIGURE 12-1. Shepard Glacier in Glacier National Park, Montana, 1913 (top) and 2005 (bottom)

Source: Top photo by W.C. Alden; bottom photo by B. Reardon.

small glaciers reflect mostly the balance between accumulation and melting, without the complex ice dynamics that may lead to an expansion in large glaciers even during a warming trend. Fortunately, the GNP was once home to many relatively small glaciers with which we can track climate changes of the past several centuries.

During the twentieth century, the extent and mass of mountain glaciers in the GNP have clearly decreased in response to warmer temperatures. In contrast, glaciers were at their largest at any time during the Holocene (which spans the past 10,000 years) at the end of the Little Ice Age (LIA), a 400-year period of markedly lower temperatures when glaciers grew in North America. The LIA ended around 1850. Numerous investigators have periodically taken both formal and informal measurements of the park's glaciers. Key and colleagues (2002) summarize these intermittent studies. More recent glacial recession has been documented by a long-term monitoring program (started in 2005) that utilizes several index glaciers.

In 1850, the GNP contained approximately 150 glaciers (Carrara 1989), but in 1966, only 37 were large enough to warrant being named on maps. By 1993, the largest glaciers in the park had shrunk by 72%, and many of the smaller glaciers had vanished or were no longer large enough to be considered glaciers (Key et al. 2002). The area within park boundaries covered by ice and permanent snow decreased from 99 km^2 to 27 km^2 by 1993 and to 17 km^2 by 1998. Many watersheds no longer contain any glaciers (Key et al. 2002) and the area covered by glaciers in any single watershed does not exceed 4%. Furthermore, glaciers have thinned by hundreds of meters, and like Shepard Glacier (Figure 12-1), may contain less than 10% of the ice volume that existed at the end of the Little Ice Age. Using a minimum size criterion of 0.1 km^2, 27 ice bodies currently qualify as glaciers; most continue to show rapid recession. Between 1993 and 1998, glaciers ranging in size from 0.15 to 1.72 km^2 became 8 to 50% smaller. The relative rate of shrinkage was greatest for the smaller glaciers. Hall and Fagre (2003) created a geospatial model of glacier recession for the Blackfoot-Jackson glacier complex. This was the largest glacier system in the park before it separated into smaller, discrete glaciers while retreating. Applying a climate scenario that assumes no major changes in the upward trend of global temperatures, Hall and Fagre estimated that all the glacial ice in the Blackfoot-Jackson area would disappear by 2030. Even if warming rates were more modest, only the tiniest remnants of glaciers would exist by 2100. We can see, then, that the glaciers of the GNP will likely be gone within our lifetimes based on the best currently available information (Hall and Fagre 2003).

The decline of glacial ice may be linked to increases in mean summer temperature, a reduction in the winter snowpack that forms and maintains glaciers, or both. Instrumental weather data from western Montana show a trend of increasing annual average temperature (Figure 12-2) for the period of record (from 1900 to 2005). A larger regional temperature analysis that includes the northern part of the CCE indicates that spring and summer minimum temperatures have increased more than other temperatures (Watson et al., forthcoming). Warmer spring temperatures, in particular, mean that the glaciers are not as cold and that melting (ablation) is more readily initiated by the onset of summer. Despite the fluctuations in temperature over the past century, many glaciers continued to shrink even during the cooler periods.

Glaciers could also be shrinking if less snow arrives during the winter. Although annual precipitation has actually increased 10% during the past cen-

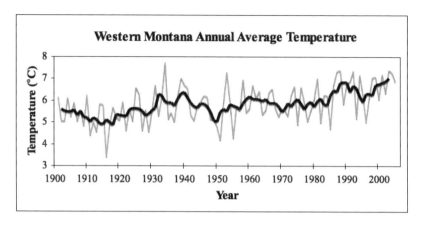

FIGURE 12-2. Summer Average Minimum Temperature, Kalispell, Montana, 1900–2005

Note: Kalispell has the longest record of weather observations in the area and is part of the Historic Climate Network.
Source: Analysis by Greg Pederson.

tury (Selkowitz et al. 2002), no similar long-term trend in snowpack exists in the vicinity of the glaciers. Snowpacks varied cyclically, but glaciers continued to shrink throughout the period, indicating that the snowpack was not adequate to counteract the temperature changes. The overall threshold for maintaining glaciers was probably exceeded sometime between 1850 and the 1920s (Selkowitz et al. 2002). This recession trend corresponds with the data on glaciers elsewhere on Earth—almost all mountain glaciers are receding as global temperatures increase.

The ecological significance of losing glaciers in the CCE is primarily through changes in hydrology. Glaciers act as a "bank" of water (stored as ice) that is released during dry periods of the year or during extended drought. This keeps a continual flow in streams that otherwise might dry up and is critical for maintaining riparian and aquatic biota. Once glaciers are gone in a watershed, many streams will become ephemeral, the overall water supply will diminish, and aquatic communities will experience a more unpredictable environment. Of equal importance to the loss of streamflow in late summer is the increased water temperature caused by the unavailability of glacial meltwater. This affects the distribution and behavior of aquatic organisms because many stream insects are sensitive to temperature and cannot complete their life cycle outside narrow temperature ranges. These insects will migrate to upper stream elevations as glacier meltwater is reduced. Some insects may become locally extinct if regional water temperatures exceed their tolerance after the glaciers are gone. Terrestrial plants and animals will eventually occupy landscapes vacated by glacial ice, and new alpine lakes often remain after the ice is gone. Although loss of glacial ice affects a small portion of the landscape, its effects are pervasive on mountain ecosystems and human communities that depend on reliable water resources.

Snowpack Trends

Snowpack trends and variability are dominant forces that limit forest growth at high elevations. The quasiperiodic Pacific Decadal Oscillation (PDO), caused by fluctuation of sea-surface temperatures in the North Pacific Ocean, affects climate throughout the western United States. The strongest climate signals associated with the PDO are seen from the north Pacific coast to the Rocky Mountains (Mantua et al. 1997). The PDO index is negatively correlated with winter precipitation in the form of snow (Selkowitz et al. 2002; refer to Figure 11-1). Growth of mountain hemlock (*Tsuga mertensiana*; Peterson and Peterson 2001) and subalpine fir (see Chapter 2 for scientific names not given in this chapter; Peterson et al. 2002) is positively correlated with the PDO at treeline (where snowpack limits growing-season length) and negatively correlated with the PDO at lower elevations (where soil-moisture storage limits summer growth) in the Olympic and North Cascade mountains Therefore, multidecadal climate phenomena such as the PDO seem likely to control growth and productivity of high-elevation forest ecosystems through trends in snow in the CCE as well. In addition, these phenomena suggest that tree-growth trends are somewhat predictable over a period of several decades. Alftine and coworkers (2003) documented that the PDO has also affected tree-seedling establishment rates at alpine treeline during the past century. Although Selkowitz and colleagues (2002) showed significant correlation of CCE snowpacks with the PDO, they found no overall changes in the moisture content of seasonal snowpacks from 1922 to 2002. Although annual mean snowpack moisture has not increased for the past 80 years, overall annual precipitation has increased by at least 10% in conjunction with warmer annual temperatures. This resulting shift in the rain-to-snow ratio implies that snowpacks are having less influence on forest growth and hydrology than they did a century ago. PDO-driven snowpack fluctuations, however, will continue to dominate high-elevation forest growth for the foreseeable future even as temperatures and the average snowline elevation rise. Eventually, future climate change will directly affect these forests as snow becomes less dominant.

Forest Responses

The response of glaciers to long-term climate change is relatively easy to measure and their rate of disappearance is a signal that more extensive changes are occurring in the mountain forests. Forests have also responded to the climate changes of the past century, but separating the climate signal of forests is more problematic because rates of disturbance by fire reflect both climate and fire-management policies. In addition, there may be forest responses to carbon dioxide (CO_2) fertilization (caused by increasing concentrations of greenhouse gases) or nitrogen inputs (resulting from atmospheric deposition) that alter growth responses to climate change.

The fact that these forests have responded to climate shifts since the end of the last major ice age (ca. 12,000 years ago) is indisputable. Elias (2002) summarizes numerous studies that document the rise of treelines; the recolonization of previously ice-covered areas; and the shifts in species dominance from grassland and juniper to pine, spruce, and fir as climates generally warmed to the present. Climate fluctuations, such as those linked to the LIA, slowed or reversed some of these changes but also underscore the dynamic relationship between climate and vegetation. In fact, the responses of forests to post-LIA warming are an historical model for the major climate changes scientists predict for the centuries ahead.

Alpine Treeline Changes

Logically, climate change should be most easily observed in the alpine treeline ecotone (ATE) where tree seedlings are established upslope under more favorable conditions, moving the ATE upslope, or where die-backs occur among the older trees under less-favorable conditions. The treeline is considered to be the upper limit of tree growth—a temperature-driven zone of tree death. Treelines likely were depressed (moved downslope) in the GNP during the LIA as shown by the distribution of very old dead trees on slopes above the current treeline (Carrara 1989). Tree-ring widths from these "fossil" trees and from high-elevation living trees together suggest increasingly harsh conditions at high elevations during the Little Ice Age, peaking in the mid-1800s when heavy snowpacks and cool temperatures suppressed alpine tree growth, and glaciers grew rapidly (Pederson et al. 2004). Chapter 6, however, makes clear that topography, geomorphology, and soils may be more dominant than climate in determining the elevation and location of the treeline or the ATE. Nonetheless, in the summer of 2003 Roush and coworkers (forthcoming) clearly showed many sites where seedlings had vigorously colonized open areas during the past 50 years at or near the ATE. Regeneration of conifers in high-elevation forest ecosystems has increased throughout western North America during the past century, particularly since the 1930s (Rochefort et al. 1994). This new establishment of trees is particularly prominent near treeline during periods of low snowpack, often associated with warm PDO regimes. Butler and colleagues (1994) did not find significant upward migration of trees from the treeline for the past century in the GNP, but they did note increased growth rates and sizes of trees, in-filling of gaps by new seedlings, and changes in growth to more upright forms. Using digital photogrammetry in the same general area, Klasner and Fagre (2002) documented a 4% increase in tree-canopy area at treeline since 1945. These researchers also corroborated the increased biomass at treeline that resulted from in-filling and more robust tree growth. Post-LIA (ca. 1850) advances of subalpine-fir krummholz into alpine tundra above treeline occurred until the early 1900s in fingers extending into the alpine tundra less than 100 m from the forest. Upright tree establishment into the GNP's alpine tundra began in 1920 when temperatures warmed and snowpack was below average. This tree-seedling establishment has been largely facilitated by microtopographic sheltering (e.g., large boulders) or the presence of a pioneer tree that provides a microclimatic envelope for subsequent seedling establishment (Resler 2006). Isolated trees that are less than 80 years old, however, have established as upright tree forms up to several hundred meters from the forest. This indicates that limitations to seedling establishment and subsequent tree growth had been ameliorated beginning in about 1920, corresponding to the rapid rise in summer average temperature that was related to the most-rapid period of glacial recession. Butler and DeChano (2001) used repeat photography to show treeline elevation changes but attributed some of these shifts to changes in fire-management policy. Overall, tree biomass at treeline is increasing, particularly as shrubby trees attain more upright forms. Increasing tree growth to the exclusion of herbaceous flora, including a more abrupt alpine ecotone (Butler et al. 1994), represents a change in biological diversity and the potential for subalpine forest ecosystems to act as a sink for carbon.

Meadows and Tundra

Repeat photography clearly shows that trees have invaded many subalpine meadows in the GNP over the past century, which is likely the result of warming temperatures and reduced snowpack persistence. The increase in growing-season length allows tree-seedling establishment to increase, but fire suppression and the reduction of grazing by horses, used more extensively before roads were built, could also have influenced the invasion. Although the same trend has been observed in high-elevation areas throughout the western United States, the direct impact on herbaceous plant abundance and diversity in meadows has not been determined, at least in the GNP. Indirectly, other investigators have focused on shifts in biodiversity in higher-elevation communities associated with forests in the GNP. For example, Lesica and Steele (1996) worked in 1990 to establish a permanent plot system in subalpine meadows and near a high-elevation wetland (the "Hanging Gardens" of Logan Pass). This wetland area supports one of the most diverse plant assemblages in the park and includes numerous species at the edges of their ranges, incorporating plants found more commonly to either the north or south along the Continental Divide. Because relatively slight changes in moisture are likely to eliminate some marginal species and favor others, this wetland will act as an indicator for climate-induced changes in alpine biodiversity occurring throughout the CCE. After ten years, Lesica and McCune (2004) found four plant species declining but the trends were not statistically significant.

A multiyear effort to create a vegetation classification for CCE alpine tundra environments has resulted in nearly 700 16-m^2 plots scattered throughout the park along 80 km of the Continental Divide (Damm 2001). During the summer of 2000, 165 of these plots were relocated and assessed for changes in species composition and dominance and then permanently marked for future assessments. After five years, no changes were seen in species presence or relative abundance. These and additional plots will be revisited at five-year intervals to assess climate change impacts. They will also be periodically compared to alpine tundra studies in other mountains.

Non-Native Plants

In the GNP, numerous non-native plant species have established populations and are considered a management problem in the effort to maintain pristine park environments. Of 1,258 vascular plants documented in the park, 10.5% are non-native. To understand and combat this problem, scientists have begun using geographic information systems (GIS) to map major non-native plant infestations, usually in disturbed areas and along road corridors. These areas are monitored annually to measure the effectiveness of the active eradication efforts. As a result of the mapping work, detailed georeferenced vegetation databases that include the distribution of non-native species are available for the GNP (see Chapter 17 for more information). A distinct elevational gradient can be seen, with most occurrences of non-native plants remaining at low elevations but relatively few existing in alpine areas. The extreme variability scenario (White et al. 1998) described in Chapter 11 indicates that lowland areas of the park—where most roads, campgrounds, and buildings have been built—are likely to undergo rapid transitions to grasses. This suggests a greater vulnerability to establishment of non-native species as a result of climate change.

Trends in Growth

Changes in rates of tree growth could reflect climate change. McKenzie and colleagues (2001) analyzed tree-growth chronologies in western North America and found that the growth rate is increasing in many high-elevation forest ecosystems below treeline. This growth increase began after 1850, concurrent with the start of industrial activity and the associated greenhouse gas emissions. Although these researchers saw no statistical evidence per se that the growth increase was caused only by increased temperature or only by precipitation changes as separate factors, the growth increase was clearly associated with atmospheric changes. McKenzie and coinvestigators cautiously infer that the growth increase is associated with elevated levels of atmospheric CO_2. If so, this is an important signal of the effects of changes in the global atmospheric environment on forest ecosystems. It also suggests that high-elevation forests are functioning as carbon sinks, a fact that will be important in regional strategies for dealing with greenhouse gas emissions.

Changes in Disturbance

Productivity, succession, and large-scale spatial patterns in the CCE are controlled by ecological disturbances, especially fire. Area burned in any particular year is at least partially related to climate variability (such as the PDO), as well as to long-term climate change. For example, years with fire areas greater than 80,000 ha in national forests of Washington and Oregon are nearly four times more common during a warm PDO than during a cool PDO (Mote et al. 1999). This regional effect does exist in the CCE, but it is moderated by synoptic-scale meteorology, particularly the effect of high-pressure ridges from eastern Washington to western Montana (Gedalof and Mantua 2002). Barrett (1986) constructed a fire history for the GNP that described the fire-return intervals for different forest types. That history showed a clear reduction in fire frequency during the height of the LIA and a period of increased fires during the early twentieth century. Both phenomena were related to climate shifts that also drove patterns of glacier recession (Pederson et al. 2006).

These historical analogs help us to estimate the potential for forest fires under future climate scenarios. We anticipate that a warmer climate will bring extended fire seasons and perhaps more large fires in much of the CCE and that the PDO will continue to influence the periodicity of fires. Forest-insect outbreaks and forest pathogens are additional sources of disturbance that may be aggravated by climate change. CCE forests have been periodically devastated by mountain pine beetle and others whose rapid population increases have been facilitated by warm, dry climate phases. With some climatologists predicting increased climate variability, the precursor conditions for insect outbreaks are likely to occur more often.

Snow avalanches disturb forests on mountain slopes and carry soil and nutrients into riparian areas during high-magnitude avalanche years. Up to 50% of some alpine watersheds show evidence of avalanche disturbance of the vegetation. Herbaceous plants that are key resources for wildlife such as bears tend to grow on avalanche paths. The conditions that create high-magnitude avalanche years are complex but are directly related to climate. Reardon and coworkers (forthcoming) have shown that more avalanches occur in high-snowpack years associated with a negative PDO. Intermediate climate change will make such conditions more common and will likely lead to more avalanches. Increased climate variability is also likely to increase avalanche frequency and disturbance in the near term. Continued warming in the CCE, however, will raise the snowline and may eventually reduce the ecological impact of snow avalanches.

Dendrochronological studies of long-lived trees indicate that multidecadal droughts have occurred several times in each century for the past 500 years (Pederson et al. 2004). Coupled with the PDO effect on winter snowpack, the pattern of these long-duration droughts helps explain the changes in forest-fire frequency, glacier mass balance, and glacier growth or shrinkage over the past several centuries. These same patterns explain other ecosystem attributes. For instance, a 70-year period of cool, moist summers and large winter snowpacks allowed for the expansion of cedar-hemlock forests that persist today. Conversely, the 1917–1942 drought, with low-to-moderate snowpacks and hot, dry summers, was a period of rapid glacier melt and the establishment of subalpine fir trees into the alpine tundra above treeline. Because snowpack is critical to regional water supplies, we must account for the effects of the PDO and lengthy droughts, as well as climate change, in comprehensive water management and forecasting. Global change effects on the strength of the PDO and the duration of droughts will have major consequences for disturbance and variability in mountain ecosystem processes.

Past and Present Observations Guide Future Projections

Global change has already had effects on the CCE during the past century, and the influence of the PDO on many ecosystem processes suggests that further changes are in store as the rate of climate change increases. In the CCE, temperature increases have been the major departure from past climates. We must be careful, though, in attributing observed changes to global change without recognizing the impacts of multidecadal droughts and pluvial periods that drive considerable variability in ecosystem dynamics. Nonetheless, visible changes are clearly occurring in the CCE. Continued documentation of these changes, coupled with ecological modeling and long-term monitoring, will give us a better scientific foundation with which to anticipate the future.

References

Alftine, K.J., G.P. Malanson, and D.B. Fagre. 2003. Feedback-Driven Response to Multidecadal Climatic Variability at an Alpine Treeline. *Physical Geography* 24: 520–35.

Barrett, S.W. 1986. *Fire History of Glacier National Park: Middle Fork Flathead River Drainage: Final Report.* West Glacier, MT: National Park Service, Glacier National Park.

Butler, D.R., and L.M. DeChano. 2001. Environmental Change in Glacier National Park, Montana: An Assessment through Repeat Photography from Fire Lookouts. *Physical Geography* 22: 291–304.

Butler, D.R., G.P. Malanson, and D.M. Cairns. 1994. Stability of Alpine Treeline in Glacier National Park, Montana, U.S.A. *Phytocoenologia* 22: 485–500.

Carrara, P.E. 1989. *Late Quaternary Glacial and Vegetative History of the Glacier National Park Region, Montana.* U.S. Geological Survey (USGS) Bulletin 1902. Denver, CO: USGS, 64.

Damm, C. 2001. A Phytosociological Study of Glacier National Park, Montana, USA, with Notes on the Syntaxonomy of Alpine Vegetation in Western North America. PhD thesis. Goettingen, Germany: Georg-August Universitaet.

Diaz, H.F., J.K. Eischeid, C. Duncan, and R.S. Bradley. 2003. Variability of Freezing Levels, Melting Season Indicators, and Snow Cover for High-Elevation and Continental Regions in the Last 50 Years. *Climatic Change* 59: 33–52.

Elias, S.A. 2002. Rocky Mountains. In *Smithsonian Natural History Series.* Washington, DC: Smithsonian Institution Press.

Gedalof, Z., and N.J. Mantua. 2002. A Multi-Century Perspective of Variability in the Pacific Decadal Oscillation: New Insights from Tree Rings and Coral. *Geophysical Research Letters* 29(4): 2204, doi:10.1029/2002GL015824.

Hall, M.P., and D.B. Fagre. 2003. Modeled Climate-Induced Glacier Change in Glacier National Park, 1850–2100. *Bioscience* 53: 131–40.

Key, C.H., D.B. Fagre, and R.K. Menicke. 2002. Glacier Retreat in Glacier National Park, Montana. In *Satellite Image Atlas of Glaciers of the World, Glaciers of North America—Glaciers of the Western United States.* Edited by R.S. Williams, Jr., and J.G. Ferrigno. Washington, DC: U.S. Government Printing Office, J365–81.

Klasner, F.L., and D.B. Fagre. 2002. A Half Century of Change in Alpine Treeline Patterns at Glacier National Park, Montana, U.S.A. *Arctic, Antarctic and Alpine Research* 34: 53–61.

Lesica, P. and B. McCune. 2004. Decline of Arctic-Alpine Plants at the Southern Margin of their Range Following a Decade of Climatic Warming. *Journal of Vegetation Science* 15: 679–90.

Lesica, P., and B.M. Steele. 1996. A Method for Monitoring Long-Term Population Trends: An Example Using Rare Arctic-Alpine Plants. *Ecological Applications* 6: 879–87.

Mantua, N.J., S.R. Hare, Y. Zhang, J.M. Wallace, and R.C. Francis. 1997. A Pacific Interdecadal Climate Oscillation with Impacts on Salmon Production. *Bulletin of the American Meteorological Society* 78: 1069–79.

McKenzie, D., A.E. Hessl, and D.L. Peterson. 2001. Recent Growth of Conifer Species of Western North America: Assessing Spatial Patterns of Radial Growth Trends. *Canadian Journal of Forest Resources* 31: 526–38.

Mote, P.W., W.S. Keeton, and J.F. Franklin. 1999. Decadal Variations in Forest Fire Activity in the Pacific Northwest. *Proceedings of the 11th Conference on Applied Climatology.* Boston: American Meteorological Society, 155–56.

NASA. 2006. 2005 Warmest Year in Over a Century. http://www.nasa.gov/vision/earth/environment/2005_warmest.html (accessed December 13, 2006).

Pederson, G.T., D.B. Fagre, S.T. Gray, and L.J. Graumlich. 2004. Decadal-Scale Climate Drivers for Glacial Dynamics in Glacier National Park, Montana, USA. *Geophysical Research Letters* 31: L12203, doi:10.1029/2004GL0197770.

Pederson, G.T., S.T. Gray, D.B. Fagre, and L.J. Graumlich. 2006. Long-Duration Drought Variability and Impacts on Ecosystem Services: A Case Study from Glacier National Park, Montana, USA. *Earth Interactions* 10: doi:10.1175/EI153.1.

Peterson, D.W., and D.L. Peterson. 2001. Mountain Hemlock Growth Responds to Climatic Variability at Annual and Decadal Scales. *Ecology* 82: 3330–45.

Peterson, D.W., D.L. Peterson, and G.J. Ettl. 2002. Growth Responses of Subalpine fire to Climatic Variability in the Pacific Northwest. *Canadian Journal of Forest Resources* 32: 1503–17.

Reardon, B.A., G.T. Pederson, C.J. Caruso, and D.B. Fagre. Forthcoming. Spatial Reconstructions and Comparisons of Historic Snow Avalanche Frequency and Extent Using Tree-Rings in Glacier National Park, Montana, USA. *Arctic, Antarctic and Alpine Research.*

Resler, L.M. 2006. Geomorphic Controls of Spatial Pattern and Process at Alpine Treeline. *Professional Geographer* 58: 124–38.

Rochefort, R.M., R.L. Little, A. Woodward, and D.L. Peterson. 1994. Changes in Subalpine Tree Distribution in Western North America: A Review of Climatic and Other Causal Factors. *The Holocene* 4: 89–100.

Roush, W., J.S. Munroe, and D.B. Fagre. Forthcoming. Using Ground-Based Repeat Photography and Spatial Analysis Tools to Detect Changes in the Alpine Treeline Ecotone, Glacier National Park, Montana. *Arctic, Antarctic and Alpine Research.*

Schimel, D., T. Kittel, S. Running, R. Monson, A. Turpinseed, and D. Anderson. 2002. Carbon Sequestration Studied in Western U.S. Mountains. *EOS Transactions* 83: 445–46.

Selkowitz, D.J., D.B. Fagre, and B.A. Reardon. 2002. Interannual Variations in Snowpack in the Crown of the Continent Ecosystem. *Hydrological Processes* 16: 3651–65.

USGCRP (U.S. Global Change Research Program). 2000. *Climate Change and America: Overview Document. A Report of the National Assessment Synthesis Team.* Washington, DC: USGCRP.

Watson, E., G.T. Pederson, B.H. Luckman, and D.B. Fagre. Forthcoming. Glacier Mass Balance in the Northern U.S. and Canadian Rockies: Paleo-Perspectives and 20th Century Change. In *Darkening Peaks: Glacial Retreat in Scientific and Social Context.* Edited by B. Orlove, E. Weigandt, and B. Luckman. Berkeley: University of California Press.

White, J.D., S.W. Running, P.E. Thornton, R.E. Keane, K.C. Ryan, D.B. Fagre, and C.H. Key. 1998. Assessing Simulated Ecosystem Processes for Climate Variability Research at Glacier National Park, USA. *Ecological Applications* 8: 805–23.

13

CCE Fire Regimes and Their Management

Robert E. Keane and Carl Key

A spectacular forest in the center of the CCE cuts a 15- by 5-km swath along the Flathead River's South Fork around Big Prairie in the middle of the Bob Marshall Wilderness Area in Montana (Figure 13-1). This wide valley bottom, which contains two patches (of about 1,000 ha each) of the last vestiges of the historic ponderosa pine ecosystem in the CCE, provides a local context and a case example for our discussion of fire dynamics in this chapter. The Big Prairie ponderosa pine (see Chapter 2 for scientific names not given in this chapter) ecosystem is a consequence of a special fire regime that has been altered during the last century. As a result, this ponderosa pine forest is declining rapidly, and the causes of its decline are similar to those in many other fire-dependent ecosystems in this diverse region. Here we discuss the many and varied fire regimes of CCE landscapes, using the Big Prairie ecosystem to demonstrate the challenges of managing fire.

Big Prairie Ponderosa Pine Forest

This forest is confined to dry river terraces along the South Fork of the Flathead River. Historically, this area was a pine savanna or an open, park-like forest where ponderosa pine grew as widely scattered trees above a grass understory (Arno et al. 2000; Figure 13-1a). The forest was maintained by frequent, low-intensity fires that occurred at intervals averaging approximately 25 years. These fires were likely started by Native Americans who used these forests seasonally as they traveled from the Flathead Valley to buffalo-hunting grounds in the Great Plains (Ostlund et al. 2005). Numerous ponderosa pine trees that are living today have distinctive large oval scars where native peoples peeled off the bark layer to harvest the underlying sap layers (Figure 13-1b). Many of these big pines also contain scars from multiple fires, which serve as documentation of fire frequency over the last three centuries. Recurrent fires would kill most of

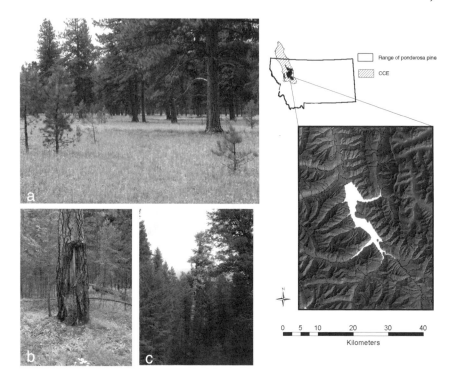

FIGURE 13-1. Ponderosa Pine Savanna Ecosystem of Big Prairie, Bob Marshall Wilderness Area, Montana

Notes: (a) pine savanna; (b) scarred ponderosa pine; (c) open, pine-dominated forest.

the encroaching saplings of competing trees species, namely Douglas fir and lodgepole pine, and maintain the open, pine-dominated forest structure (Figure 13-1c). Because ponderosa pine has thick bark, a high open crown, and deep roots, it is able to survive fires much better than its competitors.

The last 75 to 100 years have brought an increase of Douglas fir and lodgepole pine in the Big Prairie area because of the absence of aboriginal burning and an active fire-exclusion program in the wilderness area, especially prior to 1980 (Figure 13-1c). This relic forest is found near the upper elevational limit of ponderosa pine in the CCE, so the species does not reproduce and grow as well as its frost-hardy, shade-tolerant competitive tree species. Ponderosa pine is a shade-intolerant or sun-loving tree species that does not grow and regenerate in the dense forests that result from excluding fire. Without recurrent fire, large, old ponderosa pine trees become stressed and die as the forest becomes crowded with Douglas fir and lodgepole pine. This theme is repeated in many forested ecosystems in the CCE, where shade-intolerant, fire-adapted tree species are eventually replaced by more shade-tolerant species in the absence of wildland fire.

Historical Fire Regimes

The CCE is unique in that a diversity of fire regimes is represented in the region. This diversity has created a varied array of fire-dependent ecosystems, along with a distinctive landscape mosaic of plant communities that has arisen after differential burning (Figure 13-2). Because it has dictated the structure (patch distribution) and composition (plant communities) of most CCE landscapes, wildland fire has had a dominant impact on CCE ecosystems.

Fire regime is a general term that describes the temporal and spatial characteristics of fire dynamics and includes such attributes as frequency, severity, seasonality, and pattern (Agee 1993; DeBano et al. 1998). *Fire frequency*, described by the mean fire-return interval (in years), is usually defined by how often a fire burns a point on the landscape. *Severity* describes the impact of fire on the biota and soil and is often, but not always, related to fire intensity (the heat produced from the fire). Because of plant phenology, the *seasonality* of burn can produce differential effects. Finally, *pattern* refers to the size, shape, and spatial location of the burned area. For simplicity, we will confine our descriptions of CCE fire regimes to frequency and severity.

Brown and Smith (2000) describe the four major fire regime types that we consider in this chapter: (1) the nonfire (NF) regime, in which fire does not occur; (2) the understory or nonlethal surface fire (NLSF) regime, in which approximately 80% of dominant vegetation survives; (3) the stand-replacement fire (SRF) regime, in which approximately 80% of the aboveground dominant vegetation is consumed or dies; and (4) the mixed-severity (mixed) fire regime, which results in selective mortality with patches of understory and stand-replacement burns.

Historically, fire regimes in the CCE tended to be governed by the distribution of fuel moistures and loadings in space and time. Because forests at lower elevations are warmer and drier, more dry fuel tends to be available for burning over a longer period (Agee 1993). Forests at higher elevations are usually colder and damper, so they tend to be moist for most of the year and to burn only during years when the upper-elevation landscape is dry for long periods. Fires burn the most area in years of severe drought, such as 1910 and 1988, when the entire landscape is parched, ignitions (lightning strikes) are abundant, and the weather is windy (Schmoldt et al. 1999).

Fuel loadings are usually adequate to carry most fires in CCE ecosystems except for those that have recently burned (1–5 years old), the rockiest slopes (talus and scree), and some parts of the alpine tundra. Fuels can be the fallen litter (needles, leaves, cones, and buds), twigs, branches, and logs that collect on the ground; they can be "duff," which is the result of litter decomposition; and they can be living or dead plants. Fuel loadings generally increase with time since the last fire. Depending on ecosystem productivity, these loadings tend to reach equilibrium after about 100 years. Historical fire ignitions have tended to be less anthropogenic and have typically been caused by lightning with increasing elevations because of the fuel-moisture limitations (Boyd 1999). Many high-elevation fires, though, have resulted from low-elevation ignitions. The severity of CCE fires also tends to increase with time since the last fire because high fuel loadings foster more intense fires that generate higher heat and cause higher biotic mortality (DeBano et al. 1998). The invasion of shade-tolerant trees into the understory— and eventually the overstory—of a mature stand of shade-intolerant trees often increases the density and lowers the height of canopy fuels. This contributes to "torching" (when fire engulfs the overstory) and to more severe and intense fires called *crown fires* (Keane et al. 2002).

Indigenous peoples probably played an important role in fire regimes on the CCE landscape. Substantial evidence indicates that Native Americans used many portions of the Rocky Mountain landscape extensively by the early sixteenth century (Denevan 1992) and likely much earlier. John Mullan (1866) recognized that these early inhabitants had a profound bearing on forest structure and composition, resulting primarily from fires they set. They started fires for reasons including land clearing, wildlife-habitat improvement, crop cultivation, defense, signaling, and hunting (Lewis 1985; Kay 1995). There is great debate about whether lightning could have produced the same fire regimes that the Native Americans maintained (Barrett and Arno 1982; Gruell 1985), and also about whether anthropogenic burning should be considered part of the native fire regime (Arno 1985; Kilgore 1985). Fires set by Native Americans often differed from lightning fires in terms of seasonality, frequency, intensity, and ignition patterns (Kay 1995). We believe that fires set by Native Americans influenced fire dynamics in the CCE and that this factor should be recognized in the management of this vast region.

Major CCE Fire Regimes

High-elevation CCE forests were historically dominated by whitebark pine ecosystems that usually experienced infrequent fires at greater than 200-year intervals (see Tomback et al. 2000 and Figure 13-2). These rare fires were large and quite severe, killing the most trees of all species. The bird-dispersed whitebark pine, however, gained the colonization advantage because the Clark's nutcracker could plant the seeds farther into the burned area than wind could disperse seeds of the pine's major competitors. Whitebark pine is eventually supplanted by subalpine fir and sometimes by Engelmann spruce. The spread of the exotic white pine blister rust into the CCE during the last 70 years has severely reduced whitebark populations in the northern Rocky Mountains (Keane et al. 1994).

The subalpine forests below the whitebark pine zone were dominated primarily by lodgepole pine trees with infrequent SRFs racing through the forest

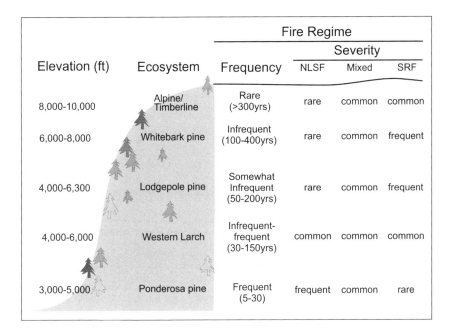

FIGURE 13-2. Characteristics of Fire Regimes along Elevation and Aspect Gradients

Notes: NLSF = nonlethal surface fire regime; SRF = stand-replacement fire regime.

canopy and killing most trees (Figure 13-2). On these sites, lodgepole pine could colonize burned areas because its cones, which remain on the tree, are sealed in wax (Tait et al. 1988). Intense fires melt the wax, open the cones, and allow the protected seeds to disperse from the fire-killed trees. This abundant "seed rain" from lodgepole pine creates dense stands of crowded trees. Subsequently, the natural mortality of pine because of crowding creates heavy fuel loads of small stems. When these heavy loads were added to the eventual dense regeneration of the shade-tolerant subalpine fir, crown fires of high intensity and severity resulted, especially in dry years, and then the cycle would repeat (Lotan et al. 1985). Similar to whitebark pine, lodgepole pine is eventually replaced by Douglas fir and subalpine fir, and sometimes by Engelmann spruce.

The montane forests below lodgepole pine were mostly composed of mixed tree species but usually dominated by western larch, Douglas fir, and sometimes western pine *(Pinus strobes;* Figure 13-2). Western larch, and, in limited areas, western white pine grow exceptionally tall and have thick bark, allowing them to survive fires of varying frequencies and severities (Arno et al. 2000). When fires were intense, the tall larches were often the only surviving trees, supplying the only source of seeds to populate burned areas. This ensured the continued presence and dominance of the western larch. Some fires, though, were low-intensity, understory surface fires that killed many smaller trees but maintained open, larch-dominated forests because the thick-barked larches could easily survive less-intense fires (Schmidt and McDonald 1995). Native Americans might have started fires in montane forests to open the forests for travel and to increase visibility, allowing them to see their enemies (Lewis 1985).

Western larch and white pine are often successively replaced by a wide variety of more shade-tolerant conifers including Douglas fir, western red cedar, subalpine fir, and western hemlock. Similar to its effect on whitebark pine, the invasion of blister rust has almost extirpated western white pine from the CCE landscape.

Last, we come to the low-elevation ponderosa pine forests (Figures 13-1, 13-2). As we mentioned previously, these forests experienced frequent fires of aboriginal origin because the fuel, primarily grass, was dry for relatively longer periods of time. Although this ecosystem has limited distribution in the CCE, where it is confined to riverine terraces, dry south slopes, and wide valley bottoms, it was often the ecosystem most heavily used by Native Americans for travel routes, camping areas, hunting grounds, and wintering areas.

Unique landscapes are created by the cumulative effects of the interactions of these diverse fire regimes across a spatial domain. In these landscapes, composition and structure are dictated by burn patterns, fire severity, vegetation development rates, and time since the last fire. The complex terrain and the availability of fuel control the pattern and extent of burned areas. Fire spread is often confined to small drainages because snow, rock, talus, and alpine landforms at the head of watersheds prevent fire spread into adjoining lands, except in severe drought years or when the fires are wind-driven and firebrands can "spot" into adjacent watersheds. The spatial pattern of fuel moisture and loading dictate the subsequent fire severity, which then governs postfire response and successional trajectories (Kessell and Fischer 1981). These diverse fire regimes create the shifting mosaics of vegetation communities that give the CCE its distinctive ecology.

Current Fire Regimes

Since the early 1930s, extensive fire-suppression programs have successfully reduced wildland fire in many portions of the CCE region (Keane et al. 2002). The absence of fire has created landscapes with atypical species compositions and accumulations of contagious fuels (i.e., fuels in close proximity that allow fire to spread) that pose a hazard to many ecosystem characteristics and human settlements. The health of many CCE landscapes appears to be declining because shade-tolerant species have been invading and sometimes replacing the forests of shade-intolerant, fire-adapted species. The continued suppression of fire has actually made it more difficult to fight fires, posing greater risks to firefighters and residents of the CCE and surrounding areas. The diverse and cascading effects of attempting to exclude fire from the northern Rocky Mountain landscapes have wide-ranging impacts, including (1) less water runoff; (2) larger, more intense fires; (3) less landscape and community diversity; (4) frequent insect and disease epidemics; (5) loss of biodiversity; and (6) loss of wildlife habitat (Keane et al. 2002). Since the 1980s, however, some land-management agencies in the United States and Canada have begun to restore fires on the landscape.

For a variety of reasons, the effects of the decades of fire exclusion are not always evident on CCE landscapes today. First, many CCE landscapes comprise ecosystems with long fire-return intervals (e.g., whitebark pine and lodgepole forests; Figure 13-2), so sufficient time may not have elapsed to force these landscapes out of the historical fire rotation. An unusually high number of old forests, however, are currently growing on many landscapes. Second, the last 20 years have seen an increase of fire on the CCE landscape. This increase results from changes in fire management that allow some lightning-caused fires to burn, increases in the loading and contagion of fuel, and severe drought that followed several decades of wetter conditions. Since the 1980s, managers have allowed some wilderness and remote fires to burn within the CCE. This has increased the burned area and started us on the long road toward restoring historical fire regimes. Once called "prescribed natural fires," this fire-management strategy is now known as "wildland fire use." Approximately 10% of the central CCE has burned during the last two decades, with the most fires occurring in the montane and subalpine ecosystems.

Fire-Management Issues

Perhaps the most important factor influencing fire management within the CCE is that the majority of the lands are protected wilderness areas, roadless areas, or national parks (see Chapter 1). Wilderness areas, de facto wilderness, and national parks comprise more than 50% of the CCE. In these wilderness settings, the set of fire and fuel treatments that can be implemented to reestablish historical fire regimes is limited. Another important factor is the extensive development on lands along the edge of the CCE. Fires occurring in and around the region now have a greater chance to burn private property and harm people. A last consideration is the introduction of exotics into the CCE ecosystem. Non-native plants, animals, insects, and diseases have wreaked havoc with native ecosystems and altered fire regimes in some local sites. Fires have actually accelerated the spread of weed in some cases.

The most immediate problem facing CCE fire managers is restoring some semblance of the historical fire regimes in the region. The decades of fire exclusion have created landscapes where fuel loadings are so high and contagious that, if a wildfire burns these areas today, it may have effects that have rarely occurred in recent history, possibly resulting in the loss of important ecosystem components. The Big Prairie ponderosa pine stand serves as an example. The first-year mortality of large, relic ponderosa pine after the fires of 2003 was approximately 34%, even though fire intensities and scorch heights were low to moderate. This was primarily because the smoldering consumption of the deep duff accumulations around the trees increased root and cambial mortality and ultimately killed the trees (Keane et al. 2006). It appears that we may need to apply proactive fuel treatments to reduce fire sizes and intensities and protect the remaining endangered CCE ecosystems, especially if these ecosystems are near homes or developments. Even though research has shown that managing fuel within 50 m of a home is more important to protecting the home than managing fuels in the forests that surround the home (Cohen 2004), the surrounding forests must nevertheless be managed to avoid losing critical ecosystem elements after they burn.

Many CCE ecosystems can benefit from reintroducing fire. Whitebark pine forests, for example, are declining rapidly because of mountain pine beetle epidemics and the spread of the exotic blister rust disease. Subalpine fir replaces the dying whitebark pine. Fires are needed to kill the subalpine fir to create openings where Clark's nutcrackers can plant whitebark pine seed caches and where new whitebark pine trees can successfully regenerate from unclaimed caches with little competition (Tomback et al. 2000). Because they were harvested from trees that survived the rust epidemic, the cached seeds are likely to produce trees with an increased level of rust resistance. Losing this keystone ecosystem would adversely affect more than 110 wildlife species that depend on this pine for food, especially the grizzly bear. Whitebark pine seed is a high-nutrition food source for grizzly and black bears and typically constitutes a major part of bear diet.

Without fire, many lodgepole pine cones do not open, limiting the species' ability to propagate. Subalpine fir will eventually replace lodgepole pine and those species dependent on lodgepole pine forests will decline. The large western larch trees still living on the montane CCE settings will eventually succumb to competing conifers or they will eventually be killed by abnormally severe wildfires. This loss will greatly reduce larch seed crops and change the composition and structure of postfire montane landscapes. In addition, ponderosa pine forests can rapidly be replaced by shade-tolerant conifers, also resulting in the loss of a unique ecosystem that many plants and animals have selectively utilized. Continuing our attempts to exclude fires in the CCE will spell the demise of these unique ecosystems.

Returning fire to these historically fire-dominated forests is not as easy as it might seem. In some stands, fire has been excluded for so long that surface and canopy fuels have accumulated to levels that, when burned, will have abnormal and undesirable consequences, such as killing the old ponderosa pine trees in Big Prairie (Brown 1985). Reducing fuel loadings using only prescribed fires—those sparked by lightning or humans—is difficult because there is a small window of opportunity to light the fire under moisture conditions that would not damage or kill the desirable ecosystem legacies and yet would still carry the fire. We can reduce these fuels using silvicultural cutting techniques, but this strategy is generally confined to scattered patches along the wildland–urban interface and near roads. These mechanical treatments are often costly and contentious and may have to be augmented with pile burning or prescribed burning for optimum efficacy. Many believe it may take two or more mechanical or fire treatments to restore some fire-excluded ecosystems.

Wilderness and Fire Management

The historical impact of anthropogenic burning in the CCE, which is now composed predominately of protected areas, presents a philosophical dilemma for management. The Wilderness Act of 1964 states that wilderness should be managed so as to be "untrammeled by man," but the very character of the landscape was probably shaped by thousands of years of burning by Native Americans (Stewart 2002). Ignoring the effect that aboriginal burning had on the flora and fauna of CCE wilderness settings would lead to the eventual creation of landscapes unlike those of the past. Yet the Wilderness Act specifically states that wilderness "not be subject to human controls and manipulations that hamper the free play of natural forces" (Hendee et al. 1978). Were aboriginal fires "natural forces" or were they "human manipulations"? Were the people who occupied North America for more than 20,000 years not a part of the natural environment? If all types of historical fires are deemed natural, the ecosystems that these fires created should indeed be conserved through frequent burning. Humans will probably need to light these fires.

Another complicating factor is the assumption in contemporary wilderness fire management that humans have not hampered the free play of natural forces. Modern humans actively suppressed most fires on wilderness landscapes for many decades before the Wilderness Act was passed. And, by definition, the act of suppressing fires and the policy of excluding fires are major human controls and manipulations. Since the 1980s, we have allowed a few fires to burn in CCE wilderness when they originated from natural ignitions (such as lightning). These fires were allowed to burn as prescribed fires under predetermined weather conditions. We have, however, suppressed the majority of wilderness fires. This exclusion of fire has led to wilderness settings where humans have hampered the most important disturbance process, wildland fire, for several decades.

In summary, two diametrically opposed anthropogenic actions have contributed to the quandary in which we find ourselves when deciding how to properly manage the CCE landscape. Humans have had their hands on wilderness ecology for a long time, yet the Wilderness Act does not fully recognize this reality or furnish any guidance for resolving this issue.

Human intervention is probably needed to save some of the remaining fire-dependent ecosystems such as the Big Prairie ponderosa pine forest. Proactive treatments might include raking around the base of the pines to minimize heating of the stem and roots, igniting prescribed burns during periods when lower duff is still moist to reduce fire intensities, and cutting encroaching conifers to reduce their potential for crown fires. These treatments may compromise the wildness character of the CCE and reduce the quality of the wilderness experience for some people.

A Crossroads in Wilderness Fire Management

We find ourselves at a crossroads in the management of wilderness and large natural areas such as those that comprise the CCE. Land management practices in the recent past have resulted in a buildup of surface fuels and a thickening of crown fuels, factors that lead to large, more-severe, and more-intense fires. This has put forth yet another dilemma for wilderness managers. If we decide that conserving historical CCE fire regimes is important, we must accept a small loss of wildness so that historical ecological processes and vegetation types can be restored. This can also protect the valuable ecosystem elements, such as the old-growth ponderosa pines, from future wildfires. Conversely, if we decide that wildness is more important, we must accept the consequences of the new fire regime, which will probably create landscapes that do not resemble those of the last 10,000 years. But important CCE wilderness legacies, such as large, cone-bearing whitebark pine forests, will dwindle to nothing if we do nothing, especially in light of the concomitant adverse effects of exotics, climate change, and human development in the CCE.

References

Agee, J.K. 1993. *Fire Ecology of Pacific Northwest Forests.* Washington, DC: Island Press.
Arno, S.F. 1985. Ecological Effects and Management Implications of Indian fires. In *Proceedings of the Symposium and Workshop on Wilderness Fire.* Edited by J.E. Lotan, B.M. Kilgore, W.C. Fischer, and R.W. Mutch. General Technical Report INT-182. Washington, DC: U.S. Department of Agriculture (USDA) Forest Service, 81–89.
Arno, S. F., D.J. Parsons, and R.E. Keane. 2000. Mixed-Severity Fire Regimes in the Northern Rocky Mountains: Consequences of Fire Exclusion and Options for the Future. In *Wilderness Science in a Time of Change Conference, Volume 5: Wilderness Ecosystems, Threat, and Management.* Fort Collins, CO: USDA Forest Service Rocky Mountain Research Station, 225–32.
Barrett, S.W., and S.F. Arno. 1982. Indian Fires as an Ecological Influence in the Northern Rockies. *Journal of Forestry* 80: 647–50.

Boyd, R. 1999. *Indians, Fire and the Land in the Pacific Northwest.* Corvallis, OR: Oregon State University Press.

Brown, J.K. 1985. The "Unnatural Fuel Buildup" issue. In *Symposium and Workshop on Wilderness Fire.* Edited by J.E. Lotan, B.M. Kilgore, W.C. Fischer, and R.W. Mutch. Missoula, MT: USDA Forest Service Intermountain Forest and Range Experiment Station, 127–28.

Brown, J. K., and J.K. Smith (eds.). 2000. *Wildland Fire in Ecosystems: Effects of Fire on Flora.* General Technical Report RMRS-GTR-42-vol. 2. Ogden, UT: USDA Forest Service Rocky Mountain Research Station.

Cohen, J.D. 2004. Relating Flame Radiation to Home Ignition Using Modeling and Experimental Crown Fires. *Canadian Journal of Forest Research* 34: 1616–26.

DeBano, L.F., D.G. Neary, and P.F. Folliot. 1998. *Fire's Effect on Ecosystems.* New York: John Wiley and Sons.

Denevan, W.M. 1992. The Pristine Myth: The Landscape of the Americas in 1492. *Annals of the Association of American Geographers* 8: 369–85.

Gruell, G.E. 1985. *Indian Fires in the Interior West: A Widespread Influence.* General Technical Report INT-182. Washington, DC: USDA Forest Service.

Hendee, J.C., G.H. Stankey, and R.C. Lucas. 1978. *Fire in Wilderness Ecosystems.* USDA Forest Service Miscellaneous Publication 1365: 249–78.

Kay, C.E. 1995. Aboriginal Overkill and Native Burning: Implications for Modern Ecosystem Management. *Western Journal of Applied Forestry* 10: 121–26.

Keane, R.E., S. Arno, and L.J. Dickinson. 2006. The Complexity of Managing Fire-Dependent Ecosystems in Wilderness: Relict Ponderosa Pine in the Bob Marshall Wilderness. *Ecological Restoration* 21: 71–78.

Keane, R.E., P. Morgan, and J.P. Menakis. 1994. Landscape Assessment of the Decline of Whitebark Pine (*Pinus albicaulis*) in the Bob Marshall Wilderness Complex, Montana, USA. *Northwest Science* 68: 213–29.

Keane, R.E., T. Veblen, K.C. Ryan, J. Logan, C. Allen, and B. Hawkes. 2002. The Cascading Effects of Fire Exclusion in the Rocky Mountains. In *Rocky Mountain Futures: An Ecological Perspective.* Edited by J.S. Baron. Washington, DC: Island Press, 133–53.

Kessell, S.R., and W.C. Fischer. 1981. Predicting Postfire Plant Succession for Fire Management Planning. General Technical Report INT-94. Washington, DC: USDA Forest Service.

Kilgore, B.M. 1985. What Is "Natural" in Wilderness Fire Management? In *Symposium and Workshop on Wilderness Fire.* Edited by J.E. Lotan, B.M. Kilgore, W.C. Fischer, and R.W. Mutch. Missoula, MT: USDA Forest Service Intermountain Forest and Range Experiment Station, 57–67.

Lewis, H.T. 1985. Why Indians Burned: Specific Versus General Reasons. General Technical Report INT-182. Washington, DC: USDA Forest Service.

Lotan, J.E., J.K. Brown, and L.F. Neuenschwander. 1985. Role of Fire in Lodgepole Pine Forests. In *Lodgepole Pine: The Species and Its Management, Symposium Proceedings.* Edited by D.M. Baumgartner, D.M. Krebill, J.T. Arnott, and G.F. Weetman. Pullman: Washington State University Cooperative Extension Service, 133–52.

Mullan, J. 1866. *Report on Military Roads.* House of Representatives Executive Document #44.

Ostlund, L., R.E. Keane, S.F. Arno, and R. Andersson. 2005. Culturally Scarred Trees in the Bob Marshall Wilderness, Montana, USA—Interpreting Native American Historical Land Use in a Wilderness Area. *Natural Areas Journal* 25(4): 315–25.

Schmidt, W.C., and K.J. McDonald. 1995. *Ecology and Management of Larix Forests: A Look Ahead.* General Technical Report GTR-INT-319. Ogden, UT: USDA Forest Service Intermountain Research Station.

Schmoldt, D.L., D.L. Peterson, R.E. Keane, J.M. Lenihan, D. McKenzie, D.R. Weise, and D.V. Sandberg. 1999. *Assessing the Effects of Fire Disturbance on Ecosystems: A Scientific Agenda for Research and Management.* General Technical Report GTR-PNW-455. Washington, DC: USDA Forest Service.

Stewart, O. 2002. *Forgotten Fires: Native Americans and the Transient Wilderness.* Norman, OK: University of Oklahoma Press.

Tait, D.E., C.J. Cieszewski, and I.E. Bella. 1988. The Stand Dynamics of Lodgepole Pine. *Canadian Journal of Forest Research* 18: 1255–60.

Tomback, D., S.F. Arno, and R.E. Keane. 2000. *Whitebark Pine Communities: Ecology and Restoration.* Washington, DC: Island Press.

PART V
Management Issues and Challenges

14

Cumulative Effects Analysis and the Crown Managers Partnership

Michael Quinn, Danah Duke, and Guy Greenaway

> ... [I]f the right people come together in constructive forums with the best available information, they are likely to shape effective solutions to shared problems.
>
> —McKinney et al. 2002, *102*

One of the greatest threats to the sustainability of the CCE, including the impacts of climate change, loss of biodiversity, and declining quality and quantity of freshwater, is the insidious erosion of ecological integrity resulting from the cumulative effects of individual decisions and actions. Even though the ecosystem impacts of individual actions may seem insignificant, the additive and synergistic effects in time and space of repeated and multiple actions have the potential to impair the structure and function of the CCE.

In this chapter, we describe the work of a group of public land managers called the Crown Managers Partnership (CMP). The members of the partnership are exploring innovative and collaborative approaches to evaluate the cumulative effects of human activities on natural resources in the transboundary CCE.

The Nature and Significance of Cumulative Effects

Our discussion illustrates transboundary cumulative effects in the CCE. We illustrate the meaning of cumulative effects with an example of the ecologic and hydrologic conditions of a river with its headwaters on the Continental Divide (e.g., the Milk, Belly, Waterton, and Flathead rivers). If we were to sample one of the rivers as it egresses from the CCE, the quality, quantity, and flow regime, along with the associated aquatic biota, would be a complex function of numerous biophysical factors. Although any single factor could have measurable impacts on river conditions, the interaction of these effects over time and space results in highly complex, potentially large, and often unpredictable changes to aquatic ecosystems.

The proliferation of cumulative effects has been attributed to the incremental and disjointed nature of decisionmaking (the so-called "tyranny of small decisions") that characterizes many contemporary institutional structures and regulatory approval processes (Creasey 2002). The CCE is characterized by jurisdictional fragmentation, with each administrative authority making decisions within its jurisdiction, independently of other authorities. Furthermore, within each administrative boundary, individual resource sectors are regulated in a piecemeal fashion with a focus on mitigation of local, short-term effects. Each new subdivision, gas well site, logging cutblock, or recreational trail is subject to an approval process that contains few requirements for consideration of cumulative effects.

In northern Alberta, a striking example of the negative effects of jurisdictional fragmentation pertains to two resource sectors, petroleum and timber harvesting. A cumulative effects study (AXYS Environmental Consulting Limited 2000) demonstrated that the activities of the petroleum industry (i.e., construction and installation of seismic lines, roads, well pads, and pipelines) were causing more timber to be removed than the forest industry was harvesting in the same area. Furthermore, the petroleum industry is not subject to the same planning and regulatory processes for timber harvesting as the forest industry. Modeling of business-as-usual scenarios predicted an eventual exhaustion of the wood supply and associated negative ecological effects (Schneider et al. 2003). In the same modeling effort, Schneider and colleagues demonstrated that implementation of best management practices and better integration between the activities of the petroleum and forest industries could significantly improve future prospects.

Sustainable development and adaptive ecosystem-based management, along with emerging holistic regional planning, policy, and management paradigms, echo the need for novel approaches that strategically address cumulative effects. These approaches should be designed to achieve three goals: ecological integrity, economic sustainability, and social equity (Stinchcombe and Gibson 2001). Unfortunately, political, financial, and technical barriers impede landscape-scale data collection and the development of frameworks necessary for transjurisdictional assessment of cumulative effects. These barriers are magnified when political borders divide a landscape, as in this case. In the CCE, no single agency has the mandate or the resources to focus on the entire transjurisdictional region. Recognizing this situation, a group of resource-agency managers launched a new collaborative initiative, the CMP.

Crown Managers Partnership

In February 2001, government representatives from more than 20 agencies gathered in Cranbrook, British Columbia, to explore ecosystem-based collaboration on shared issues in the CCE. Participation included federal, aboriginal, provincial, and state agencies or organizations with significant land- or resource-management responsibility in the CCE. The aim of the meeting was to increase interactions among senior and middle managers and technical and professional staff (e.g., conservation biologists and land-use planners) that have management responsibilities at the ecosystem scale. The Miistakis Institute for the Rockies, a nonprofit research organization affiliated with the University of Calgary, was invited to facilitate the process and act as a neutral third party. The highly successful founding workshop, hosted by managers of the Waterton Glacier International Peace Park, resulted in a commitment by all participants to move forward collaboratively on regional communication and management.

At the founding workshop, participants reached consensus on the need to address five strategic or priority regional-scale issues: (1) cumulative effects of human activity across the region; (2) increased public interest in how lands are managed and how decisions are reached; (3) increased recreational demands and visitation; (4) collaborative data sharing and standardizing assessment and monitoring methodologies; and (5) maintenance and sustainability of shared (transboundary) wildlife populations.

To advance progress, the CMP established a steering committee composed of volunteer representatives from participating agencies to develop a work plan that addresses the priority issues identified by the CMP. The steering committee meets on a regular basis and some progress has been made on all priority issues. In this chapter, we describe CMP progress on the first-priority issue—evaluating cumulative effects of human activity across the CCE.

Approaches to Assessing Cumulative Effects

Cumulative effects of human activities on the environment are identified and assessed through cumulative effects assessment (CEA), which is the antithesis of site-specific, linear, cause-and-effect analysis. CEA gives proponents and stakeholders a set of tools with which to systematically evaluate the impact of proposed actions on the environment and socioeconomic system (Griffiths 1998). CEA allows stakeholders to address environmental impacts from a single activity as well as from the additive and synergistic impacts of multiple activities. As defined by the Canadian Environmental Assessment Act of 1992 and summarized by the Cumulative Effects Assessment Working Group (1999), CEA has four purposes: (1) to assess effects over a larger ("regional") area that may cross jurisdictional boundaries, including effects resulting from natural perturbations that affect environmental components and human actions; (2) to evaluate effects over a longer period of time, reaching into the past and future; (3) to consider effects on valued ecosystem components (VECs) that result from interactions with other human activities (not just the effects of the single action under review); and (4) to account for other past, existing, and future (reasonably foreseeable) actions.

The theory and practice of CEA of individual project proposals have advanced significantly in the past two decades. Project-based CEAs are generally conducted under statutory environmental assessment processes required by provincial/territorial, state, or federal legislation. Project-based CEAs tend to be proponent-driven and focus on the incremental impacts of proposed projects within a limited area. As project-based CEAs have evolved and become more comprehensive, project proponents have become increasingly dissatisfied with the requirements to address cumulative effects over larger temporal and spatial scales. In particular, insufficient information about other projects and a lack of management control have proven to be significant obstacles (Davey et al. 2002).

The experience and insight gained from project-based CEA points to the need for assessment of regional cumulative effects, which will provide more and better strategic information for planning and comprehensive assessment of individual project proposals. Strategic environmental assessment has been defined as

> ... a systematic on-going process for evaluating, at the earliest appropriate stage of publicly accountable decision making, the environmental quality, and consequences of alternative visions and development intentions incorporated in policy, planning, program initiatives, ensuring full integration of relevant biophysical, economic, social and political considerations. (Partidário and Clark 2000, *4*)

The principles that make the approach "strategic" include (1) focusing on strategies for action; (2) using backcasting and forecasting; (3) considering alternative means of achieving goals; (4) clearly articulating goals and objectives; (5) being proactive; (6) integrating ecological, social, and economic factors; (7) being broadly focused; and (8) using a tiered approach. A strategic CEA is considered to be one of the most promising approaches for advancing national and international commitments to sustainable development (Stinchcombe and Gibson 2001).

Strategic regional CEAs typically occur over large spatial and temporal scales and assess human activities across jurisdictional boundaries. Generally, regional CEAs are initiated outside the statutory environmental assessment process, although they can play an important role in project-based CEA. By definition, regional approaches require significant interagency collaboration. According to

AXYS Environmental Consulting Limited (2000, 22) "A critical component in developing a framework for the assessment of regional cumulative effects is government agencies working in partnership to develop a management strategy in areas where future development will likely occur."

Regional CEAs are used in planning processes to assess current conditions, thresholds, and environmental and social capacities. A strategic approach to addressing cumulative effects over large geographic areas overlaps conceptually with the long-standing field of regional planning. In a review of regional approaches to public policy and natural-resource management in the American West, McKinney and colleagues (2002, 102) suggest that "regionalism looks beyond political and jurisdictional boundaries, embracing a distinctly transboundary approach that recognizes the natural territory of public issues, such as watersheds, ecosystems, bioregions or other organic regions." Emerging approaches to ecoregional planning are a response to the failure of existing jurisdictional infrastructure to effectively manage complex transboundary issues.

In pursuing its stated priority of addressing cumulative effects in the CCE, the CMP sought a strategic, regional, and highly collaborative approach that would support decisionmaking at the higher planning and policy levels.

Cumulative Effects Modeling and Framework Development

The framework for a transboundary, collaborative approach to assessing cumulative effects in the CCE was first discussed at the CMP's inaugural meeting in 2001. As mentioned previously, the ensuing work plan identified a CEA for the region as a top priority. In 2002, the Miistakis Institute was retained as project manager and tasked with developing a framework for the regional CEA, which eventually became known as the Regional Landscape Analysis Project (RLAP; Quinn et al. 2005).

ALCES® Model

Concluding that CEA would require robust landscape-scale modeling, the CMP began searching for an appropriate model. Although modeling software cannot drive the strategic CEA, the model is a fundamental component of the framework, significantly influencing the type of data to be collected and the scenarios generated. After two exploratory workshops, the CMP selected a model developed by Forem Technologies (www.foremtech.com) called A Landscape Cumulative Effects Simulator (ALCES®). The CMP chose this model because of its proven applicability for regional CEA in nearby areas (such as Alberta), its validation and verification by independent experts (Hudson 2002), and the familiarity of several CMP participants with the model.

ALCES is a stock-and-flow (systems dynamics) model constructed in a STELLA® (Richmond et al. 1987) modeling environment. The model operates by establishing relationships (pathways and rates of flow) between entities (stocks) of interest (e.g., land-use and land-cover types). Next, it simulates changes in the entities over time. ALCES is spatially stratified rather than spatially explicit, meaning that it simulates the area of the landscape in a particular land-cover type (e.g., lodgepole pine forest), but does not simulate the spatial distribution of that land-cover type on the landscape.

The model contributes to a strategic CEA framework by making available an exploratory tool that helps to identify emerging regional issues and opportunities. In addition, it examines the potential implications of trends and policy choices under a range of future scenarios. Finally, the model simulates the cumulative effects of all selected landscape activities and processes.

RLAP Management

The CMP steering committee directs the management of the RLAP. Specifically, the CMP steering committee (1) works with upper management/higher political levels to encourage agency participation; (2) collaborates with the project team to identify and prioritize land uses to be modeled; (3) assists in defining scenarios for cumulative effects modeling; and (4) communicates modeling results to the CMP.

The project team is responsible for conducting the regional cumulative effects modeling and analysis, and ensuring that the overall vision for the project is realized. The Miistakis Institute reports to the steering committee and directs the work of the project, data collection, and modeling teams.

Data Collection

Before ALCES can be used to generate scenarios, spatial and nonspatial descriptive and trend projection data must be incorporated into the model. Specific data requirements are (1) current land uses, such as energy development (e.g., well sites, coal mines, and seismic lines) and agriculture (e.g., feedlots, pasture, and crops); (2) natural processes, such as fire (return interval and suppression), vegetation growth and yield, and insect outbreak; (3) existing land cover, such as forests (e.g., lodgepole pine and white spruce) and water (e.g., lakes and rivers); and (4) projections of specific land uses (e.g., future forest harvesting and energy production).

Working under the direction of the project team, the data collection team coordinates data collection, which accounts for the greatest proportion of time and effort devoted to the project. To facilitate acceptance of modeling results, the most appropriate data for each land-use sector are used. At least one representative is recruited from each sector to serve as a liaison between the RLAP and that sector (e.g., representatives from the petroleum industry in each major jurisdiction would be recruited).

The first step in the modeling process is to establish a baseline landscape description. The focal area for the project is the CCE (see CCE boundary map on p. xviii). The "initial landscape" description for the CCE includes the various land-cover types (e.g., vegetation, waterways, rock, and ice) overlain by various land-use "footprint" types (e.g., transportation infrastructure, residential development, and well pads). Because data produced by different jurisdictions have different spatial resolutions, degrees of completeness, and vegetation and base feature classification schemes, the spatial data must be standardized and converted to a form that the model can use.

Perhaps the largest and most challenging data collection task is acquiring nonspatial and trend data related to the full spectrum of land uses. In the CCE, human activity on the landscape is classified into the following land-use types: (1) forestry; (2) energy and mining; (3) crop and livestock production; (4) transportation; (5) human settlements; (6) protected areas; (7) general industry and electrical power equipment and services; and (8) tourism, recreation, hunting, and trapping. Although some of this information exists in published reports and internal databases, much of it does not exist or is in dispute. Published data and reports are consulted first, and then expert opinion (via consultation or consensus-based workshops) is used to fill in data gaps.

Natural processes that affect landscape dynamics (e.g., meteorological and hydrological influences, fire, and insects) are tracked and incorporated into the cumulative effects modeling. Analysts choose a selected suite of wildlife species (also called "guilds" or "communities") to act as indicators of ecological health in the region. In these cases, data are derived in much the same manner as they are for land uses—from a combination of readily available sources and expert workshops.

Base-Case Modeling

The role of modeling is to generate plausible representations of future land-use dynamics and impacts based on model inputs and assumptions. These representations allow the user to explore scenarios for mitigating undesirable impacts. The main outputs of the model are a series of user-defined graphs that show various parameters and relationships over the simulation time period, as well as spatially stratified descriptions of potential future landscapes.

Once the data have been collected, validated, and entered into the model, the modeling team tests the model and establishes a baseline against which potential scenarios are assessed. Individuals from the CMP make up the modeling team, which holds responsibility for coordinating the ALCES model runs based on input from the CMP. The team initially conducts multiple runs of the model, and identifies and rectifies apparent inconsistencies in the data or model relationships. The result is a base case, which yields output based only on the initial trends and metrics gathered for the model. The team then compares results of proposed mitigations and alternative scenarios to results for the base case.

Individual agencies or subregions of the CCE can run simulations on smaller geographic areas. This requires that spatial data be collected in such a way that it can be "clipped" to the desired subregional boundaries. In addition, trend and nonspatial data, normally determined for the entire region, must be collected at the subregional scale. Moving forward with a subregional approach would make testing the utility of the modeling exercise less onerous for the CMP. The CMP is currently evaluating which approach is best given the current level of commitment, time, and resources.

Scenario Modeling and Communication for Decision Support

Model outputs (graphs and tables) do not immediately demonstrate the regional, strategic-level issues of interest to managers that might need to be addressed through policy and management actions. For this reason, the modeling team converts the model outputs into issue statements written in plain language. These summary statements make no judgments about the changes in management action that may be required; they simply identify areas where the model has indicated that conflicts may occur in the future.

When scenario modeling begins, the modeling team communicates base-case issue statements to the CMP. The CMP then collectively identifies the highest priority issue statements that need to be investigated through further modeling.

A CMP subcommittee works with the modeling team to identify mitigation strategies that address the identified issues. The modeling team converts the subcommittee's feedback into workable individual mitigation scenarios or multifaceted management scenarios that are modeled in ALCES, then compares the modeling results for these scenarios against the base case. The CMP then receives the team's reports on the results of the scenario investigations, its assessment of their potential for success, and its description of any new issues that have arisen.

This iterative process continues as new scenarios are created, tested, and analyzed. Individual agencies extract and incorporate information from the scenario exercises into policy and management actions as they see fit. The power of this modeling approach lies in its ability to examine cumulative effects for multiple sectors and land uses at a regional scale.

Lessons Learned

The team, the subcommittee, and the CMP have learned several key lessons in the process of developing a framework for assessing strategic cumulative effects in the CCE. These lessons, discussed in the sections that follow, pertain to agency engagement, modeling, project coordination, and agency decisionmaking.

Agency Engagement

Because the regional CEA was conceived as a strategic tool for land-management agencies in the CCE, obtaining "buy-in" from those agencies is critical if they are ever to employ the modeling results in their decisionmaking. Additionally, data collection would be next to impossible without the participation of these agencies. As a result, it is necessary to accommodate multiple-agency circumstances, develop a shared understanding of goals, gain buy-in for the process in general, and acquire the right political support at the right time.

Securing agency commitments to a strategic regional CEA has been a significant challenge. The process did not begin with a concrete goal or specific objectives. Instead, the CMP started by discussing needs and possibilities for collaborative research and management, which resulted in the partnership eventually taking on the cumulative effects project.

Maintaining multiple-agency involvement in a complex, multiyear project has been difficult despite the high level of support expressed repeatedly by those agencies for a regional CEA. Agencies are operating under difficult conditions in terms of sparse budgets, shifting priorities, staff changes, and limited human resources. As a result, many participants are hard-pressed to provide funding, time, and personnel, which has had a detrimental effect on the ability to maintain engagement. For example, agency personnel changes and departmental reorganizations have presented constant challenges.

Factors that cause agencies to maintain support for the project include an annual forum that gives participating agencies formal access and input to the process, a steering committee of peers that helps generate support, a handful of champions who maintain the process when it falters, and the ability to leverage finances to fund the partnership. Two key challenges to success are a lack of an agency mandate to pursue a regional CEA, and a lack of legislative or policy mechanisms that support the initiative. Fortunately, many middle- to high-level managers have interpreted their policy environment to include support for transboundary assessment initiatives. CMP participants have recognized and communicated that a key precursor to higher levels of engagement is the unambiguous articulation that a regional CEA makes valuable contributions to each agency.

Participants need to have a shared understanding of the goals of the regional CEA and the process used to implement it. Because most agency experience with CEA is project-based, there is a range of beliefs about the goals of the project. Among the diverse agency perspectives on the value of a regional CEA project are that it would (1) create a compelling vision of the future, (2) serve as grist for a scenario-building exercise, (3) act as a strategic planning tool, (4) serve as another operational-level tool, (5) be a multiyear effort, and (6) prove its worth within a few months.

Middle- to high-level managers participating in the CMP continually expressed interest in the CEA's completion. Support at higher, more political levels, however, continues to be unenthusiastic or absent, largely because of the agency circumstances we described previously and political reservations about pursuing activities outside existing mandates and jurisdictional boundaries. Ultimately, this lack of support at higher levels has caused significant delays in advancing the project.

Modeling

It was clear at the outset that a robust modeling process should be able to handle the complex land-use and natural process data that need to be incorporated into a CEA. ALCES was familiar to a number of CMP participants and has a proven record of supporting CEAs within the region. Before ALCES became available, computer models for regional cumulative assessment had two significant and related limitations—existing models did not incorporate the complete spectrum of land uses and natural processes (Hudson 2002; Stelfox 2003), and, until recently, desktop computers could not execute highly complex CEA models.

Significant misunderstanding and concern surrounded the ALCES model itself. Because it is the central integrating element of the project, participating agencies must accept the model as a legitimate tool. Furthermore, participants and other stakeholders must be confident that the model adequately reflects relevant land uses and sectors. Although understanding of the model's capabilities and limitations varied significantly, a few perspectives were common—concern that the model would usurp agency decisionmaking ability; excitement that it would predict the future; and fear that the public would view model output as predictive, pressuring managers into ill-advised actions based on false interpretations of model outputs.

Project Coordination

Ross (1994) proposes that project proponents are not able and should not be expected to address issues at the temporal and spatial scales necessary in regional CEAs. Dubé (2003) suggested that the lack of a "responsible owner" has been a strong limiting factor in strategic CEA versus project-based CEA. In the case of the CMP regional CEA, this issue was resolved by establishing a steering committee within the CMP and retaining a third party (the Miistakis Institute) to manage the project. As a nonprofit research institute with CEA experience, the Miistakis Institute has the advantage of not representing a single agency or land use.

The logistical challenges faced by the Miistakis Institute were considerable, but not unpredictable. These challenges included coordinating logistics for data gathering and meetings across the 42,000-km^2 CCE and accommodating the schedules, time frames, and budgets of more than 20 government agencies over multiple years. Developing and promoting the framework described in this chapter is helping to address these challenges.

Securing stable funding for the project is a primary challenge to implementing CEA in the CCE. Agencies initially committed to the project with the understanding that funding would be highly leveraged: each agency would contribute a relatively small amount, funds furnished by different agencies would be pooled, and total funding would be leveraged through the Miistakis Institute's charitable sources. Agencies have repeatedly been unable to fulfill their commitments for several reasons, namely (1) misaligned agency-budgeting processes (e.g., the state of Montana budgets on a two-year cycle and jurisdictions have different fiscal years); (2) ever-present agency downsizing and budget trimming; (3) the lack of agency mandates for having regional cumulative effects drive budget priorities; and (4) weak support at higher levels of administration.

The long-term future of the program is predicated on a regular budgetary commitment from all participating agencies to support member involvement (e.g., out-of-jurisdiction travel and time allocation). In addition, a mechanism that allows agencies to secure and transfer funds to the program is essential. Such budgetary commitment requires higher-level agency support, as well as inclusion of the CMP in agency work plans.

Agency Decisionmaking

A detailed process for incorporating results of the CCE regional CEA into agency decisionmaking is not a part of the framework we have described. Because establishing this process is the ultimate goal of the program, it is worth describing some of the findings on how to design a framework that incorporates this process.

If agency representatives are to be comfortable using model outputs, they must understand those outputs. To date, no set process for converting model (ALCES) outputs to policy actions has been developed. The modeling framework does, however, include a step in which the modeling team converts graphical outputs into issue statements that are presented to participating agencies. Issue statements form the basis of the mitigation scenarios that participants want ALCES to examine.

Involving agencies in analyzing and communicating output and selecting simulations and scenarios was an intentional step. Crafting management strategies based on model outputs, however, was not designed into the process. Both agency involvement and the tenets underlying the model rely on agencies having the flexibility to develop management plans and policies by combining modeling results with the myriad other ecological, social, and political factors as they see fit. The downside of this approach, which we believe is outweighed by the risks of a more aggressive approach, is that agencies might decide not to base decisionmaking on model outputs.

In summary, cumulative effects arising from jurisdictional fragmentation and incremental decisionmaking constitute a significant threat to sustainable management of landscapes in the CCE. The CMP was born of the recognition that innovative and collaborative approaches are required to realize more integrated regional management of human and natural resources. A major goal of the CMP is a collaborative effort to assess the cumulative effects of human activities at a regional scale, which has become known as the RLAP. The project participants are using a simulation software package called ALCES as a primary analytical tool. Significant challenges have presented themselves in this complex, multiyear project. For example, despite the high level of agency support repeatedly expressed for a regional CEA, maintaining multiple-agency involvement has proven problematic. Agencies are operating under challenging conditions characterized by tight budgets, shifting priorities, changing staffs, and limited human resources.

Despite the challenges it faces, the CMP and the ensuing cumulative effects project are world-class examples of transboundary management. Sustaining the CCE depends on the future of this and similar initiatives.

Seven Suggestions for Success

To enhance the long-term viability of the CMP initiative, we recommend a set of seven actions:

- The CMP should continue to explore and identify mechanisms and appropriate timing for attaining higher-level recognition and support for the partnership and the cumulative effects project.
- Project goals and objectives should be periodically revisited with CMP members through a series of small group meetings.
- CMP members should work to secure adequate long-term financial and human resources for the project and take advantage of the benefits of leveraging.
- The CMP should develop mechanisms for better and more frequent internal communication.
- Activities and benefits of the CMP should be communicated to the public and relevant interest groups.
- The CMP should be committed to monitoring, feedback, and continuous improvement of the project.

- Participating agencies should identify opportunities to explicitly incorporate the data, tools, and model outputs from the cumulative effects project into existing agency programs, thereby demonstrating the benefits of participation.

References

AXYS Environmental Consulting Limited. 2000. *Regional Approaches to Managing Cumulative Effects in Canada's North.* Prepared for the Department of the Environment, Government of Canada. Calgary, AB: AXYS.

Cumulative Effects Assessment Working Group. 1999. *Cumulative Effects Assessment Practitioners Guide.* Hull, Quebec, Canada: AXYS Environmental Consulting and CEA Working Group for the Canadian Environmental Assessment Agency.

Creasey, J. R. 2002. Moving from Project-Based Cumulative Effects Assessment to Regional Environmental Management. In *Cumulative Environmental Management, Tools and Approaches.* Edited by A. Kennedy. Calgary, AB: Alberta Society of Professional Biologists, 3–16.

Davey, L.H., J.L. Barnes, C.L. Horvath, and A. Griffiths. 2002. Addressing Cumulative Environmental Effects: Sectoral and Regional Environmental Assessment. In *Cumulative Environmental Effects Management: Tools and Approaches.* Edited by A. Kennedy. Calgary, AB: Alberta Society of Professional Biologists, 187–205.

Dubé, M.G. 2003. Cumulative Effect Assessment in Canada: A Regional Framework for Aquatic Ecosystems. *Environmental Impact Assessment Review* 23: 723–45.

Griffiths, A. 1998. *Cumulative Effects Assessment: Current Practices and Future Options.* Calgary, AB: Macleod Institute.

Hudson, R. 2002. *An Evaluation of ALCES, A Landscape Cumulative Effects Simulator for Use in Integrated Resource Management in Alberta.* Unpublished discussion paper circulated for review by the A Landscape Cumulative Effects Simulator (ALCES) review team.

McKinney, M., C. Fitch, and W. Harmon. 2002. Regionalism in the West: An Inventory and Assessment. *Public Land and Resources Law Review* 23: 101–91.

Partidário, M.R., and R. Clark. 2000. *Perspectives on Strategic Environmental Assessment.* New York: Lewis Publishers.

Quinn, M.S., G. Greenaway, D. Duke, and T. Lee. 2005. *A Collaborative Approach to Assessing Regional Cumulative Effects in the Transboundary Crown of the Continent.* Research and Development Monograph Series. Ottawa, ON: Canadian Environmental Assessment Agency Research and Development Program.

Richmond, B., S. Peterson, and P. Vescusco. 1987. *An Academic Users Guide to STELLA.* Lyme, NJ: High Performance Systems, Inc.

Ross, W.A. 1994. Assessing Cumulative Environmental Effects: Both Impossible and Essential. In *Cumulative Effects Assessment in Canada: From Concept to Practice.* Edited by A.J. Kennedy. Calgary, AB: Alberta Society of Professional Biologists, 3–9.

Schneider, R. R., J.B. Stelfox, S. Boutin, and S. Wasel. 2003. Managing the Cumulative Impacts of Land Uses in the Western Canadian Sedimentary Basin: A Modeling

Approach. *Conservation Ecology* 7: 8. Available at http://www.consecol.org/vol7/iss1/ (accessed November 25, 2005).

Stelfox, J.B. 2003. Personal communication with the authors, February 2.

Stinchcombe, K., and R. Gibson. 2001. Strategic Environmental Assessment as a Means of Pursuing Sustainability: Ten Advantages and Ten Challenges. *Journal of Environmental Assessment Policy and Management* 3: 343–72.

15
Transboundary Conservation and the Yellowstone to Yukon Conservation Initiative

Marguerite H. Mahr

For millennia, when glacial ice covered most of Canada and the northern tier of the United States, many plants and animals found refuge in ice-free areas between the Cordilleran and Laurentide ice sheets. As the climate warmed, a long, north- to south-trending corridor of life emerged between the margins of the receding ice sheets just east of the Rocky Mountains (Pielou 1991). This ice-free opening allowed organisms to move between southern temperate and northern boreal landscapes. Ten thousand years ago as the ice sheets were melting, the corridor started to widen, plants migrated northward and colonized ice-free areas, and animals adapted to the changing habitats along the Rockies' eastern slopes.

In the twenty-first century, wildlife still move along ancient pathways, and ecological connectivity between temperate and boreal landscapes remains central to sustaining diversity of life in the Rockies. Descendants of early colonizers are now being rapidly displaced by habitat alteration and loss from urban sprawl, competition with domestic livestock, and hydropower and energy development. In 1993, the Yellowstone to Yukon (Y2Y) Conservation Initiative (www.y2y.net) was formed in response to losses in biological diversity caused by these habitat changes. The initiative is a network of American and Canadian conservationists, scientists, land trusts, communities, businesses, and other collaborators working to maintain and restore biological diversity and habitat connectivity within a 1.2-million km^2 area that stretches from Wyoming's Wind River Range south of Yellowstone National Park to the Peel River Basin in north–central Yukon (Figure 15-1). The Y2Y region straddles the Rockies with the Great Plains' grasslands and dry forests on the east side, and a patchwork of moister forests blanketing the mountains, plateaus, and valleys on the west side. The area includes the ancient corridor of life that critically sustained biological productivity during recurring ice ages. In this chapter, I describe how the

Y2Y Conservation Initiative is working to maintain and restore biological diversity and habitat connectivity in the Y2Y region.

Background

One of the most biologically intact areas in North America, the Y2Y corridor contains some of the largest mountains and river systems on the continent, including the Columbia, Missouri, Fraser, and Yukon rivers. Large predators still inhabit the region (Laliberte and Ripple 2004), and the Central and Pacific Flyways converge within Y2Y, serving as continental thoroughfares for migrating birds.

Field studies document the often unimaginable distances across which wildlife roam in the Y2Y region and the vital role of landscape connectivity for preserving their existence. The cycles of birds and fish are some of the great connecting agents in Rocky Mountain ecology. For example, migration of Rocky Mountain trumpeter swans (*Cygnus buccinator*) annually traces the 3,200-km-long Y2Y corridor from the Yellowstone region to breed on remote lakes in the Yukon Territory. Similarly, every spring and fall tens of thousands of golden eagles (see Chapter 2 for scientific names not given in this chapter) fly the "eagle highway" high above the spine of the Rockies. The migratory corridor for golden eagles covers a phenomenal distance that extends from as far north as interior Alaska for summer breeding, then south to Utah, Nevada, and northern Mexico for the winter (Sherrington 2003).

Another spectacular migration occurs when millions of sockeye salmon (*Oncorhynchus nerka*) endure a hazardous journey that takes them as far as 10,000 km from their natal freshwater streams in British Columbia's Fraser River in the mountains of the Y2Y region to the middle of the north Pacific Ocean. After two to three years in the ocean, the Fraser sockeye return to spawn in the same remote tributaries of the river where they hatched. Recorded movements for these and other species give clear and convincing evidence for the importance of conserving large, secure, and connected habitat to enhance adaptability and long-term survival.

For more than a decade, the Y2Y Conservation Initiative has been exploring how best to conserve wildlife and wild places over very large landscapes. One of the initiative's principal messages is that the Y2Y's contiguous landscapes should be treated as one large ecoregion in which nature continues to reign as unimpaired as possible, and where human communities coexist with nature. The scientists working on the initiative—who number more than 100—have been inspired by its vision of landscape connectivity and preservation of big landscapes. This approach represents a very different way of perceiving and thinking about nature in Rocky Mountain landscapes. In recognition of this emerging paradigm shift, in 1997, the International Union for the Conservation of Nature (IUCN) declared the Y2Y Conservation Initiative at the forefront of the world's

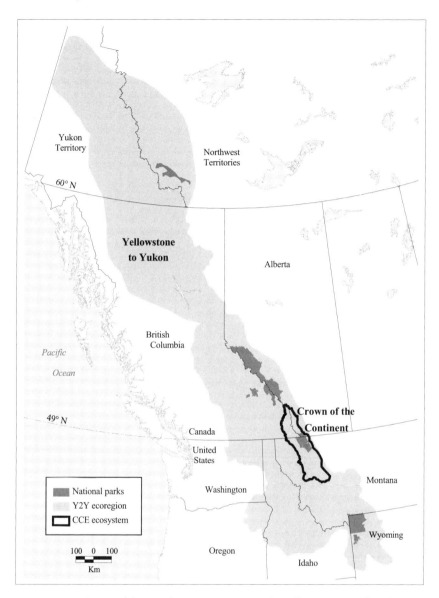

FIGURE 15-1. Crown of the Continent Ecosystem on the Yellowstone to Yukon (Y2Y) Corridor

large, landscape-scale conservation strategies. In 2004, the IUCN recognized it as one of the leading conservation initiatives on the planet (Bennett 2004).

The Y2Y Conservation Initiative has underpinned its conservation efforts with science to improve our accuracy and insights about the expression of the natural world. Although conservation is intertwined with and ultimately dependent on our values, ethics, and sense of place, factual and intimate biological knowledge is especially critical if humans are to coexist with nature and human and natural systems are to be sustained. Employing this rationale, the Y2Y Conservation Initiative has created opportunities and incentives for scientists to work within a common framework that has encouraged sharing data sets and analytical tools to study the region's ecological patterns and processes. By weaving together examples from studies on large carnivores, birds, and fish in the sections that follow, I illustrate what has been learned from research conducted on a continental scale in the context of the Y2Y region. The following wildlife portraits lay a foundation for shaping the challenging and imperative need for coexistence.

A Remarkable Journey

Beginning in 1991, a radio-collared gray wolf (*Canis lupus*) named Pluie embarked on an extraordinary four-year journey through the CCE. Her journey, recorded via satellite, crossed more than 30 different Canadian and U.S. government jurisdictions encompassing more than 100,000 km^2. Sadly, in December of 1995, Pluie, her mate Orion, and one of their pups were shot legally by a hunter south of Kootenay National Park in British Columbia. Pluie's remarkable journey mirrors the story told by daily, seasonal, and lifetime movements of other wildlife that crisscross the narrow swath cut in the forest that demarks the Canada–U.S. border at the heart of the CCE (Weaver 2001). Pluie's travels and those of many other species illustrate that the CCE's international border has no real biological or ecological significance for wildlife. The entire CCE is their home.

In the larger context of the Y2Y region, the CCE is significant because it is slightly south of the midpoint of the region and represents an important transitional area in terms of human impact, which decreases with increasing latitude. The portion of the Y2Y region from Banff National Park south is undergoing an exponential increase in the number of humans whose activities are fragmenting and destroying wildlife habitat at unprecedented levels. Human intolerance has prevented wolves from recolonizing all of their former range. Pluie's journey may be interpreted in two ways—as an indication that wildlife may still have enough habitat to roam wild, or as a potential harbinger of the unraveling of the ecological tapestry that can occur when critical, high-quality habitat is not conserved and mortality rates are not managed.

Enough Room to Roam

Parks and wilderness areas in the Rockies are much celebrated for their scenic beauty and cultural heritage, but they are not fully protecting nature's diversity.

In a sobering realization, biologists agree that protected areas are not large enough and connected enough to (1) support healthy populations of large carnivores, (2) maintain native species diversity, (3) recover threatened species, and (4) sustain the natural phenomena that maintain healthy ecosystems (Noss and Cooperrider 1994). Although more than 30% of the CCE is already in protected areas (i.e., parks, wilderness, and wildlife reserves), individual areas may be too small in themselves to adequately conserve species near the top of the food web. These species travel long distances in search of mates and food. Moreover, protected areas in many places in the CCE consist of disconnected habitat islands surrounded by developed areas that are inhospitable to wolves and other large carnivores. Lack of habitat connectivity limits the exchange of individuals and genes, putting small, isolated populations at greater risk of extinction. Establishing habitat "networks" is an ecological solution. These networks connect portions of the landscape by way of local, linear pathways that allow wildlife to move from one place to another. Landscape connectivity is also achieved through broadly connected swaths of habitat that extend across landscapes and permit interaction between small groups of wide-ranging animals. This solution is backed by science indicating that both contiguous habitat in the form of secure, concentrated habitat areas (commonly referred to as core areas) and linkages between core areas for populations of wide-ranging large carnivores are needed to achieve interconnectedness of individual, family, and subpopulation units (Carroll et al. 2003, 2004). Wolves like Pluie that roam vast distances in search of food and companionship illustrate how regional networks of connected reserves benefit wide-ranging species that exist in a metapopulation—defined as a large population composed of connected smaller subpopulations (Hanski 1997)—and need to interact with the larger ecosystem to satisfy their needs.

Wide-ranging wildlife would benefit from managing the CCE as one ecosystem across multiple jurisdictions in the true sense of an international biosphere reserve. The fate of many threatened transboundary species begs a cooperative transboundary effort to keep shared biological diversity abundant. Table 15-1 lists species at risk of extinction in the transboundary CCE along with their protected status. Not surprisingly, many of the region's wide-ranging species that have large area requirements and depend on connectivity are listed as "species at risk" in at least a portion of their geographic range. This status designation is conferred by the Committee on the Status of Endangered Wildlife in Canada (COSEWIC) or the U.S. Endangered Species Act of 1973 (ESA), or both.

In the rest of this chapter, I consider research that investigates how grizzly bears, birds, and fish use the Y2Y region's landscapes. I also identify critically important habitat for these species. Each section offers perspectives on threats and opportunities to conserving habitat, species diversity, and ecological integrity at both the North American and CCE landscape scales.

TABLE 15-1. Transboundary CCE Species at Risk in Canada and the United States

Species name	Protected status in Canada	Protected status in United States
Canada lynx (*Felis lynx canadensis*)	—*	Threatened
Gray wolf (*Canis lupus*)	—	Endangered: Experimental/nonessential
Grizzly bear (*Ursus arctos horribilis*)	Species of special concern	Threatened
Wolverine (*Gulo gulo*)	Species of special concern	—
Woodland caribou (*Rangifer tarandus caribou*)	Threatened and species of special concern	Endangered: Experimental/nonessential
Bald eagle (*Haliaeetus leucocephalus*)	—	Endangered and threatened, but proposed for delisting
Peregrine falcon (*Falco peregrinus anatum*)	Threatened	
Northern goshawk (*Accipiter gentilis*)	Threatened	—
Flammulated owl (*Otus flammeolus*)	Species of special concern	—
Lewis's woodpecker (*Melanerpes lewis*)	Species of special concern	—
Trumpeter swan (*Cygnus buccinator*)	—	Threatened
Bull trout (*Salvelinus confluentus*)	—	Threatened
Chinook salmon (*Oncorhynchus tshawytscha*)	—	Threatened
Sockeye salmon (*Oncorhynchus nerka*)	—	Endangered
White sturgeon (*Acipenser transmontanus*)	Endangered	Endangered

*The long dash (—) indicates that no designation has been assigned.
Notes: These species are listed under Canada's Species at Risk Act of 2003 by the COSEWIC (http://www.cosewic.gc.ca), the U.S. Endangered Species Act of 1973 (http://endangered.fws.gov), or both. A species may have a double listing depending on how populations are classified in different portions of a species range. For example, the southern mountain population of woodland caribou in British Columbia and Alberta is listed as "threatened" while the northern mountain population in northern British Columbia and the Yukon is listed as "species of special concern." A native gray wolf in the United States is listed as "endangered" and a gray wolf introduced as part of a recovery planning effort in central Idaho or Yellowstone National Park is listed as "experimental/nonessential."

Grizzly Bear Essentials

Grizzly bears, a type of brown bear, require large, secure, and diverse habitat to meet their life needs. They must be able to "live their way" across a landscape in broad, residential linkages between core habitat areas. Grizzly bears are considered an important "umbrella species," which means that protecting grizzly bears also protects large areas of habitat as well as many other species and ecosystem functions. Almost everywhere south of 60° N latitude, populations of grizzly bears are smaller than historical levels and face potential extirpation because their habitat is being degraded by mining, logging, housing developments, recreational resorts, off-road motorized recreation, and other types of human encroachment. In the lower 48 states, 1,100 to 1,200 grizzlies occupy less than 2% of their former range (Mattson and Merrill 2002). U.S. border populations inhabit the last remaining stepping stone habitats in what was once a continuous distribution of grizzly bears from northern Mexico to coastal Alaska. Grizzly bears have been extirpated from U.S. and Canadian prairies, listed as a "threatened species" in the United States, and recommended for "threatened" status in Alberta. They persist to varying degrees in British Columbia, and the Yukon and Northwest territories.

Because the habitat requirements of grizzly bears generally conflict with most human activities, it is common for research to focus on identifying the location, size, and productivity of critical grizzly habitat. For example, information about grizzly bear foods and landscape characteristics occurring in different areas of the Y2Y can be used to determine habitat productivity and estimate the density (i.e., number of bears per area) a region could support over time. Merrill and Mattson (2003) estimated annual bear mortality caused by humans. Estimating annual bear mortality allowed Merrill (2005) to determine the area required to meet two levels of population security indicated by demographic or evolutionary vigor. For example, Merrill evaluated areas based on whether they could support a healthy population of 500–700 individuals for several hundred years and called these "demographically robust populations." Merrill reasons that this number would likely provide enough genetic and behavioral diversity to withstand environmental variation over several hundred years. If these populations could be linked to allow genetic and behavioral interchange among 2,000 or more individuals for thousands of years, Merrill called them "evolutionary robust populations."

Of the four existing and one potential grizzly bear populations south of 60° N latitude (i.e., Northern British Columbia, Central Canadian Rockies, Central Idaho [potential], Northern Continental Divide, and Greater Yellowstone; see Figure 15-2 for additional details), only the first three met Merrill's criterion of having enough secure habitat to support estimated populations of 500–700 grizzlies. According to Merrill, although the Northern Continental Divide and Greater Yellowstone populations are likely to contain more than 500 individu-

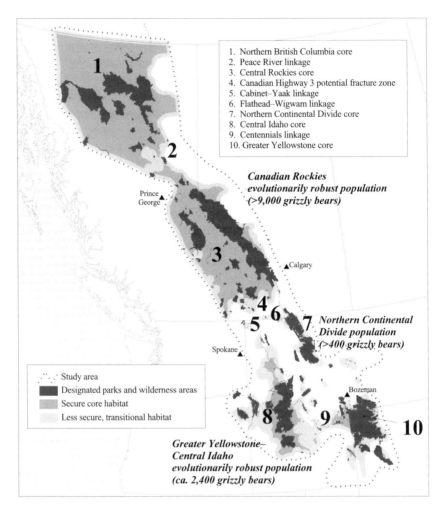

FIGURE 15-2. Grizzly Bear Populations from Northern Wyoming to Northern British Columbia

Source: Merrill 2005.

als, a significant portion of those bears reside in less-secure habitat and do not contribute to overall population security because they frequently come into conflict with humans and are killed or removed.

A closer look at grizzly bear populations in the U.S. portion of the CCE reveals that the Northern Continental Divide population may be at risk because of the relatively long and peninsular shape of its habitat core area. This linear shape gives the Northern Continental Divide Ecosystem (NCDE) a high edge-to-area ratio. Such a ratio decreases the security of the interior wilderness habitat and increases the likelihood that a large portion of the grizzly population living along the perimeter will come into regular contact (and conflict) with humans. Local accounts confirm an increasing risk of lethal contact with humans as grizzlies disperse from the NCDE's wilderness core habitat into lands occupied by human activity or settlements. We have no clear explanation for what is motivating grizzly bears to move out of the secure core into inhospitable areas. Local biologists speculate that movements may result from several consecutive years of drought, huckleberry crop failures, an increasing number of grizzly bears, or other factors. Unfortunately, more bears on the move have translated into more bears getting into trouble and being killed. More Northern Continental Divide bears have died on an annual basis from 1997 to 2004 than is believed acceptable to sustain or grow a healthy grizzly bear population. For example, in 2004, 31 bears died in the NCDE, more than double the total allowable number. Even more significant, 18 of the 31 mortalities were female bears.

Implications for Grizzly Bear Conservation

In their study of large carnivores in theY2Y region, Carroll and colleagues (2004) predict that the highest risk of extinction for grizzly bears exists along the border between Canada and the United States. The modeling these scientists conducted indicated that, based on the fast pace of development and degradation of habitat that straddles the border, long-term persistence of large carnivores is precarious without major conservation efforts. Peek and coworkers (2003) note that the phenomenon of habitat fragmentation followed by extirpations of small grizzly bear populations in the United States—described by Mattson and Merrill (2002)—is also occurring in southern British Columbia.

Scientists recommend four types of action to enhance the conservation of transboundary grizzly bear populations. First, conservation measures should be applied at the broad landscape and population scales. As an example, viable north–south connections should be maintained between the NCDE and Canadian Rockies through the CCE. Grizzly bear connectivity in southwestern Canada requires bears to cross Highway 3. This two-lane highway, which slices east–west through the transboundary mountain ranges of the Rockies, Cabinets, Purcells, and Selkirks poses perhaps the most insidious threat to grizzly bear habitat and population connectivity in the transboundary region. The Highway 3 corridor is creating what carnivore biologists refer to as a fracture zone. This zone has the potential to act as a barrier to movement and ultimately to completely separate members of a population (Proctor et al. 2002; Apps et al. 2006).

Second, to conserve existing, secure habitat in known linkage areas, conservation efforts should focus on enhancing the size and security of existing linkages that straddle the international border. For example, Waterton Lakes National Park (WLNP) should be expanded to encompass and protect additional riparian areas key for carnivores in the transboundary Flathead River Basin (see Weaver 2001) that adjoin the habitat core of Glacier National Park (GNP) in the United States. Other key actions should ensure protection of wildlife movement corridors in southwestern British Columbia's Elk and Flathead river watersheds and the Alberta Foothills before any further development of oil, gas, and coal-bed methane takes place.

Third, the density and public access of government roads on Crown lands in Alberta and British Columbia and on U.S. Department of Agriculture (USDA) Forest Service lands in Montana should be reduced to enhance wildlife security and connectivity in critical habitat bounded by Canada's Highway 3 in the north and U.S. Highway 2 in the south. For example, bear mortality has been exacerbated by the high number of open roads along the South, Middle, and North forks of the Flathead River. These roads permit motorized access into remote areas and displace bears from essential seasonal habitats. Flathead National Forest managers should comply with standards mandated by Amendment 19 of the Flathead National Forest Plan (USDA Forest Service 1995) to reduce the number of and public accessibility to agency roads in bear-management units.

Fourth, conservation measures that directly reduce grizzly bear mortality should be enacted in the NCDE's urban–wildland interface and security along the perimeter of the peninsular core habitat should be increased. Human-caused grizzly bear mortality represents approximately 90% of the total grizzly bear mortality in the NCDE. Most of these mortalities are attributed to government management control of "nuisance" bears that go looking for food in garbage dumpsters, bird feeders, pet food containers, bee boxes, and seasonal cabins. Illegal poaching and mistaken identity by black bear hunters add to the annual death toll of grizzly bears, as do frequent collisions with trains, trucks, and automobiles. Examples of local conservation initiatives include efforts to decrease human-caused mortality within the Montana portion of the U.S. Highway 2 and Burlington Northern Santa Fe railroad corridor adjacent to the GNP; reduce the number of deadly conflicts on private agricultural lands that are associated with cattle boneyards, calving and lambing areas, stored grain, beehives, and fruit orchards; and improve rural and urban sanitation practices to reduce waste that attracts bears.

The Bird's Eye View

Migratory birds flying the Central Flyway through the Y2Y corridor elegantly illustrate the concept of ecological connectivity. Some bird species temporarily congregate on wetlands for rest and food; other species set up residence in grasslands and forests. To prioritize habitats for the nearly 300 bird species found in the Y2Y region, we must understand bird species distribution and abundance. Using bird field surveys collected throughout western North America by the North American Breeding Bird Survey program, Hansen and Jones (2003) measured bird species richness, which is simply a count of the total number of species at a given location. They found a continental avian hotspot in western North America—a region of exceptionally high bird species diversity—that extends along the international border from the Okanagan Valley of Washington and British Columbia to the Flathead Valley of Montana and British Columbia.

Hansen and Jones (2003) investigated bird conservation priority areas in western Montana in more detail. Using a large data set containing field observations of resident and migrant birds, they calculated the total number of observations of native landbirds. Next, they separately summarized the occurrences of sensitive landbird species that were prioritized for conservation action and monitoring by Montana Partners in Flight in each transect. Partners in Flight (PIF) (2006) was established in 1990 to conserve bird populations in North and South America. Analyses indicated that areas relatively lower in elevation and higher in temperature, precipitation, and productive vegetation are the major hotspots for both overall landbird and sensitive species richness in the Montana portion of the Y2Y corridor (Figure 15-3).

Figure 15-3 depicts the average number of PIF Montana priority species recorded on each field transect. Shades of gray represent values based on the predicted number of priority bird species in an area (i.e., darker areas on the map represent places of high bird species richness, whereas lighter areas represent places of low bird species richness). Hansen and Jones (2003) did not model regions that are predominately nonforested (e.g., urban areas and grasslands) and these areas have been clipped from Figure 15-3.

Within the Y2Y portion of Montana, maps derived from both the continental North American Breeding Birds Survey (2006) and local field observations show bird richness to be very high in northwestern Montana. The North American Breeding Birds Survey is a program that monitors the continent's breeding bird populations. This pattern likely applies equally across the border in southern British Columbia's Creston wetlands, a well-known bird refuge and flyway, and the Elk River, although Hansen and Jones (2003) did not model these areas.

The Flathead region is important for bird biodiversity because it has higher levels of precipitation, more growing degree days, greater soil nutrients, and more plant species and habitat types than other areas of Montana (Hansen and Jones 2003). Hansen and Jones found that productive river valley bottoms with a significant forest component, especially forested riparian habitats, tend to feature high bird species richness. Forested riparian habitats are local hotspots for Montana PIF (2006) priority species, such as the brown creeper (*Certhia americana*), common loon (*Gavia immer*), harlequin duck (*Histrionicus histrionicus*), and northern goshawk.

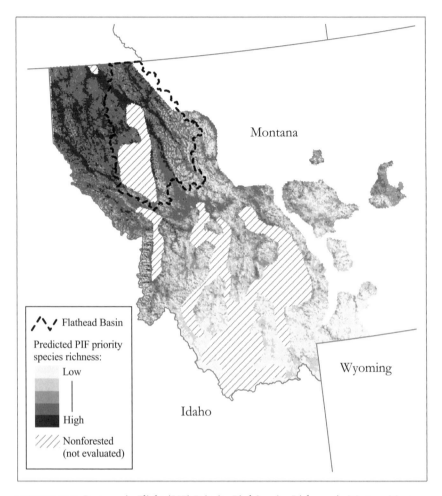

FIGURE 15-3. Partners in Flight (PIF) Priority Bird Species Richness in Western Montana

Source: Based on data from the U.S. Forest Service Northern Region Landbird Monitoring Project.

Implications for Bird Conservation

Conserving bird biodiversity in the Montana portion of the CCE will benefit birds regionally and throughout North America. Four conclusions can be drawn from Hansen and Jones's (2003) results that inform bird conservation efforts (see Mahr and Jones 2005). First, bird diversity and abundance are strongly associated with productive, lush habitat that is structurally diverse with different canopy levels and various ages and species of trees and shrubs. These habitat characteristics are very common in the Flathead region and should be maintained and, wherever possible, enhanced to benefit bird species that depend on a variety of plant communities.

Second, bird diversity is evenly distributed between privately and publicly owned forest land in the Flathead. The hotspot river valley bottoms of the North, Middle, and South forks of the Flathead River, along with the Stillwater and Swan rivers and Swift Creek, occur primarily in unprotected areas outside of the GNP and surrounding designated wilderness areas. Given that Montana's Flathead Valley is undergoing some of the highest rates of human population growth in the state, we must identify remaining undeveloped, highly productive areas and assess their importance for resident and migratory birds seeking summer breeding residence or stopover habitat. To date, conservation easements have been used to protect several high-priority wetlands on private land along the mainstem of the Flathead River. In addition, Montana Audubon has leased more than 178 ha of undisturbed riparian habitat along the confluence of the Stillwater and Flathead rivers on the outskirts of Kalispell from the state of Montana. We should expand such efforts to include other quality wetland and riparian areas with the goal of achieving avian habitat networks along the Flathead's valley bottoms.

Third, large stands of riparian forests of black cottonwood (*Populus trichocarpa*) are particularly important landscape elements to conserve in both the U.S. and Canadian portions of the Flathead Basin. These stands contribute nesting and perching habitat for cavity-nesting species, such as woodpeckers and owls, great blue herons (*Ardea herodias*), bald eagles (*Haliaeetus leucocephalus*), and red-tailed hawks (*Buteo jamaicensis*). Conservation and restoration strategies should be based on new empirical studies that, among other themes, consider black cottonwood's enormous benefit to birds throughout the entire basin. We can evaluate this benefit by comparing current and historical distribution of cottonwood galleries to establish how much of this habitat type has been changed or lost over time and by describing how migrant and resident birds depend on black cottonwood habitat.

Finally, and perhaps most importantly, we should design avian conservation and restoration activities from the start to benefit as many bird species as possible. Approaches that emphasize all-bird concepts of species diversity and abundance can complement more traditional management actions aimed at waterfowl, raptors, or a specific threatened species. In terms of the Flathead region, increased bird conservation in riparian hotspot areas would benefit suites of landbird, shorebird, and waterbird species. In addition, existing species-specific strategies that local conservationists are already pursuing for the common loon and harlequin duck would be supported by the larger effort.

A Big Fish Story

Mighty rivers such as the Yukon, Fraser, Athabasca, Columbia, Missouri, and Snake are fed from thousands of headwater streams that drain from the Continental Divide north, south, east, and west ultimately into the Pacific and Atlantic Oceans, and Hudson Bay. Alongside images of cold, pristine Rocky Mountain creeks are those of massive dams, sediment-choked and channelized streams, contaminated water, exotic species invasions, and overfishing. In the U.S. portion of the Y2Y corridor, the Yellowstone River is the only major river that is not dammed. The Canadian portion hosts several rivers without dams, with the Fraser being the most southerly. Given that the Y2Y region still contains the best cold-water fisheries in North America, scientists have called for new approaches designed to support conservation planning for freshwater species diversity and ecosystem health. Research by aquatic scientists associated with the Y2Y Conservation Initiative gives us a basis for understanding conservation potential within river basins and across multiple basins.

Of the 152 species and sub-species of fish that now occur or once occurred in the 23 major drainage basins of the Y2Y corridor, 99 are native species that still occupy some part of the corridor, and 48 species are exotic to the region (Mayhood 2004). Five species are now extinct in the region, including two endemics—the Banff longnose dace (*Rhinichthys cataractae smithi*) and the Snake River sucker (*Chasmistes muriei*). Rainbow trout (*Oncorhynchus mykiss*) is the most widespread species, occurring in all 23 major drainages (although it is native only to 13 of these drainages). Brook trout is by far the most widespread exotic, having been introduced into 17 drainages. The most widespread wholly native species is mountain whitefish (*Prosopium williamsoni*), which is native to all 20 major drainages in which it occurs. Another widespread native is the bull trout, which occurs naturally in 18 drainages and has been extirpated in 1 drainage. Although most native species are widely distributed in the Y2Y region's river systems, a remarkable 26 (25%) native fishes are restricted to living within a single drainage.

To understand the current distribution of fish diversity and habitat conditions of Y2Y's river systems, we must consider the long geographic isolation that occurred during glacial and postglacial history. As mentioned at the beginning of the chapter, where fish species live relates to where their ancestors lived during the last ice age, the immigration routes their ancestors used after the glaciers retreated, and their ability to disperse. Over millennia, many of these populations have evolved to create genetic diversity within each species. Mayhood's (2004) review of fish databases for all of the Y2Y region's river basins allowed him to distinguish four natural faunal groups of fishes (Figure 15-4a) that reflect their geographical and glacial isolation, along with their postglacial history. Mayhood called these groups (1) the Arctic Interior glaciated, (2) the Pacific glaciated, (3) the Pacific Columbian refugium, and (4) the Interior unglaciated. Historical and current fish faunal groups are illustrated in Figures 15-4a and 15-4b, respectively. In his work, Mayhood (2004) evaluated changes in species composition in each basin using lists of fish

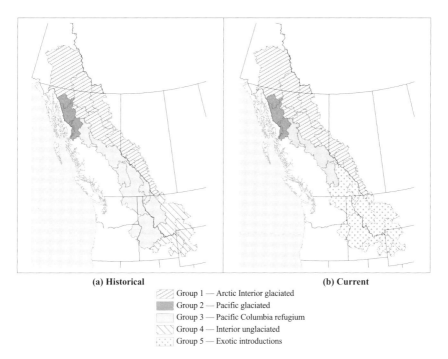

FIGURE 15-4. Historical and Current Fish Faunal Groups in the Y2Y Corridor

species occurrences in 297 sub-basins of the 23 major drainage basins that comprise the Y2Y region.

Mayhood (2004) found that native fish species richness was highest in the northern third of the Y2Y region, which includes all of the Pacific glaciated drainages and the northernmost Arctic Interior glaciated drainages. The southern basins of the Pacific Columbia refugium faunal group (i.e., Upper and Middle Columbia and Upper Kootenay) and all of the Interior unglaciated basins are currently occupied by a fifth distinct and artificial faunal group, referred to as "exotic introductions." This group is characterized by widely introduced exotic and transplanted native sportfish, notably salmonids (e.g., species of trout and whitefish), cyprinids (e.g., species of chub and minnow), and centrarchids (e.g., species of bass and crappie).

Analysis of historical and current fish composition illustrates increased homogenization of once-distinct fish faunas within faunal group 5 (exotic introductions). The greatest need and immediate challenges lie in the drainages of group 5 in the southern one-third of the Y2Y region. According to Mayhood (2004), these severely modified faunas have suffered habitat damage, especially from dams that commonly favor introduced generalist species at the expense of locally adapted native species.

Mayhood looked at fish species distribution from a zoographical perspective, whereas Hitt and Frissell (1999) assessed the aquatic health of thousands of watersheds. In particular, Hitt and Frissell assigned each watershed a conservation score based on characteristics such as the density of roads, the number of native fish species, the number of sensitive nonfish species, and the history of fish stocking for every watershed in major river basins in the U.S. portion of the Y2Y region. Corresponding data were not available for the Canadian portion of the region.

Hitt and Frissell's results revealed a patchwork of pristine, semipristine, degraded, and highly degraded watersheds throughout the Upper Columbia and Missouri river basins of the U.S. portion of the CCE. On the west side of the Continental Divide, Hitt and Frissell (1999) determined that fish faunas in watersheds with designated wilderness areas were the least modified. More than half of wilderness-containing watersheds scored within the highest category of aquatic integrity. Clear patterns of high-scoring watersheds follow the boundaries of the large, contiguous Great Bear/Bob Marshall/Scapegoat wilderness complex. A large, contiguous block of watersheds with high aquatic conservation value extends several hundred kilometers from the international border south to tributaries to the renowned Blackfoot River near Lincoln, Montana.

On the east side of the Continental Divide in the Montana portion of the Upper Missouri River Basin, Oechsli and Frissell (2002) determined that the headwater streams in the Sun and Dearborn drainages contain most of the best remaining watersheds with native aquatic diversity. Similar to the west side of the Continental Divide, the large block of wilderness that straddles the Rocky Mountains offers important protection to the eastside aquatic habitat and fisheries.

Implications for Fish Conservation

Because natural aquatic ecosystems and their fish communities usually develop over a long time period, efforts to conserve the natural faunas of the Y2Y corridor should adopt the credo "protect the best, restore the rest." This means that we must design management strategies to work simultaneously on protection and restoration. We should assign top priority to preventing further degradation of the most ecologically valuable drainages and rivers, and protecting the most genetically and demographically important native fish populations. Hitt and Frissell's (1999) broad systematic assessment of aquatic integrity lends itself to systematic landscape planning to preserve pristine watersheds, avoid continued deterioration of healthy aquatic systems, and prevent further listings of imperiled fish and other aquatic organisms under the U.S. ESA. West of the Continental Divide, the South, Middle, and North forks of the Flathead River are the best drainages in which to protect native fisheries. Transboundary rivers such as the Flathead and the Wigwam, which hydrologically stitch together the border, should be protectively managed as important examples of the best of the best.

"Restore the rest" applies to the majority of unprotected watersheds in the CCE, which are moderately to highly degraded. Management efforts in these watersheds could attempt to recover fisheries by analyzing the remnant native fish faunas with sensitive species, such as bull trout and westslope cutthroat trout and their relationships to introduced fishes and modified habitats. Successful restoration in the CCE depends on major ecological improvements at the basin scale. In addition, local remediation efforts in small catchments must continue. We should focus conservation activities on restoring modified drainages in the Upper Missouri and Columbia river basins, which have been substantially altered by introductions of non-native species and stocks. Restoring strongholds of native fish populations may also require restoration of riparian and upland vegetation. Ultimately, we must focus watershed restoration plans on maintaining the ecological integrity of landscapes where people live and work. Scientific evidence indicates that protected areas—although important—are insufficient by themselves to conserve aquatic systems.

Many of the existing problems in CCE drainages are caused by the destruction of habitat resulting from dams, water contamination, fish-management practices, and overfishing. These damaging activities often occur outside of the Y2Y region. Aquatic conservation planners should consider how flowing water contributes to connectivity in terms of downstream–upstream, upstream–downstream, and groundwater–surface water connections. Barriers to connectivity, such as dams, not only block upstream and downstream movements of numerous migratory species, but their associated reservoirs provide habitats that favor introduced species. Maintaining the natural connectedness of transboundary river basins of the U.S. and Canadian Rocky Mountains is among the most critical issues for aquatic conservation.

Sustaining Biodiversity in the CCE

Scientific findings presented in this and other chapters indicate that the CCE contains significant pristine habitats for North American wildlife. Conserving and connecting these habitats is essential for achieving viable populations of large carnivores, birds, and fish. The Y2Y Conservation Initiative's continental perspective offers an excellent landscape scale from which to investigate the functioning of biological and ecological integrity in the U.S. and Canadian Rocky Mountains. The Y2Y initiative's vision of habitat networks calls for new ways to manage land and wildlife. In addition, we must rethink human systems such as transportation, resource extraction, and recreation, because ultimately many of the solutions must be considered in a larger human social context of systems, values, and behaviors.

Ongoing speculation in the scientific community centers on whether species that inhabit the CCE and have endured on a geologic time scale (such as grizzly bear, northern goshawk, and bull trout) will survive the impacts of modern civilization. Throughout the ages, humans have faced questions about how to live. Today, we are questioning how we should live so as to reduce the negative impacts of population growth and all our activities on all life on Earth. Taking a holistic view of our place in the world, which acknowledges that we live in and are surrounded by wildlife habitat, will allow us to design and implement the most effective land-use planning and management strategies. This is especially true in a place like the CCE where rural and wild landscapes predominate, and where people drink water and eat fish from local rivers and lakes, harvest wild huckleberries, eat wild game, and raise free-range livestock. As long as sandhill cranes return in the spring to dance in wetlands and elk continue to bugle in the mountains, most people feel that life is good. If people in the transboundary area continue to point to the distant mountains and say "that's where the habitat is," however, our daily decisions and living patterns will become further disengaged from the natural world and have greater impacts on that world.

If, as humans, we do not have continual rapport with nature, our eyes will no longer observe subtle changes and our actions will become more self-serving. When the connections between human ecology and natural systems are diminished, which occurs in an urbanizing environment, we too easily fall out of rhythm and stop paying attention to the ecological conditions and prerequisites for sustaining life. From scientists, we continually learn how human actions affect ecosystems and the biosphere. Yet we all must become more ecologically astute and imaginative, allowing us to make more informed decisions about managing natural resources and wildlife; using water and energy resources wisely; transporting ourselves with minimal impact; managing waste effectively; growing and distributing food to all who need it; designing sustainable communities; building minimal-impact businesses; and, ultimately, caring for each other.

References

Apps, C.D., J.L. Weaver, B. Bateman, and P.C. Paquet. 2006. *Planning Tools for the Conservation of Wide-Ranging Carnivores in the Southern Canadian Rocky Mountains and Across the Crowsnest Highway: Research Phases I & II*. New York: Wildlife Conservation Society.

Bennett, G. (ed.). 2004. *Integrating Biodiversity Conservation and Sustainable Use*. Gland, Switzerland: International Union for the Conservation of Nature (IUCN).

Carroll, C., R.F. Noss, P.C. Paquet, and N.H. Schumaker. 2003. Use of Population Viability Analysis and Reserve Selection Algorithms in Regional Conservation Plans. *Ecological Applications* 13: 1773–89.

———. 2004. Extinction Debt of Protected Areas in Developing Landscapes. *Conservation Biology* 18: 1110–20.

Hansen, A., and K. Jones. 2003. Spatial Patterns of Bird Diversity in the Yellowstone to Yukon Region: Drivers and Implications for Conservation. Technical Report No. 2. Bozeman, MT: Montana State University Landscape Biodiversity Lab. http://www.homepage.montana.edu/~hansen/ (accessed October 24, 2006).

Hanski, I. 1997. Metapopulation Dynamics: From Concepts to Predictive Models. In *Metapopulation Biology: Ecology, Genetics and Evolution.* Edited by I. Hanski and M.E. Gilpin. London: Academic Press, 69–92.

Hitt, N.P., and C.A. Frissell. 1999. *Wilderness in a Landscape Context: An Evaluation of Wilderness and Aquatic Biointegrity in Western Montana.* Missoula, MT: Wilderness Science Conference.

Laliberte, A.S., and W.J. Ripple. 2004. Range Contractions of North American Carnivores and Ungulates. *Bioscience* 54: 123–38.

Mahr, M.H., and K. Jones. 2005. *Bird Biodiversity in the Flathead River Basin: A Conservation Hotspot in the Yellowstone to Yukon Corridor.* Technical Report No. 5. Canmore, AB: Yellowstone to Yukon Conservation Initiative. http://www.y2y.net (accessed October 24, 2006).

Mattson, D.J., and T. Merrill. 2002. Extirpations of Grizzly Bears in the Contiguous United States, 1850–2000. *Conservation Biology* 16: 1123–36.

Mayhood, D. 2004. *Fishes of Yellowstone to Yukon: Overview.* Canmore, AB: Yellowstone to Yukon Conservation Initiative. http://fwresearch.ca/ (accessed October 24, 2006).

Merrill, T. 2005. *Grizzly Bear Conservation in the Yellowstone to Yukon Region.* Technical Report No. 6. Canmore, AB: Yellowstone to Yukon Conservation Initiative. http://www.y2y.net (accessed October 24, 2006).

Merrill, T., and D.J. Mattson. 2003. The Extent and Location of Habitat Biophysically Suitable Habitat for Grizzly Bears in the Yellowstone Region. *Ursus* 14: 171–87.

Montana PIF. 2006. http://biology.umt.edu/landbird/mbcp/mtpif/mtpif.htm (accessed December 4, 2006).

North American Breeding Birds Survey. 2006. http://www.absc.usgs.gov/research/bpif/Monitor/bbs.html (accessed December 4, 2006).

Noss, R.F., and A. Cooperrider. 1994. *Saving Nature's Legacy: Protecting and Restoring Biodiversity.* Washington, DC: Island Press.

Oechsli, L.M., and C.A. Frissell. 2002. *Aquatic Integrity Areas: Upper Missouri River Basin.* Canmore, AB: Yellowstone to Yukon Conservation Initiative.

Partners in Flight. http://www.pwrc.usgs.gov/pif/ (accessed December 4, 2006).

Peek, J., J. Beecham, D. Garshelis, F. Messier, S. Miller, and D. Strickland. 2003. *Management of Grizzly Bears in British Columbia: A Review by an Independent Scientific Panel.* Unpublished report submitted to Minister of Water, Land and Air Protection, Government of British Columbia, Victoria.

Pielou, E.C. 1991. *After the Ice Age: the Return of Life to Glaciated North America.* Chicago, IL: The University of Chicago Press, 147–63.

Proctor, M.F., B.N. McLellan, and C. Strobeck. 2002. Population Fragmentation of Grizzly Bears in Southeastern British Columbia, Canada. *Ursus* 13: 153–60.

Sherrington, P. 2003. *Trends in a Migratory Population of Golden Eagle in the Canadian Rocky Mountains Bird Trends. A Report on National Ornithological Surveys in Canada.* Ottawa, ON: Canadian Wildlife Service, 24–39.

USDA (U.S. Department of Agriculture) Forest Service. 1995. *Flathead Forest Plan Amendment #19, Decision Notice and Environmental Assessment.* Kalispell, MT. March 1.

Weaver, J.L. 2001. *The Transboundary Flathead: A Critical Landscape for Carnivores in the Rocky Mountains.* New York: Wildlife Conservation Society.

16
Adaptive Ecosystem Management
Tony Prato

> ... [I]t is uncertain how well we can sustain ecosystem values while providing commodity resources. Prudence calls for an adaptive ecosystematic approach that spreads environmental risks across management strategies....
> —Kohm and Franklin (1997, *27*)

The CCE furnishes significant social, economic, and ecological values that can be sustained only through effective planning and management. Although resource management agencies in the CCE have different goals and responsibilities and employ different planning and management methods, all resource management agencies face a common challenge—how to handle uncertainty in evaluating the effectiveness of management actions for resolving particular management problems. Adaptive ecosystem management (AEM) is an approach for handling uncertainty that has significant potential. Because the potential applications of AEM in the CCE are too numerous to address here, I focus on describing the AEM approach and explaining how it can be used to assess the suitability of grizzly bear habitat in a portion of the CCE called the Northern Continental Divide Ecosystem (NCDE). AEM integrates the concepts of ecosystem management, adaptive management, and stakeholder collaboration, and Bayesian statistical inference is used to implement the approach.

Ecosystem Management

Traditionally, managers have evaluated and selected management actions based on "... casual observations and unreplicated case studies [both of which tend] to be biased and misleading" (Wilhere 2002, *20*). Such approaches assume that managers know with certainty the social, economic, and ecological consequences of alternative management actions before they are implemented. This assumption is invalid for two reasons. First, our incomplete understanding of ecosystem complexity prevents us from accurately predicting ecosystem responses to future management actions (Costanza et al. 1993; Conroy 2000). Second, sampling and measurement errors make it difficult to accurately measure ecosystem responses to implemented management actions.

Most traditional resource planning and management approaches focus on only one part of the ecosystem and often cover a relatively short period of time. For example, general management plans for U.S. national parks and national forests identify and compare the social, economic, and environmental impacts of proposed alternatives. Planning is typically done as though the unit being managed is independent of the larger ecosystem to which it contributes ecosystem goods and services (e.g., habitat for threatened and endangered species; see Chapter 1 for a definition of "ecosystem goods and services" after Daily 1997) and by which it is affected (e.g., adverse impacts of landscape fragmentation on wildlife habitat; see Chapter 4). In other words, we typically give insufficient consideration to the social and ecological connectivity between areas inside and outside the management unit. This deficiency is likely to impede successful planning and management in the CCE.

Traditional approaches have other weakness in that they usually (1) focus on short-term ecosystem goals, such as reducing visitor congestion in a campsite, at the expense of long-term ecosystem goals, such as improving the habitat for a species; (2) are inefficient, particularly when stakeholders are unaware of one another's efforts; (3) ignore the diversity of human interests in resource management; and (4) often lead to conflicts and impasses, such as the northern spotted owl (*Strix occidentalis caurina*) controversy in the Pacific Northwest (Prato and Fagre 2005).

In contrast, ecosystem management focuses on larger spatial scales, longer time periods, and more variables, and actively involves stakeholders in identifying and evaluating management actions (Diaz and Bell 1997; Franklin 1997; Schowalter et al. 1997; Thomas 1997). Several federal land management agencies have changed their management philosophies, moving from extracting and exploiting timber, minerals, and forage to managing ecosystem goods and services in an ecologically sustainable way. This shift in philosophy and action is at the heart of ecosystem management (NRC 1992; Haeuber and Franklin 1996; Williams et al. 1997).

Although ecosystem management has been defined as a concept, we have paid less attention to the *goal* of ecosystem management. The goal is to sustain the productive capacity of an ecosystem over time as measured by either the stock of reproducible capital and natural capital, or the flow of goods and services produced by both forms of capital. Reproducible capital includes human knowledge and manufactured capital such as automobiles, buildings, and equipment. Natural capital includes natural resources such as soil, water, plants, energy, and wildlife. Simply sustaining the amounts of reproducible capital and natural capital in an ecosystem allows us to achieve only weak sustainability in that ecosystem (Pearce et al. 1990; Prato 1999).

We find ecosystem management more challenging to implement because it (1) encounters all the problems that normally accompany group decisionmaking; (2) requires personnel in resource-management agencies to abandon their top-down, technocratic management approach; (3) necessitates considerable time to achieve meaningful interaction among stakeholders; (4) demands considerable time, money, and effort to accumulate, organize, and interpret data and information; (5) requires scientists from different disciplines to share ideas and information and collaborate with managers and other stakeholders; (6) is generally not supported by current management policies (although this is changing); and (7) is viewed by some as a threat to private property rights (Prato and Fagre 2005). Despite these challenges, we are making overall progress in implementing ecosystem management (Yaffee et al. 1996), and recommendations specific to advancing ecosystem management in the CCE have been put forth (Pedynowski 2003).

Adaptive Management

Traditional management methods assume that managers know how ecosystems function and respond to management actions. Some actions have highly predictable ecosystem impacts. Most do not. For example, the impacts of forest roads on stream sedimentation can be estimated with greater accuracy than the impacts of roads on grizzly bear habitat and mortality. Our imperfect knowledge about variability in the many factors influencing ecosystem responses (e.g., changes in land use, climate, social attitudes, and economic growth), coupled with our incomplete understanding of ecosystem processes, give rise to uncertainty about how an ecosystem will respond to management actions (Conroy 2000).

Adaptive management is a way to manage an ecosystem when this uncertainty exists. Some, in fact, view adaptive management as the only logical approach to management under uncertainty (Holling 1978; Walters 1986; Irwin and Wigley 1993; Kohm and Franklin 1997; Parma and the NCEAS Working Group on Population Management 1998; Woodley 2002). Other advantages of adaptive management are that it increases the ability of policymakers and resource managers to acquire knowledge about ecological relationships, enhances information flows among policymakers, and creates shared understandings among scientists, policymakers, and managers (Peterman 1977; Holling 1978; Clark et al. 1979; McLain and Lee 1996; Wondolleck and Yaffee 2000). AEM allows stakeholders to maximize their capacity to learn about ecosystem responses to management actions.

Adaptive management can be passive or active. In applying its passive form, we develop predictions of an ecosystem's likely responses to management actions, select and implement actions based on those predictions, monitor ecosystem responses to implemented actions, and modify actions and subsequent predictions based on those responses (Walters and Hilborn 1978; Hilborn 1992). Passive adaptive management has been used to (1) implement a timber, fish, and wildlife agreement for the state of Washington (Halbert 1993); (2) implement a program for controlling spruce budworm (*Choristoneura fumiferana*) in New Brunswick, Canada (Gunderson et al. 1995); (3) develop an elk management strategy for Banff National Park (Parks Canada 2006); (4) monitor and evaluate the short- and long-term impacts of snowmobile use on resources and values in Grand Teton and Yellowstone National Parks and adjust snowmobile use and management based on those impacts (Sacklin et al. 2000); and (5) manage the Missouri River System (USACE 2001). Although passive adaptive management is relatively simple and inexpensive to implement, it does not give us reliable information with which to make decisions (Hurlbert 1984; Wilhere 2002). The National Research Council (NRC) recommended using passive management to implement the natural regulation policy for the northern range in Yellowstone National Park, and more generally "to the degree possible, [that] all management at YNP should be done as adaptive management" (NRC 2002, *135*).

In choosing active adaptive management, we design experiments to collect data and test hypotheses about ecosystem responses to management actions. Each action is evaluated in a "statistically valid" experiment (Halbert 1993), which incorporates replication and randomization of management actions (treatments). Active adaptive management allows stakeholders to assess ecosystem responses to management actions. Unlike passive adaptive management and trial-and-error methods, adaptive management experiments yield reliable information with which we can make informed decisions (Lee 1993).

Active adaptive management is being used in the salmon recovery program for the Columbia River Basin (Lee 1993, 1995; McLain and Lee 1996). In the lower Colorado River, the technique is being used to improve our understanding of the influence of water releases from Glen Canyon Dam on cultural, economic, and environmental resources (Glen Canyon Adaptive Management Program 2002). Managers in British Columbia used active adaptive management to develop biophysical and economic models, experiment with management actions, and test hypotheses about factors affecting salmon recovery (McClain and Lee 1996). Active adaptive management does have several limitations, discussed by Prato and Fagre (2005).

Stakeholder Collaboration

Collaboration between scientists, managers, and other stakeholders is essential for successful implementation of AEM. To illustrate how this might be accomplished in the CCE, consider the hypothetical formation of an Adaptive Management Council (AMC). The council's goals are to identify desired ecosystem outcomes for the CCE (or a portion of it), assess the extent to which threats and impacts prevent achievement of those outcomes, and use AEM to evaluate management actions for achieving desired outcomes. The latter goal is relevant when current ecosystem outcomes are undesirable. An example of a desired ecosystem outcome is maintaining sufficient habitat to sustain viable populations of threatened and endangered species that have all or a large portion of their designated habitat in the CCE.

At the outset, the AMC should recognize that it is powerless to alleviate certain kinds of threats and impacts to the CCE. For example, the AMC can take no management actions to mitigate the causes and consequences of global warming. The AMC can, however, develop and evaluate adaptation strategies designed to compensate for adverse impacts of global warming. For example, restricting road use or eliminating roads in multiple-use areas, such as national forests, could counteract adverse impacts of global warming on grizzly bear habitat and population.

Implementing AEM

Management actions can have positive and negative ecosystem impacts. For example, restricting logging and residential development in or near wildlife migration corridors in the CCE would not only benefit wide-ranging carnivores, but also reduce sediment delivery to streams, which benefits fish. Possible adverse impacts of this action are lower regional income and employment.

Suppose that the AMC can distinguish between management actions or elements of such actions for which outcomes can be predicted with reasonable accuracy (first type) and those that have relatively uncertain outcomes (second type) in terms of alleviating particular threats and impacts to the CCE. The council would evaluate the first type of action using passive adaptive management and, if feasible, would use active adaptive management to evaluate the second type.

An example of passive adaptive management in the CCE involves deciding whether to ban sightseeing air tours in the GNP. Although this option was considered in the last update of the park's general management plan, it was not implemented because the park lacks jurisdiction over use of airspace. Such a ban would presumably eliminate visual and auditory disturbances to on-the-ground visitors, the option of viewing the park from the air, and the income and employment generated by air tours. Because most elements of this decision have relatively certain outcomes, it can be evaluated using passive adaptive management.

The Bayesian Approach: An Example Using Grizzly Bear Habitat

In this section, I describe applying active adaptive management to the problem of assessing the suitability of grizzly bear habitat in the NCDE under various management actions. Grizzly bear is a threatened species in the United States and the NCDE is the primary recovery area for the species in the U.S. portion of the CCE. The boundaries of the NCDE and U.S. portion of the CCE are very similar.

In the hypothetical application, I will describe how Bayesian statistics can be used to test hypotheses about whether particular management actions improve grizzly bear habitat in the NCDE. Here, I give a nonquantitative explanation of the procedure. A quantitative explanation can be found in Prato (2005).

Bayesian statistics is well suited for testing hypotheses about the impacts of management actions on grizzly bear habitat in the NCDE (Prato 2005). To assess habitat in this example, I use two competing hypotheses: H_U states that the NCDE contains unsuitable habitat, and H_S states that the NCDE contains suitable habitat. If H_U is true, the AMC is justified in changing management actions to improve habitat. If H_S is true, the AMC is not justified in changing management actions, at least for the time being. H_U and H_S are tested using data collected from an experiment that selects a random sample of n equally sized areas in the NCDE. Each area is classified as having suitable or unsuitable habitat based on a given set of criteria. The number of areas with unsuitable habitat, designated X_0, is tallied. Table 16-1 indicates the most likely habitat condition and the optimal decision (whether or not to change management of bear habitat) for all possible values of X_0 for a sample size of 12 areas. Note that a sample size of 12 is small; it is used here only for illustrative purposes.

For these hypothetical results, H_S is true and changing bear habitat management is not justified for $X_0 \leq 7$. Conversely, H_U is true and changing bear habitat management is justified for $X_0 \geq 11$. Because the most likely habitat condition is inconclusive, the decision about changing bear habitat management is ambiguous for $8 \leq X_0 \leq 10$.

If the AMC can specify the expected losses or expected gains for alternative management actions, ambiguous decisions can be eliminated by determining Bayes action. I illustrate the latter for two management actions: (1) closing and/or decommissioning selected roads in the NCDE and (2) no change in

TABLE 16-1. Hypothetical Most Likely Bear Habitat Condition and Optimal Bear Management Decision

Number of areas with unsuitable bear habitat (X_0)	Most likely habitat condition[a]	Decision: Should management of bear habitat be changed?
0	Sustainable	No
1	Sustainable	No
2	Sustainable	No
3	Sustainable	No
4	Sustainable	No
5	Sustainable	No
6	Sustainable	No
7	Sustainable	No
8	Inconclusive	Ambiguous
9	Inconclusive	Ambiguous
10	Inconclusive	Ambiguous
11	Unsustainable	Yes
12	Unsustainable	Yes

a. Based on hypothetical posterior probabilities in Prato (2005).

management. Road closures and/or decommissioning are reasonable actions for improving bear habitat because, with other factors being equal, more roads increase bear mortality (Harris and Gallagher 1989; Mace and Manley 1993; Mace et al. 1996).

Two habitat conditions and two actions make four expected losses or expected gains possible. First, closing and/or decommissioning roads when the NCDE does not provide suitable habitat could result in a gain (G_2) or a loss (L_3), depending on the magnitude of the benefits of improved habitat relative to two costs—the cost of implementing road restrictions and the cost of reducing road density in the NCDE (in terms of decreases in motorized recreational opportunities). Second, closing and/or decommissioning roads when the NCDE does provide suitable habitat is expected to result in a loss (L_1) that stems from reduced recreational opportunities and lessened accessibility for timber harvesting and fire fighting.

Third, making no change in management when the NCDE provides unsuitable habitat is expected to result in a loss (L_2) because it perpetuates the unsuitability of habitat and possibly increases the risk of species extinction. Fourth, not changing management when the NCDE provides suitable habitat is expected to result in a gain (G_1) equal to the savings in cost from not imposing road restrictions. Expected gains, expected losses, and Bayes action are likely to vary with management actions.

For the hypothetical expected gains and losses specified in Table 16-2, Bayes action is to leave management unchanged (H_S is true) for $X_0 = 8$, and to close and/or decommission roads (H_U is true) for $X_0 = 9, 10$.

TABLE 16-2. Hypothetical Expected Gains and Losses with Two Road Management Actions and Two Habitat Conditions

Habitat condition	Action	
	Road closing/ decommissioning	No change in road management
Unsuitable	$G_2 = \$500/ha$	$L_2 = -\$1,500/ha$
Suitable	$L_1 = -\$1,000/ha$	$G_1 = \$800/ha$

Factors other than roads can change habitat over time. For instance, extensive fires can reduce bear food sources, global warming can shift bear range to the north, and residential development can fragment habitat. In the example, then, if feasible, adaptive management experiments should be repeated periodically. To illustrate this procedure, suppose the first experiment indicates that $X_0 = 11$ without any change in management. For this value of X_0, H_U is true and action should be taken to improve habitat. Suppose road restrictions are implemented and—after a sufficient period of time enables those restrictions to take effect—a second experiment is conducted and H_U and H_S are retested. If results from the second experiment indicate that H_U is true, the road restrictions do not improve habitat, and the road restrictions should be increased or another management action should be implemented. If results from the second experiment indicate that H_S is true, the road restrictions improve habitat. In this manner, sequential testing of H_S and H_U over time allows the AMC to identify management actions that significantly improve habitat. Impacts of management actions on social and economic values should be evaluated as well, and multiple-attribute evaluation is useful for this purpose (Prato and Fagre 2005).

Active Adaptive Management Results in Statistically Reliable Information

Applying AEM to the CCE allows stakeholders to evaluate the impacts of management actions on social, economic, and environmental values. Passive adaptive management develops predictions about how an ecosystem is likely to respond to new management actions, selects actions based on those predictions, implements actions, monitors ecosystem responses to those actions, and, if warranted, alters actions.

Active adaptive management employs sequential experiments to test hypotheses about ecosystem conditions with various management actions. Unlike passive adaptive management and the trial-and-error approach, active adaptive management yields statistically reliable information for decisionmaking. Note, however, that active adaptive management is more expensive and difficult to apply than passive management and that some of the prerequisites for successful implementation may not be in place. One possible strategy is using passive adaptive management to evaluate actions for which outcomes can be predicted with reasonable accuracy and, if feasible, drawing on active adaptive management techniques to evaluate actions that have uncertain outcomes.

References

Clark, W.C., D.D. Jones, and C.S. Holling. 1979. Lessons for Ecological Policy Design: A Case Study of Ecosystem Management. *Ecological Modeling* 7: 1–53.
Conroy, M.J. 2000. Mapping Biodiversity for Conservation and Land Use Decisions. In *Spatial Information for Land Use Management.* Edited by M.J. Hill and J. Aspinall. Amsterdam: Gordon and Breach Science Publishers, 145–58.
Costanza, R., L. Wainger, C. Folke, and K.-G. Maler. 1993. Modeling Complex Ecological Economic Systems: Toward an Evolutionary, Dynamic Understanding of People and Nature. *BioScience* 43: 545–55.
Daily, G.C. 1997. Introduction: What Are Ecosystem Services? In *Nature's Services: Societal Dependence on Natural Ecosystems.* Edited by G.C. Daily. Washington, DC: Island Press, 1–10.
Diaz, N.M., and S. Bell. 1997. Landscape Analysis and Design. In *Creating Forestry for the 21st Century: The Science of Ecosystem Management.* Edited by K.A. Kohm and J.F. Franklin. Washington, DC: Island Press, 255–70.
Franklin, J.F. 1997. Ecosystem Management: An Overview. In *Ecosystem Management: Applications for Sustainable Forest and Wildlife Resources.* Edited by M.S. Boyce and A. Haney. New Haven, CT: Yale University Press, 21–53.
Glen Canyon Adaptive Management Program, Adaptive Management Work Group. 2002. http://www.pn.usbr.gov/ (accessed December 5, 2006).
Gunderson, L.H., C.S. Holling, and S.S. Light (eds.). 1995. *Barriers and Bridges to the Renewal of Ecosystems and Institutions.* New York: Columbia University Press.
Haeuber, R., and J. Franklin. 1996. Forum: Perspectives on Ecosystem Management. *Ecological Applications* 6: 692–93.
Halbert, C. 1993. How Adaptive Is Adaptive Management? Implementing Adaptive Management in Washington State and British Columbia. *Reviews in Fisheries Science* 1: 261–63.
Harris, L.D., and P.B. Gallagher. 1989. New Initiatives for Wildlife Conservation: The Need for Movement Corridors. In *Preserving Communities and Corridors.* Edited by G. Mackintosh. Washington, DC: Defenders of Wildlife, 11–34.
Hilborn, R. 1992. Can Fisheries Agencies Learn from Experience? *Fisheries* 17: 6–14.
Holling, C.S. 1978. *Adaptive Environmental Assessment and Management.* Chichester, UK: John Wiley & Sons.
Hurlbert, S.H. 1984. Pseudoreplication and the Design of Ecological Field Experiments. *Ecological Monographs* 54: 157–211.
Irwin, L.L., and T.B. Wigley. 1993. Toward an Experimental Basis for Protected Forest Wildlife. *Ecological Applications* 3: 213–17.
Kohm, K.A., and J.F. Franklin. 1997. Introduction. In *Creating Forestry for the 21st*

Century: The Science of Ecosystem Management. Edited by K.A. Kohm and J.F. Franklin. Washington, DC: Island Press, 1–5.

Lee, K.N. 1993. *Compass and Gyroscope: Integrating Science and Politics for the Environment.* Washington, DC: Island Press.

———. 1995. Deliberately Seeking Sustainability in the Columbia River Basin. In *Barriers and Bridges to the Renewal of Ecosystems and Institutions.* Edited by L.H. Gunderson, C.S. Holling, and S.S. Light. New York: Columbia University Press, 214–38.

Mace, R.K., and T. Manley. 1993. *The Effects of Roads on Grizzly Bears: Scientific Supplement. South Fork Flathead River Grizzly Bear Project Report for 1992.* Helena, MT: Montana Department of Fish, Wildlife & Parks.

Mace, R.K., J. Waller, T. Manley, L.J. Lyon, and H. Zuring. 1996. Relationships among Grizzly Bears, Roads, and Habitat in the Swan Mountains, Montana. *Journal of Applied Ecology* 33: 1395–1404.

McLain, R.J., and R.G. Lee. 1996. Adaptive Management: Promises and Pitfalls. *Environmental Management* 20: 437–48.

NRC (National Research Council). 1992. *Restoration of Aquatic Ecosystems: Science, Technology, and Public Policy.* Washington, DC: The National Academies Press.

———. 2002. *Ecological Dynamics on Yellowstone's Northern Range Committee on Ungulate Management in Yellowstone National Park.* Washington, DC: The National Academies Press.

Parks Canada. 2006. Elk Management in Banff National Park. http://www.pc.gc.ca/pn-np/ab/banff/plan/plan23a_e.asp (accessed December 5, 2006).

Parma, A.M., and the NCEAS (National Center for Ecological Analysis and Synthesis) Working Group on Population Management. 1998. What Can Adaptive Management Do for our Fish, Forest, Food, and Biodiversity? *Integrative Biology* 1: 16–26.

Pearce, D., E. Barbier, and A. Markandya. 1990. *Sustainable Development: Economics and Environment in the Third World.* London: Earthscan Publications, 57–66.

Pedynowski, D. 2003. Prospects for Ecosystem Management in the Crown of the Continent Ecosystem, Canada–United States: Survey and Recommendations. *Conservation Biology* 17: 1261–69.

Peterman, R.M. 1977. Graphical Evaluation of Environmental Management Options: Examples from a Forest-Insect Pest System. *Ecological Modeling* 3: 133–48.

Prato, T. 1999. Multiple Attribute Decision Analysis for Ecosystem Management. *Ecological Economics* 30: 207–22.

———. 2005. Bayesian Adaptive Management of Ecosystems. *Ecological Modelling* 183: 147–56.

Prato, T., and D. Fagre. 2005. *National Parks and Protected Areas: Approaches for Balancing Social, Economic and Ecological Values.* Ames, IA: Blackwell Publishing.

Sacklin, J.A., K.L. Legg, M.S. Creachbaum, C.L. Hawkes, and G. Helfrich. 2000. Winter Visitor Use Planning in Yellowstone and Grand Teton National Parks. *Proceedings RMRS-P-15-VOL-4.* Ogden, UT: U.S. Department of Agriculture (USDA) Forest Service, Rocky Mountain Research Station.

Schowalter, T., E. Hansen, R. Molina, and Y. Zhang. 1997. Integrating the Ecological Roles of Phytophagous Insects, Plant Pathogens, and Mycorrihzae in Managed Forests. In *Creating Forestry for the 21st Century: The Science of Ecosystem*

Management, edited by K.A. Kohm and J.F. Franklin. Washington, DC: Island Press, 171–89.

Thomas, J.W. 1997. Foreword. In *Creating Forestry for the 21st Century: The Science of Ecosystem Management,* edited by K.A. Kohm and J.F. Franklin. Washington, DC: Island Press, ix–xii.

USACE (U.S. Army Corps of Engineers). 2001. *Summary: Missouri River Revised Draft Environmental Impact Statement, Master Water Control Manual, Review and Update.* Omaha, NE: USACE Northwestern Division.

Walters, C. 1986. *Adaptive Management of Renewable Resources.* New York: Macmillan and Co.

Walters, C.J., and R. Hilborn. 1978. Ecological Optimization and Adaptive Management. *Annual Review of Ecology and Systematics* 9: 157–88.

Wilhere, G.F. 2002. Adaptive Management in Habitat Conservation Plans. *Conservation Biology* 16: 20–29.

Williams, J.E., C.A. Wood, and M.P. Dombeck (eds.). 1997. *Watershed Restoration: Principles and Practices.* Bethesda, MD: American Fisheries Society.

Wondolleck, J.M., and S.L. Yaffee. 2000. *Making Collaboration Work: Lessons from Innovation in Natural Resource Management.* Washington, DC: Island Press.

Woodley, S. 2002. Planning and Managing for Ecological Integrity in Canada's National Parks. In *Parks and Protected Areas in Canada: Planning and Management* (2nd ed.), edited by D. Dearden and R. Rollins. Don Mills, ON: Oxford University Press, 97–114.

Yaffee, S.L., A.F. Phillips, I.C. Frenz, P.W. Hardy, S.M. Maleki, and B.E. Thorpe. 1996. *Ecosystem Management in the United States: An Assessment of Current Experience.* Washington, DC: Island Press.

17
Challenges of Managing Glacier National Park in a Regional Context

Tara Carolin, Steve Gniadek, Sallie Hejl, Dawn LaFleur, Joyce Lapp, Leo Marnell, Richard Menicke, and Jack Potter

Glacier National Park (GNP), along with Waterton Lakes National Park (WLNP) and the Bob Marshall Wilderness Complex, are protected areas that comprise the relatively ecologically intact core of the CCE. Although these protected areas are arguably the least trammeled lands in the CCE, they are not isolated from the rest of the ecosystem or from other ecosystems. Compared to other areas in the coterminous 48 states, the GNP is unique in that it has a relatively pristine environment and is home to the full complement of post-Pleistocene carnivore species. The GNP also contains headwaters for three major river systems in North America (Columbia, Missouri, and South Saskatchewan). Many, but not all, natural processes (avalanches, wildlife migration, and sometimes fire) occur fairly unimpeded. Since it was established in 1910, however, the GNP has not been immune to human impacts. These impacts include park management actions (e.g., predator control), air pollution, landscape change on lands within and around the park, global climate change, and the effects of visitors on the park's resources. The GNP is the focal point for most visitors to the CCE, receiving approximately 2 million visitors per year, primarily in the peak summer season.

Resource management, as a distinct discipline in the GNP, originated with the first specialists hired by the park in the 1970s. Suggestions for science-based management had been highlighted by the "Leopold Report" (Leopold et al. 1963) and the "Robbins Report" (NRC 1963). The initial focus for resource management in the GNP was on internal issues because there was a well-documented history of resource manipulation in the park including fish stocking, predator and fire control, and wildlife feeding, in addition to growing concerns about recreational overuse, construction rehabilitation needs, wildlife–human conflicts, and sewage disposal.

During the next three decades, three main factors led to a broader focus for park managers and resource specialists. First, the listing of several species under the U.S. Endangered Species Act (ESA) of 1973 emphasized the need to manage from an ecosystem perspective. Second, energy exploration and development started in the early 1970s with the proposed Cabin Creek coal mine in British Columbia (see Chapter 18) and continued with gas, oil, and coal-bed methane in various locations. This underscored the impacts of external threats to park resources. Third, increasing influxes of people brought land-use changes and associated concerns about habitat loss, breaks in connectivity corridors, spread of exotic species, and noise and light pollution. The connectivity to a larger, regional landscape became a key issue in park management.

Additional emphasis on the need for more research to inform park managers came from a National Research Council (NRC) committee report on improving the science and technology programs of the National Park Service (NPS; NRC 1992). In addition, a former NPS employee published a book that criticized managers of national parks for making resource decisions without scientific knowledge (Sellars 1997). In 2000, Congress funded the Natural Resource Challenge (NPS 1999a) to reinvigorate natural resource programs and outline a new management style that would emphasize and benefit park resources. The challenge aims to bring more information to park managers so that decisions can be active and informed. As a result of this new strategy, inventory programs have been expanded and monitoring programs initiated through the development of a nationwide inventory and monitoring program (I&M); Cooperative Ecosystem Study Units (CESUs) were created to encourage more research in national parks; and Research Learning Centers were established to facilitate research activities, engage partners, and share knowledge with park staff, educational institutions, and the public. The GNP is a member of the Rocky Mountain I&M network, works primarily with the Rocky Mountain CESU, and is the home park for the Crown of the Continent Research Learning Center.

In this chapter, we describe several current major natural resource issues, management concerns, and conflicts that resource managers deal with in the GNP. We illustrate how these issues relate to resource concerns in the greater CCE and how forces or inputs from the greater ecosystem result in different management decisions than would be made if the park were a wholly self-contained, isolated landscape. Highlighted issues include health of aquatic systems, integrity of wildlife communities, the threat of invasive non-native plant species, importance of native plant restoration for long-term integrity, and whitebark pine restoration. We end the chapter with the role of geographic information systems (GIS) in managing transboundary resources.

Health of Aquatic Systems

The GNP contains more than 250 lakes, as well as numerous ponds, marshes, and wetlands interconnected by a network of first- through fourth-order streams and rivers. Meltwaters drain to lowland ponds and marshes from glaciers and permanent snowfields at high elevations on both sides of the Continental Divide.

The park's aquatic biota has affinities with all three continental drainages that arise from within the park. The dominant species assemblage of native fish includes the Rocky Mountain whitefish (*Prosopium williamsoni*), the westslope cutthroat trout, and the bull trout (see Chapter 2 for scientific names not given in this chapter). Several native species of minnows, suckers, and sculpins are also present.

Historically, the park's precipitous mountain topography limited the distribution of fish mainly to large valley lakes below 1,371 m (4,500 ft), predominantly on the west side of the Continental Divide. Native fish were able to reach a few headwater lakes in drainages where upstream barriers to fish movements were absent (e.g., Quartz, Cerulean, and Upper Kintla lakes in the Flathead River North Fork drainage).

Several large valley lakes on the park's east side, however, such as Saint Mary Lake and Waterton lakes, were colonized by an entirely different species assemblage. This assemblage, which included lake trout and burbot (*Lota lota*) as the dominant piscivores, has affinity with Great Lakes fisheries and probably had an independent origin. Lake whitefish (*Coregonus clupeaformis*) are believed to be native to the St. Mary lakes, but the historical status of westslope cutthroat trout and bull trout in these waters is poorly understood. It appears that both species colonized some rivers in the Hudson Bay drainage (i.e., Belly River, upper Red Eagle Valley, Divide Creek), but failed to attain dominance as they did west of the Continental Divide. Small populations of bull trout are widely scattered among several streams throughout the Hudson Bay drainage, but this species evidently failed to establish resident populations in the large valley lakes. Native fish were unable to reach many interior lakes on the park's east side because of physical barriers. As a result, most of these waters were historically depauperate.

Disturbances to Aquatic Systems

The physical nature of aquatic habitats in the GNP remains relatively undisturbed. Aquatic ecosystems are affected by some localized developments, including damage to stream banks and channels near developed areas, local water supply infrastructure (such as pipelines and tanks at Rubideau Springs), and culvert placement that affects drainage from roads. External dams raise the level of Lower Two Medicine Lake into the GNP and have created Lake Sherburne, which is predominantly in the park. Natural impacts include wildfires that periodically cause short-term (i.e., two- to five-year) changes to the character of park streams and creeks such as elevated nitrates and other nutrients (Hauer and Spencer 1998). Greater burn intensity and spatial extent will increase effects on streams, as will postburn precipitation patterns.

During the past 100 years the most noticeable and far-reaching disturbance to the aquatic resources of the GNP has been the introduction of non-native fish through stocking in the park and by invasion from aquatic systems outside the park. Between 1912 and 1971, an estimated 55 million non-native fry, fingerlings, and juvenile fish were planted in park waters (Marnell 1988). At least eight species were introduced or otherwise gained entry into the GNP, including Yellowstone cutthroat trout (*Oncorhynchus c. bouveri*), golden trout (*O. aguabonita*), grayling (*Thymallus arcticus*), rainbow trout, kokanee, brook trout, lake trout, and lake whitefish (Marnell et al. 1987; Marnell 1988). The latter two species, native to some park waters east of the Continental Divide, are now widely dispersed in many of the large valley lakes in the North and Middle Fork drainages. Tables 17-1 and 17-2 summarize the current status of native and non-native fish species in GNP waters.

Two kinds of impacts are typically associated with the stocking or invasion of non-native fish into pristine waters, including waters harboring indigenous fisheries. In cases where a native fishery existed, genetic corruption of native fish through hybridization and introgression is of paramount concern. For example, because both rainbow trout and Yellowstone cutthroat trout have been widely stocked in park waters, the potential exists for hybridization with native westslope cutthroat trout.

The other concern is the unpredictable ecological disturbances following the introduction of non-native fish. Introduced fish may have an impact on the native fishery through intense competition for food and habitat, predation, behavioral stresses, and population expansion. In extreme cases these impacts can lead to local extirpation of native populations. As a prime example, in the GNP the lake trout invasion has had a catastrophic impact on native bull trout and westslope cutthroat trout populations in several large lakes on the western slope of the Continental Divide, including Lake McDonald (Marnell 1988). Invasive lake whitefish have tied up a huge amount of biomass at the expense of native fish in many of these same waters.

Native fish have not been the only victims of fish introductions in park waters. The stocking of planktivorous species such as Yellowstone cutthroat trout, rainbow trout, and kokanee into historically fishless waters has profoundly disrupted indigenous aquatic microbiota, notably zooplankton communities. Macroinvertebrates have probably also been affected in some waters (Verschuren and Marnell 1997). Similarly, amphibian populations adapted to waters without fish predators have been negatively affected in some park drainages (Marnell 1997). Cascading effects of non-native fish introduc-

TABLE 17-1. Native Fish Species in Glacier National Park

Salmonids			
Westslope cutthroat trout (*Oncorhynchus clarki lewisi*)[a]	C	M	H
Bull trout (*Salvelinus confluentus*)[a]	C		H
Mountain whitefish (*Prosopium williamsoni*)	C	M	H
Pygmy whitefish (*Prosopium coulteri*)[b]	C		
Lake trout (*Salvelinus namaycush*)			H
Minnows			
Redside shiner (*Richardsonius balteatus*)	C		
Peamouth (*Mylocheilus caurinus*)	C		
Northern Pikeminnow (*Ptychocheilus oregonensis*)	C		
Fathead minnow (*Pimephales promelas*)[c]		M	
Suckers			
Longnose sucker (*Catostomus catostomus*)	C	M	H
Largescale sucker (*Catostomus marcrocheilus*)	C		
Sculpins			
Slimy sculpin (*Cottus cognatus*)	C		
Shorthead sculpin (*Cottus confusus*)[a]	C		
Spoonhead sculpin (*Cottus ricei*)[a]			H
Mottled Sculpin (*Cottus bairdi*)		M	H
Others			
Burbot (*Lota lota*)			H
Northern pike (*Esox lucius*)			H
Trout-perch (*Percopsis omiscomaycus*)[a,b]			H

a. Denotes "species of special concern" in the state of Montana.
b. Denotes "rare" in the GNP.
c. Fathead minnow status is uncertain (may have been stocked into Three Bears Lake in the GNP).
Notes: C = Columbia River drainage; M = Missouri River drainage; H = Hudson Bay drainage.

tions are likely to be significant on terrestrial vertebrate predators such as bald eagles, great blue herons, ospreys (*Pandion haliaetus*), belted kingfishers (*Megaceryle alcyon*), and river otters.

Notwithstanding the threat posed by coal-bed methane gas and surface coal mining operations in British Columbia (see Chapter 18), the greatest long-term threat to aquatic habitats in the GNP is climate change. Current models suggest that climate shifts could cause significant changes to thermal and hydrologic characteristics of park waters (see Chapter 12). Climate-driven changes could dramatically alter the frequency and intensity of wildfires, leading to changes in stream hydraulics, aquatic nutrient and sediment dynamics, and riparian habitats. The combined effects of these perturbations have the potential to radically alter aquatic systems and their associated biotas across all trophic levels and at all elevations throughout the park.

TABLE 17-2. Non-Native Fish Species in Glacier National Park

Salmonids			
Rainbow trout (*Oncorhyncus mykiss*)	C	M	H
Brook trout (*Salvelinus fontinalis*)	C	M	H
Kokanee (*Oncorhynchus nerka*)	C		H
Yellowstone cutthroat trout (*Oncorhynchus clarki bouvieri*)[a]	C	M	H
Lake whitefish (*Coregonus clupeaformis*)[b]	C		H
Arctic grayling (*Thymallus arcticus*)[a]			H
Lake trout (*Salvelinus namaycush*)[c]	C		
Minnows			
None known			
Suckers			
None known			
Sculpins			
None known			
Others			
None known			

a. Denotes species of special concern in the state of Montana.
b. Status uncertain (possibly native to St. Mary Lake in the Hudson Bay drainage).
c. Denotes non-native in Columbia River Basin (lake trout are native to the Hudson Bay drainage).
Notes: C = Columbia River drainage; M = Missouri River drainage; H = Hudson Bay drainage.

Management Considerations

NPS policy requires that park managers protect, maintain, and restore native species and natural processes. The first line of defense is preventing further degradation of the resource. One major step toward this goal was the cessation of all fish stocking in GNP waters in 1971. This did not address the problem of continuing invasion of non-native fish from outside waters, but GNP managers have incrementally reduced the creel limit of native fish while concurrently allowing the liberal harvest of non-native species as a partial solution.

In addition, NPS policy stipulates that natural resources that have been disturbed or significantly altered must, to the extent possible, be rehabilitated or restored (NPS 2000). In many instances, however, this may be impossible or economically impractical. Invasive lake trout in several of the GNP's large westside lakes, once established, cannot be eliminated unless new technology is developed. Management's goal is now limited to suppressing non-native lake trout populations to enable native fish to coexist more successfully.

Recent research, in which lake trout implanted with sonic tags were acoustically tracked, has yielded useful information about the movements and spawning behavior of lake trout in Lake McDonald (Dux and Guy 2004). This information may aid park biologists in their efforts to reduce the lake trout population by allowing the placement of gill nets and other capture devices in targeted areas such as key spawning shoals.

Blocking structures strategically placed at key locations show some promise for preventing the invasion or continued entry of invasive fish into park waters. An experimental fish barrier has been constructed below the outlet of Quartz Lake in the North Fork drainage of the park to protect the native fishery residing in upper Quartz Creek, which was thought to be genetically and ecologically intact. In 2005, however, anglers caught two lake trout in the creek.

GNP waters harbor several introduced populations of Yellowstone cutthroat trout and at least one park lake contains non-native grayling. Both species have been petitioned for listing under the ESA. If one or both of these species becomes listed, an interesting predicament will be created because the NPS would be the custodian of a listed but non-native species. As more species become listed, the NPS will likely face similar dilemmas in other parks.

Despite the daunting problems faced by park managers, restoration of the GNP's aquatic resources is not a hopeless task because several native fish populations in the park's interior remain substantially intact. The highest priority must be given to rigorously protecting these resources for their inherent value and as a future source for restoration of fisheries elsewhere in the CCE. Research related to fisheries and other aquatic populations has been conducted in the GNP during the past two decades. This research will greatly facilitate the task of recovering selected aquatic resources. For example, Verschuren and Marnell (1997) documented the survival of a westslope cutthroat trout population in Avalanche Lake that appears to represent the ancestral stock native to the Lake McDonald Valley. If the lake trout population in Lake McDonald can be reduced, biologists will be able to use this surviving population for recovery of the native cutthroat fishery in that basin.

Extensive limnological research has been completed in Ptarmigan Lake, where non-native trout were introduced during the last century, as well as in similar lakes that have not been disturbed by non-native fish. If managers remove the fish population at Ptarmigan Lake at some future date, this limnological research will serve as a baseline that will enable biologists to document the recovery of the historical microbiotic community in this water. In addition to being useful for CCE fish management and restoration, this work could act as a useful benchmark for restoring other historically fishless park waters that have been disturbed by non-native fish introductions. Much work remains to be done, however, to document the condition of aquatic resources in the GNP and determine the degree to which they can be used as a reference point for restoration throughout the CCE. As new information becomes available, park managers will be in a stronger position to carry out actions to restore and recover damaged aquatic systems.

Integrity of Wildlife Communities

Although some wildlife species help to define the CCE by their distribution and movement patterns (e.g., mountain goat and bighorn sheep), many other wildlife species challenge our concept of an ecosystem with discrete boundaries. Many of the bald eagles that migrate through the GNP nest in the subarctic and winter in Utah and California (McClelland et al. 1996). Many golden eagles migrate through the GNP (Yates et al. 2001), flying to and from the boreal forest in the north and to winter haunts as far south as Mexico. Several pairs of bald and golden eagles nest in the GNP. Juvenile bald eagles from nests in the park migrate to the coast of Washington and British Columbia for the winter (McClelland et al. 1996). Harlequin ducks (*Histrionicus histrionicus*) also nest in the park and winter along the Pacific Coast of British Columbia and Washington (Reichel et al. 1997). Many of the bird species found in the GNP are neotropical migrants, spending only four or five months in the park during the nesting season and the rest of the year in transition or in wintering areas in Mexico, Central, or South America. Many other bird species are short-distance migrants that also nest in the park, and then migrate relatively short distances to their wintering grounds. These migratory movements may range from very short, such as birds that move downslope for the winter, to as far as 800 km (500 mi), such as birds that winter in Arizona, New Mexico, and Texas. All these feathered bundles of energy represent a piece of the CCE that connects with other parts of the continent. The CCE is a destination as well as a place of transition and genetic interchange, where many birds come to nest and raise their young, and then disperse to more hospitable climes.

Many mammal species are year-round residents or move locally across varying distances. Moose, elk, mule deer, and white-tailed deer that summer in the GNP often move downslope and out of the park during winter. For example, most elk on the east side of the park typically move into patches of mixed forest and grassland on the Blackfeet Indian Reservation during the winter, returning to mountain valleys in the park in the spring. Most elk probably calve in the park. The pattern in the North Fork of the Flathead River drainage is similar with respect to topography, but has a twist. Most elk that spend winters in the small prairie patches in the park and adjacent private lands migrate north to spend their summers in southeastern British Columbia (Kunkel 1997). Elk that use winter ranges on the south-facing slopes along the southern boundary of the GNP probably summer in the park as well as on U.S. Department of Agriculture (USDA) Forest Service lands to the south. Moose and deer probably exhibit similar migratory patterns. Bighorn sheep and mountain goats use year-round habitats that are almost entirely within the park. Goats using the Walton Lick along the Middle Fork of the Flathead River come from higher ranges in both the GNP and the Flathead National Forest (Singer and Doherty 1985).

Carnivores exhibit home ranges that vary from a few hundred hectares for weasels and pine marten to hundreds of square kilometers for grizzly bears. Wolves from packs in the GNP's North Fork drainage have dispersed more than 320 km (200 mi) into Alberta to the north (Ream et al. 1991), and as far south as the Ninemile Valley near Missoula, Montana. A young male wolverine trapped and equipped with a radio transmitter in the Many Glacier Valley was legally trapped near Libby, 200 km (125 mi) to the west after 11 months of movement. The GNP and the CCE may harbor important source populations of some carnivores that disperse into surrounding but less-well-protected landscapes.

Because of the wide-ranging nature of many species, managing wildlife in the CCE requires a cooperative interagency approach, into which diverse management objectives must also be integrated. The Montana portion of the CCE is termed the Northern Continental Divide Ecosystem (NCDE) in the U.S. Fish and Wildlife Service (FWS) *Grizzly Bear Recovery Plan* (FWS 1993). The NCDE Managers Subcommittee considers grizzly bear management and conservation in the U.S. portion of the CCE; the Rocky Mountain Grizzly Bear Planning Committee integrates management in the CCE across the international boundary into Alberta and British Columbia. Various interagency groups, including the Montana Bat Working Group, the Montana Common Loon Working Group, the Montana Harlequin Duck Working Group, the Montana Bald Eagle Working Group, the Montana Bird Conservation Partnership, the Interagency Wolf Working Group, and the Western Forest Carnivore Committee direct the management of other types of wildlife in the CCE. GNP managers also work extensively and cooperatively with managers of the WLNP. Cooperation across administrative boundaries and among biologists from diverse agencies, the private sector (Plum Creek Timber Company), and nongovernmental organizations (NGOs) is essential for effective management and conservation of wildlife in the CCE.

The U.S. portion of the CCE is considered to be among the most ecologically intact landscapes in the coterminous United States, and has therefore qualified for World Heritage and Biosphere Reserve designations. The assumption of intactness, however, has not been tested, and some have questioned the ability of the GNP and other western national parks to protect ecosystem components (Newmark 1995). Because of the connections between the GNP and diverse regions of the continent, a concern for ecological integrity in the park requires a concern for activities elsewhere. The protection of wildlife in the GNP may be affected by, for example, mining and road-building activities in adjacent Canadian provinces; hunting and trapping regulations on surrounding lands in the United States or Canada; forest-management practices in the Canadian boreal forest (where golden eagles nest); water management practices in the Great Basin (where bald eagles winter); oil spills along the Pacific coast (where common loons winter); or habitat conversion in Mexico (where most of our songbirds winter). A single catastrophic event, such as a coastal oil spill, could eliminate nearly the entire nesting population of the park's common loons or harlequin ducks.

The assumption of ecological intactness or resilience for ecosystems in the GNP, and the complementary idea of species persistence, is based on relatively limited information because we do not monitor most components of these ecosystems, including wildlife populations, for long-term change. Also, we have no practical prescriptions for an approach to ecosystem management that will guarantee species persistence. The GNP is establishing a monitoring framework that will monitor a few selected "vital signs." Even with the advent of the new monitoring program, we will continue to base the management of most species of wildlife on anecdotal information. Our ability to detect significant biological perturbations, let alone predict future changes, is and will continue to be woefully inadequate.

Dramatic alterations of less-well-protected areas of the North American continent may help inform us of impending changes in the CCE, reiterate the need to assess impacts at a landscape scale, and develop strategies for predicting and addressing those impacts. For example, an assessment of environmental change in the province of Nova Scotia found that incremental and cascading effects over the past century (e.g., conversion of forest composition, loss of wolves and lynx, decline in moose populations, and increase in white-tailed deer populations) were probably not predictable because of the complexity of the interactions among species (Telfer 2004). Could incipient and undetected events be occurring within the CCE that could lead to similar cascading and perhaps irreversible effects? Evidence of unintended impacts to terrestrial wildlife from non-native fish introductions is still emerging decades after those introductions ceased. Will there be unintended consequences of current fire or vegetation management practices? Could feeding of wildlife on private lands adjacent to the GNP, hunting practices outside the park, or poaching within the park result in loss of genetic or behavioral diversity within the CCE? Without applying an intensive and extensive monitoring program, we may be increasingly unlikely to detect significant alteration of the natural environment in time to intervene successfully and reverse unfavorable trends similar to those documented in Nova Scotia. Our lack of information on the distribution, status, and trend for nearly every species of wildlife in the park makes it nearly impossible to prevent an undesirable outcome. Indicators of environmental change, based largely on anecdotal information, include the following:

- The apparent extirpation of the porcupine (once considered abundant but rarely sighted over the last decade)
- The decline in mountain goat use of a popular mineral lick
- Apparent declines in other species (great blue heron, Clark's nutcracker, Cassin's finch [*Carpodacus cassinii*])
- Increases in non-native wildlife (European starling [*Sturnus vulgaris*], turkey [*Meleagris gallopavo*], house finch [*Carpodacus mexicanus*]) and plants

- Unknown effects of West Nile virus and other exotic diseases
- Concerns for the persistence of pikas (with declines documented in the Great Basin)
- Known impacts to grizzly bears from climbers (White et al. 1999)
- Uncertain impacts of climate change
- The increasing isolation of the GNP because of development around the boundaries (e.g., transportation corridors, evidenced by fragmentation and invasive/exotic introductions)
- Construction activities within the park.

All these realized and potential changes have negative consequences for the GNP directly and in its role as a source for biodiversity for the rest of the CCE. In turn, these changes will directly influence the pace of change throughout the CCE.

Threat of Invasive Non-Native Plant Species

The perspectives we have described to this point extend to flora as well as fauna, and threats become dramatically evident when we consider the great diversity of vascular plants found in the GNP. A major threat to the integrity of native plant communities is the invasion, establishment, and spread of non-native species. The park's enabling legislation in 1910, along with the 1999 general management plan, mandated that natural resources will be managed to "prevent degradation of high quality wildlife habitat, protect the wild character of the area, protect the pristine character of the area, and the integrity of biologic communities" (NPS 1999b, *26*).

Invasive non-native plants initially became established in the GNP toward the end of the nineteenth century, when EuroAmericans settled the area. Sources of invasion included agricultural practices on private and federal land; horse and livestock operations; seeds carried in mud and snow falling from vehicles, blowing off railroad cars, or transported by wind and animals; contaminated gravel used in road construction; sand applied to roads in winter; and ornamental plants, topsoil, seed, and sod brought in for landscaping (Tyser and Key 1988).

About 1,100 native species of plants exist in the park. At least 126 non-native plant species have been reported in the GNP (Lesica 2002), primarily in the aster, grass, mustard, and snapdragon families. A number of these species are aggressive and competitive and are increasing in quantity, area, and density. These invasive non-native plants alter native plant communities that GNP managers are charged with protecting and also decrease the quality of wildlife habitat. The spread of invasive non-native plants displaces forage species for wildlife and modifies habitat structure (e.g., changing grassland to a forb-dominated community) (Duncan 2005). These non-native plants displace native vegetation by robbing them of moisture, nutrients, and sunlight. As native habitat is lost, soil erosion increases and leads to long-term changes in plant communities and loss of biodiversity.

Most invasive non-natives in the GNP have a strong association with disturbed areas such as roadsides, construction projects, old homesteads, grazed fields, floodplains, and utility sites. Removal of topsoil and vegetative cover creates favorable microhabitats for non-native colonization. Spread beyond centers of infestation occurs by transport of seeds on construction equipment, visitors, animals, wind, and water.

Once established, invasive non-native plants can be quite difficult to eradicate because they often produce great numbers of seeds or have extensive underground root systems or rhizomes. In addition, because they are growing away from their native environments, they may be immune to natural control agents such as insects, disease, or competition. These species often have low nutritional value or palatability for wildlife and successfully compete with preferred grassland species.

Current Status

Of the 126 species of non-native plants that have become established, most are considered innocuous. Park managers, however, are particularly concerned about 15 species because they can spread rapidly. The state of Montana maintains a "noxious weed list," which divides invasive non-native plants already found in the GNP into two categories (see http://www.agr.state.mt.us/weedpest/noxiousweedslist.asp). Category 1 weeds are weeds that are currently established and generally widespread in many counties of the state. These weeds are capable of rapid spread and render land unfit or greatly limit beneficial uses. As of June 27, 2003, Category 1 weeds in the GNP included the following:

- Spotted knapweed (*Centaurea maculosa*)
- Canada thistle (*Cirsium arvense*)
- Leafy spurge (*Euphorbia esula*)
- Dalmatian toadflax (*Linaria dalmatica*)
- Sulfur cinquefoil (*Potentilla recta*)
- St. Johnswort (*Hypericum perforatum*)
- Oxeye daisy (*Chrysanthemum leucanthemum* L.)
- Houndstongue (*Cynoglossum officinale* L.)
- Common tansy (*Tanacetum vulgare*)
- Field bindweed (*Convolvulus arvensis*)
- Yellow toadflax (*Linaria vulgaris*).

Category 2 contains noxious weeds that have recently been introduced into the state or are rapidly spreading from their current infestation sites. These weeds are capable of rapid spread and invasion of lands, rendering lands unfit for beneficial uses. As of June 27, 2003, Category 2 weeds in the GNP included the following:

- Orange hawkweed (*Hieracium aurantiacum* L.)
- Meadow hawkweed complex (*Hieracium pratense, H. floribundum, H. caespitosum*)
- Tall buttercup (*Ranunculus acris* L.)
- Tansy ragwort (*Senecio jacobea* L.).

Some of these non-native plants are being carried beyond initial establishment sites into wilderness and backcountry areas. The expansion of invasive non-native plants in backcountry and proposed wilderness areas at the expense of native plants is particularly alarming because (1) these areas represent the park's natural conditions and contain the greatest reservoir of the biodiversity that makes this park unique; (2) native plants in the backcountry provide critical wildlife habitat; (3) backcountry areas furnish seed stock for native plant production and restoration efforts; and (4) non-native plants are difficult and expensive to control once they have become established, particularly in remote locations.

The GNP's Integrated Weed Management Program

To combat the establishment and spread of invasive non-native plants, GNP managers initiated an Integrated Weed Management (IWM) program in 1991. This program is based on NPS policy, which states that invasive non-native plant species should be controlled if they displace native flora, interrupt ecological processes, or interfere with interpretation of natural scenes. Control techniques include the use of herbicides, as well as manual, mechanical, and biological treatments. NPS guidelines (NPS 2000) further direct parks to place the highest priority on species that have potential to cause major impacts and are feasible to manage.

In 2004, 468 ha (1,157 acres) of invasive non-native plant establishment were recorded throughout the park. The park's IWM program treats approximately 60.7 ha (150 acres) each year by using herbicides, cutting, mowing, pruning, hand pulling, and weed whipping.

Weed Prevention

Given the difficulty of eradicating established populations of non-native plants, a high priority is placed on prevention. Local prevention priorities are set according to the abundance of a particular non-native species within the park and the resource values at risk. The regional scale of the entire CCE scale takes into account the non-native invaders that are widespread in the state, giving higher priority to those that are well-established elsewhere. Once non-native infestations have been documented by inventory activities, treatment actions are prioritized based on the ability to eliminate a new or small infestation and/or on the ability to contain larger infestations that are associated with habitats that have significant resource values (e.g., grasslands, wilderness, rare species).

GNP managers have adopted a number of prevention measures in recent years. The park participates in a communication network about new and potential invaders among surrounding states, Canada, and within Montana. Managers follow existing weed and seed laws enacted in Montana. Park personnel conduct education programs on newly invading non-native species, and inventory and monitor high-risk sites such as roadsides, recreation areas, and trails for new invaders. Non-native seed-free hay is required for livestock used on all trails. Disturbed sites are rehabilitated with native plant materials, and all construction and rehabilitation projects must use non-native weed-free seed, topsoil, sand, and gravel. Before any construction activities begin in the park, contract specifications mandate that equipment must be cleaned and inspected to prevent non-native plant establishment. Managers coordinate revegetation projects with invasive non-native plant program activities.

Partners in the CCE

The GNP has developed partnerships with park neighbors to address the non-native invasive plant species issue. Current partners include Flathead National Forest, Flathead and Glacier counties, the WLNP (Parks Canada), Alberta and British Columbia provinces, the Blackfeet Nation, and NGOs such as BNSF Railway and the Montana Department of Transportation. These partnerships exist primarily to share knowledge, management practices, and educational tools among the many entities that have a connection or an obligation to the CCE. Upcoming plans call for the members of the Crown Managers Partnership (CMP) to work together on a weed brochure specific to the CCE. Up until now, managers have used many different field guides to monitor weed invasions. Because non-native plants do not adhere to boundaries, these partnerships are extremely important for facilitating communication among all land managers.

Challenges

The GNP's current challenges in terms of invasive species follow:
- Livestock: Trespass livestock carry seeds and create increased areas of impact within the park, increasing the potential for weed invasion.
- Fire: Fires have resulted in areas of disturbed ground with an increased potential for invasive non-native plant recruitment.

- Riparian/river corridors: These areas are difficult to treat because they are constantly disturbed by high water flooding. In addition, their proximity to water, terrain, and access limits management options.
- Grasslands: The fescue grasslands are highly susceptible to weed encroachment. Many are located along road corridors that serve as vectors for weed invasions.
- Effectiveness of current management efforts: Are treatments successful or are niches being created for more-invasive non-natives?
- Climate change: Global change favors new species and more severe fires.

Importance of Native Plant Restoration to Long-Term Integrity of the GNP

Eliminating or preventing the establishment of non-native plant populations is clearly a difficult task for the GNP when there is constant pressure from invasive plants from the surrounding CCE (among other factors). This pressure means that native plant populations are regularly affected and need to be restored. The remarkable diversity of native flora in the GNP, which results from maritime and continental climate influences in combination with a variety of topographic gradients, is an asset to the CCE. At least seven rare plant species, preserved in the alpine zones of the GNP, are at the southern edge of the arctic boreal climate. Pacific Coast and Great Plains plants reach their eastern and western limits in the park. In all, the GNP supports approximately 1,200 native species of vascular plants and encompasses four distinct floristic provinces—boreal, Cordilleran, arctic alpine, and Great Plains (Lesica 2002).

Many areas of fragile meadows, riparian and wetland areas, grasslands, forests, and alpine tundra are accessible by foot or car (or both) in the GNP. This access gives the 1.5 to 2 million visitors that travel through the park each year a rare opportunity to experience the spectacular landscapes and native vegetation of this mountain ecosystem. A diverse mosaic of plant communities is relatively intact throughout the park and provides critical habitat to several state-listed sensitive species and one candidate species for federal listing, slender moonwort (*Botrychium lineare*). These communities are extremely sensitive to trampling, soil erosion, and species loss, particularly in areas of concentrated visitor use such as campgrounds, scenic vistas, stream banks, and lakeshores.

As these impacts proliferate they manifest as habitat and species loss, disruption of foraging and nesting opportunities for wildlife, and diminished ecological productivity. Without management intervention, significant negative visual and ecological impacts caused by visitors will only increase. Intervention examples include educational outreach, signs, seasonal and long-term area closures, trail improvement to confine use, and natural and human-made barriers. Moving facilities out of more fragile areas is an additional option.

The GNP has developed a comprehensive restoration program to address these impacts within the park, along with methodologies that can be useful throughout the larger CCE. The rehabilitation strategies seek to restore structure, function, and plant diversity to areas affected by overuse, construction, maintenance, and trail- and road-building activities. As part of the restoration program, park managers conduct site evaluations and vegetation inventories to describe the existing site conditions. Next, a restoration plan is developed to address project-specific, measurable objectives. These objectives are designed to accomplish the following:

- Preserve the genetic integrity of the parks' native floral populations and encourage optimum survival and vigor of plant materials by using species collected at or near a disturbed site.
- Replace eroded soil and lay down rapid vegetation cover to stabilize soil and eliminate erosion.
- Inhibit the invasion and establishment of exotic plant species and introduce native plants after weed eradication.
- Restore species composition and structure of disturbed sites with vegetation that is ecologically and visually compatible with the adjacent undisturbed plant communities.
- Make use of vegetation that is sustainable, low-maintenance, and durable.
- Initiate opportunities for research and technology transfer.

Specifically, park managers use indigenous plant material to maintain genetic integrity and salvage and store native soils and plants for replanting whenever possible. Park personnel collect seeds and cuttings each year and propagate them in the GNP's native plant nursery for use in restoration projects. Seasonal crews of biological technicians operate the native plant nursery and engage in site restoration, monitoring, and educational stewardship activities.

Large quantities of native seeds and plants are required to revegetate the park each year. Up to three years advance notice is needed for the collection and production of plant materials for major rehabilitation projects such as building and road reconstruction efforts. Park managers maintain a stock of genetically appropriate native plant and seed materials from all the major drainages in the park to meet restoration needs. Seeds are sent to the Natural Resources Conservation Service (NRCS) Bridger Plant Materials Center in Montana for cleaning and storage designed to ultimately increase production. Annual production is roughly 30,000 to 50,000 plants.

Revegetation work includes decompacting and amending the soil, installing native plant and seed materials, obliterating unnecessary social trails or other extraneous pedestrian travel paths, blocking with large organic debris to discourage continued impacts, and putting up signs to inform visitors about the revegetation work and the need to stay off revegetated areas.

Monitoring of completed work is the critical last step, and strategies range from taking simple measurements of canopy and ground cover to establishing permanent transects and microplots. Park personnel document all work done by nursery, revegetation, and monitoring crews by establishing photo points and preparing implementation, completion, and annual reports.

The GNP has worked with partners in the planning, design, construction, and revegetation of eight major road rehabilitation projects along the historic Going-to-the-Sun Road (GTSR). Partnerships have been forged with the Federal Lands Highway Program, local school districts, the USDA Forest Service, the Montana Conservation Corps, and other national parks. Using these relationships as a model, partnerships have been expanded to other parts of the CCE to facilitate successful restoration programs.

Innovative partnerships have also proved critical in sustaining the GNP restoration program. A partnership with the USDA Forest Service has allowed the park to include propagation protocols for 225 species native to the GNP in a Native Plants Network Protocol Database (see http://www.nativeplantnetwork.org). In addition, restoration planning, seed collection, plant materials production, and rehabilitation work are done collaboratively in the Bob Marshall Wilderness and other forest lands adjacent to the park. Restoration plans have been completed for wetland mitigation projects on Salish Kootenai tribal lands in conjunction with the Federal Lands Highway Program and the Montana Department of Transportation. GNP personnel are also working on restoration plans for Rocky Mountain, Carlsbad Cavern, and Crater Lakes national parks in the United States and the WLNP in Canada.

This process of creative partnering enhances both park preservation and visitor enjoyment. Integration of resources management across the regional ecosystem is facilitated by involving numerous groups, including students, community members, volunteers, visitors, and other agencies in hands-on restoration work. Finally, drawing on the collaborative exchange of ideas, energies, and resources has made a positive and lasting improvement across real and imagined boundaries of public programs and lands.

Whitebark Pine Restoration

The story of whitebark pine clearly illustrates the challenges of managing the GNP in a regional context. Whitebark pine communities were once a significant component of high-elevation forests throughout the CCE. The tree is considered a keystone species because it plays important roles in the ecosystems where it lives. A wide variety of wildlife depend on the high-fat seeds for food, including species that range far across boundary lines such as the grizzly bear and

Clark's nutcracker, as well as closer-ranging red squirrels and many others. The pines influence vegetation patterns across the landscape by pioneering dry, cold, exposed sites where no other tree species can establish. They create microclimates favorable to other plant species, as well as to subalpine fir and Engelmann spruce. Whitebark pines have a penchant for growing on windy ridges, where they accumulate and retain snow, thus lengthening the snowmelt period. The unique growth form of the whitepark pine adds beauty to the subalpine landscape. Whitebark pine stands are also present in areas of cultural and spiritual importance to neighboring indigenous peoples.

The existence of whitebark pine, unfortunately, has been affected both by white pine blister rust and fire exclusion. The rust was inadvertently introduced to the United States early in the twentieth century and eventually decimated whitebark pine stands in the GNP. Since then, fire exclusion has hampered regeneration. A thorough inventory of whitebark pine in the GNP conducted from 1995 to 1997 showed that 44% of the trees were dead (Kendall et al. 1996). Of the remaining live trees, 78% are lethally infected with white pine blister rust and will likely die within a couple of decades. A quarter of their cone-bearing crowns were already dead. Data collected in 2003 and 2004 indicate that average crown kill has risen to 30% (Carolin 2006a).

White pine blister rust has been, is, and will continue to be a widespread problem throughout the Northwest. Whitebark stands surveyed in Washington, northern Idaho, northwest Montana, and southern Alberta and British Columbia were found to have a 50–100% infection rate in live trees, with 40–100% of trees in each stand already dead (Kendall and Keane 2001). Scientists' and managers' greatest concern is that whitebark pine will be functionally lost in the GNP and the CCE without active management intervention. As a result, a whitebark pine restoration program was initiated in 1998.

The goal of this restoration program is to maintain whitebark populations in the park by planting stock that appear to be genetically resistant to white pine blister rust within appropriate habitats in the park. The program's objectives are to (1) develop cooperative agreements with neighboring agencies to meet the common restoration goal; (2) survey for healthy trees in blister-rust-decimated stands and collect seeds from these apparently rust-resistant trees; (3) develop or refine propagation protocols as needed and propagate seed into whitebark pine stock for future outplanting; (4) select appropriate habitats and plant tree stock in the field; and (5) monitor plantings to determine successes, failures, and needed improvements.

Results of Cooperative Partnerships

True to the mission statement of the Whitebark Pine Ecosystem Foundation (WPEF; see http://www.whitebarkfound.org/mission-stmt.html)—"Through the cooperative organization of WPEF, and networking with other organizations, government agencies, and individuals, we will be able to do together what we cannot do alone"—restoration of the tree in the GNP has involved cooperation among a number of entities. The U.S. Geological Survey (USGS) Glacier Field Station first conducted surveys on the status of whitebark pine and researched the history of blister rust treatment in the region. Next, the NPS sought funding to initiate a restoration program with input from a variety of federal agencies and universities. The U.S. Fish and Wildlife Service (FWS) Coeur d'Alene Nursery supplied whitebark pine tree stock to the Blackfeet Nation. In addition, a small amount of whitebark pine stock was raised in cooperative greenhouses at high schools in Browning and Columbia Falls, Montana. Students from these schools planted more than 600 whitebark pines while learning about the ecological and cultural significance of the tree. Seeds, seedlings, and information have been shared with the GNP's sister park in Canada, the WLNP. The WLNP in turn has shared funding for additional blister rust surveys using standardized methodology developed by the WPEF.

Germination Results

Between 1997 and 2000, more than 22,500 whitebark pine seeds were collected. Germination results varied from year to year. The biggest cone crop during the project was in 1998, and percentage fill of the lots in this batch of seeds ranged from 72 to 96%, with the seeds averaging 67% germination. In contrast, in 1999 the whitebark cones were very late in maturing, most of the seed molded in stratification, and there was virtually no germination. It is likely that the seed never fully matured that year. In addition, scientists discovered that 86% germination could be obtained from using a majority of 2.5-year-old seed.

Seedling Survival

During monitoring for survival in 2005, we found that 34% of the seedlings planted in 2001 and 2002 had survived. Forty-seven percent of the three-year-old trees were still living, most in healthy condition (Carolin 2006b). Because the trees enjoyed a wet fall in 2004 and a very wet spring in 2005, these results are highly encouraging considering the 2001–2003 drought.

Conclusions

Developing cooperative efforts with the USDA Forest Service, the Blackfeet Nation, and the WLNP allowed us to achieve our common restoration goal for whitebark pine ecosystems. We developed successful seed collection and propagation methods, and a number of healthy, cone-producing trees (sometimes called "plus" trees) have been located in heavily hit blister rust stands. Trees have been planted and the monitoring of success rates will highlight the strategies that work well.

Role of Geographic Information Systems (GIS) in Management of Transboundary Resources

The natural resources issues discussed so far all have a common element—the spatial relationship between GNP and the CCE. A GIS allows us to understand spatial relationships by connecting visual map information with powerful relational database management software. This results in a very functional decision-support tool for land managers. Multiple layers of physical, biological, cultural, and socioeconomic information can be integrated within the GIS, leading to synergistic map-based results that guide land-management decisions. Common GIS terms such as overlay, spatial query, and spatial modeling speak to the powerful capability of GIS to quickly reference multiple data sources and extract information that enables more informed decisionmaking. We use the GIS as a "dynamic map database" to meet park management and research needs.

The GNP initiated its GIS program in the early 1980s, becoming one of the earliest park units within the NPS with a functional GIS program. Park GIS applications have generally focused on natural resource issues, with great attention paid to fire ecology and wildlife management issues. In more recent years, other aspects of park management have utilized GIS as both a mapping and spatial information management tool. A detailed spatial database exists for road-related assets along the GTSR, for example, which gives road rehabilitation contractors information necessary for their work.

The park's GIS database consists of hundreds of spatial data themes, continually updated as improved or more recent information becomes available. Park managers commonly use global positioning systems (GPS) technology and remote sensing data sources (i.e., satellite images or aerial photography) to refine the mapping accuracy of existing spatial data and to map previously unrecorded features.

GPS data collection in recent years has allowed the GNP staff to improve data layers. Park personnel have used recent GPS data to update deficient mapping of trails (originally compiled in the 1960s) and to accurately map fire ignitions and fire perimeters. Wildlife species of concern, such as bighorn sheep, have been fitted with GPS collars that transmit accurate location data. These data facilitate a better understanding of seasonal use of particular habitats.

Higher-resolution aerial photography can give us the scale of information required to chronicle landscape change in remote locations, which often goes undetected. Aerial photography of the GNP captured from 1999 through 2004 has guided the mapping of stream channel locations that have been affected by flooding. The North Fork of the Flathead River and some of its tributary streams alone experienced at least two major channel-altering flood events in the 1990s.

Fire managers in the GNP utilize a powerful GIS fire history data set to both plan management-ignited prescribed fire and to manage natural fire starts. The "stand age" data are based on years of field study that chronicles fire activity within mapped forest stands. Complementary information managed within the GIS database, including fire-ignition location and recent fire perimeters, informs both research and management questions that revolve around fire ecology, fuel management, and visitor safety. Recent advances in remote sensing and GIS-based mapping techniques within burned areas (i.e., fire-severity mapping) yield more timely information that describes postfire site conditions. That information, in turn, is used to assess site restoration needs and fuel conditions.

USGS-led climate change research in the GNP includes a USGS/NPS collaborative GIS-based monitoring of glacier margins. Previous mapping by USGS geographers used visible terminal moraines to establish an 1850 baseline measurement for many large park glaciers (see Chapter 12). GPS and remote sensing data, including historical and recent aerial photography, have been used to develop a compelling time series of glacier margin measurements that speak to regional climate change. All of the mapped glacier margin information is preserved within a GIS database that allows us to quantify rates of glacier recession and model future climate-related scenarios.

Transboundary GIS Applications

GIS is a potent visual tool that can assist both managers and various publics in their respective inquiries about transboundary and interjurisdictional issues. The intoxicating power of well-mapped data is compelling, even when the information is not entirely truthful. The hidden story behind the map—the spatial database—is what truly powers the information. GIS databases can help quantify regional-scale parameters, such as road densities or land-cover type, toward the end of placing a particular jurisdiction or area of interest within the context of the larger CCE. Critical to this pursuit is the ability to draw from current and accurate GIS spatial information.

Land cover and land-cover change over time are current topics of concern within the CCE, as is fire management. The relationship between fire and forest pathogens is well documented within the northern Rocky Mountains, where large-scale fire events follow large insect or disease outbreaks predictably in time and under favorable climatic conditions. Significant die-off in forest stands is typical of more serious outbreaks, such as that caused by mountain pine beetles in the U.S. portion of the North Fork drainage of the Flathead River during the 1970s. The subsequent cycle of large fire events in the North Fork drainage, beginning with the 1988 Red Bench fire, can be very closely tied to fuel conditions that result from beetle-damaged forests. Both fallen and standing dead trees, the legacy of a pine beetle population explosion, served as a source of large, ladder fuels conducive to crown fire. Mountain pine beetles, incidentally, are currently quite active in the northern and the Canadian part of the CCE.

The Landsat Thematic Mapper (Landsat™) program delivers information-rich satellite images of the earth from space, data that are not constrained by administrative or political boundaries. Landsat images are captured at least one day each month, yielding regional-scale snapshots and facilitating time-series investigations. By analyzing Landsat data, for example, we can detect forest pathogen activity across large and often remote areas. These data inform both park managers and fire ecologists, who must address broad resource protection and the characterization of forest fuels. GIS is the tool through which satellite images are processed and translated to usable management information.

Disseminating GIS Information to the Public and Facilitating Research

The U.S. federal government is increasingly making nonsensitive information easily available through a variety of venues. Such is the focus as well for GIS and spatial data. Placing well-documented spatial information on the Internet allows other agencies, agency cooperators, advocacy groups, and the public to economize their efforts and focus on information gaps. The ability of experienced GIS users to download, explore, and critically examine its spatial data will allow the NPS to improve the quality of its information.

The NPS began making GIS data available to the public via the Internet in the mid-1990s. Posting GNP data to the NPS GIS data clearinghouse has assisted many researchers and contractors, among others. At the same time, park personnel report that the ready availability of the data has simplified the increasingly time-consuming task of responding to spatial data requests as directed by the Freedom of Information Act (FOIA). The critical component in making spatial data available online is producing the documentation that supports the appropriate use of that data, otherwise known as *metadata*. Although somewhat onerous, this element is necessary in the process of making spatial information widely available.

The NPS kicked off its Natural Resource GIS Metadata and Data Store (http://science.nature.nps.gov/nrdata/) public Web site in 2004. The data store includes a competent data search function that takes advantage of current technologies, greatly simplifying the process for those searching for NPS spatial data sets. The search engines are powered by the quality of the metadata, yielding effective results when good documentation and keywords exist. In 2005, the GNP uploaded more than 100 GIS data themes to the data store, making data ranging from bedrock geology and fire history to culverts found along the GTSR publicly available.

The NPS has also created an Internet map center (http://maps.nps.gov) that allows visitors to take virtual tours of parks, including the GNP, for the purpose of vacation planning or to simply explore the park's resources. A great deal more, however, remains to be done in this arena.

Conclusions about the Challenges of Managing a National Park in a Regional Context

The biggest challenge for resource managers in the GNP is to maintain the park's native species and natural processes in a landscape where neighboring lands are changing and visitor numbers are increasing, all against a backdrop of rapidly changing climatic conditions. In all cases, we need to protect the healthy populations of native species, remove invasive non-native species, restore unhealthy populations of native species to health, and restore or perpetuate natural processes that have been disrupted. As our highlights in this chapter reveal, working cooperatively with others throughout the CCE and in far-ranging ecosystems is the only way to meet these challenges. Examples of such collaboration include sharing tools and methods; communicating needs, ideas, and goals; and working on mutual projects at every level of each of the 20 local, state, provincial, tribal, and federal agencies with an interest in the park. Educating the CCE's constituencies so that the public understands and supports agency work is equally critical. All of these activities will bolster our ability to create better, integrated regional management under which we function as a collaborative entity. The GNP cannot meet its mission and objectives in isolation.

References

Carolin, T.W. 2006a. Blister Rust Monitoring in Glacier National Park. *Nutcracker Notes* 10: 9–10.
Carolin, T.W. 2006b. Whitebark and Limber Pine Restoration in Glacier National Park: Monitoring Results. *Nutcracker Notes* 10: 14.
Duncan, C.A. 2005. *Montana Weed Management Plan* (revised). Helena, MT: Weed Summit Steering Committee and Weed Management Task Force.
Dux, A.M., and C.S. Guy. 2004. *Habitat Use and Population Characteristics of Lake Trout in Lake McDonald, Glacier National Park.* Annual report. Bozeman, MT: Montana Cooperative Fishery Research Unit.
FWS (Fish and Wildlife Service). 1993. *Grizzly Bear Recovery Plan.* Missoula, MT: FWS.
Hauer, F.R., and C.N. Spencer. 1998. Phosphorus and Nitrogen Dynamics in Streams Associated with Wildfire: A Study of Immediate and Long-Term Effects. *International Journal of Wildland Fire* 8(4): 183–98.

Kendall, K.C., and R. Keane. 2001. Whitebark Pine Decline: Infection, Mortality and Population Trend. In *Whitebark Pine Communities: Ecology and Restoration*, edited by D.F. Tomback, S.F. Arno, and R.E. Keane. Washington, DC: Island Press, 221–42.

Kendall, K.C., D. Schirokauer, E. Shanahan, R. Watt, D. Reinhart, R. Renkin, S. Cain, and G. Green. 1996. Whitebark Pine Health in Northern Rockies National Park Ecosystems: A Preliminary Report. Prepared by the U.S. Department of Agriculture (USDA) Forest Service Intermountain Research Station, Missoula, Montana. *Nutcracker Notes* 7: 16.

Kunkel, K.E. 1997. Predation by Wolves and Other Large Carnivores in Northwestern Montana and Southeastern British Columbia. Ph.D. dissertation, University of Montana, Missoula.

Leopold, S.A., S.A. Cain, C.M. Cottam, I.A. Gabrielson, and T.L. Kimball. 1963. Wildlife Management in the National Parks. In *Transactions of the Twenty-Eighth North American Wildlife and Natural Resources Conference* 28: 28–45. Edited by J.B. Tretheren.

Lesica, P. 2002. *Flora of Glacier National Park*. Corvallis, OR: Oregon State University Press.

Marnell, L.F., R.J. Behnke, and F.W. Allendorf. 1987. Genetic Identification of Cutthroat Trout (*Salmo clarki*) in Glacier National Park, Montana. *Canadian Journal of Fisheries and Aquatic Sciences* 44: 1830–39.

Marnell, L.F. 1988. Status of the Westslope Cutthroat Trout in Glacier National Park, Montana. *American Fisheries Society Symposium* 4: 61–70.

———. 1997. Herpetofauna of Glacier National Park. *Northwestern Naturalist* 78: 17–32.

McClelland, B.R., L.S. Young, P.T. McClelland, J.G. Crenshaw, H.L. Allen, and D.S. Shea. 1994. Migration Ecology of Bald Eagles from Autumn Concentrations in Glacier National Park, Montana. *Wildlife Monographs* 125: 1–61.

McClelland, B.R., P.T. McClelland, R.E. Yates, E.L. Caton, and M.E. McFadzen. 1996. Fledgling and Migration of Juvenile Bald Eagles from Glacier National Park, Montana. *Journal of Raptor Research* 30(2): 79–89.

NPS (National Park Service). 1999a. *Natural Resource Challenge: The National Park Service's Action Plan for Preserving Natural Resources*. Washington, DC: NPS.

———. 1999b. *Glacier National Park General Management Plan*. Washington, DC: NPS.

———. 2000. *Management Policies*. Washington, DC: NPS.

NRC (National Research Council). 1963. *A Report: Advisory Committee to the National Park Service*. Washington, DC: National Academy Press.

———. 1992. *Science and the National Parks*. Washington, DC: National Academy Press.

Newmark, W.D. 1995. Extinction of Mammal Populations in Western North American National Parks. *Conservation Biology* 9(3): 512–26.

Ream, R.R., M.W. Fairchild, D.K. Boyd, and D.H. Pletscher. 1991. Population Dynamics and Home Range Changes in a Colonizing Wolf Population. In *The Greater Yellowstone Ecosystem: Redefining America's Wilderness Heritage*, edited by M. Boyce and R. Keiter. New Haven, CT: Yale University Press, 349–66.

Reichel, J.D., D.L. Genter, and D.P. Hendricks. 1997. Harlequin Duck Research and Monitoring in Montana: 1996. Helena: Montana Natural Heritage Program.

Sellars, R.W. 1997. Preserving Nature in the National Parks: A History. New Haven, CT: Yale University Press.

Singer, F.J, and J.L. Doherty. 1985. Movements and Habitat Use in an Unhunted Population of Mountain Goats, *Oreamnos americanus*. *Canadian Field-Naturalist* 99(2): 205–17.

Telfer, E.S. 2004. Continuing Environmental Change—An Example from Nova Scotia. *Canadian Field-Naturalist* 118(1): 39–44.

Tyser, R.W., and C.W. Key. 1988. Spotted Knapweed in Natural Area Fescue Grasslands: An Ecological Assessment. *Northwest Science* 62: 151–60.

Verschuren, D., and L.F. Marnell. 1997. Fossil Zooplankton and the Historical Status of Westslope Cutthroat Trout in a Headwater Lake of Glacier National Park, Montana. *Transactions of the American Fisheries Society* 126: 21–34.

White, D. Jr., K.C. Kendall, and H.D. Picton. 1999. Potential Energetic Effects of Mountain Climbers on Foraging Grizzly Bears. *Wildlife Society Bulletin* 27(1): 146–51.

Yates, R.E., B.R. McClelland, P.T. McClelland, C.E. Key, and R.E. Bennetts. 2001. The Influence of Weather on Golden Eagle Migration in Northwestern Montana. *Journal of Raptor Research* 35(2): 81–90.

18

Resolving Transboundary Conflicts
The Role of Community-Based Advocacy

Steve Thompson and David Thomas

The mountains and alpine meadows of the Rocky Mountain region feed three major continental rivers—the Columbia, the Missouri, and the Saskatchewan. These are the headwaters of the Pacific Ocean, the Atlantic Ocean, and Hudson Bay, respectively. Long before these waters reach their destinations, they are likely to flow across the U.S.–Canadian border, sometimes again and again. The Wigwam River, for example, originates in Montana's Ten Lakes Scenic Area and flows north into the Elk River in British Columbia before splashing into border-straddling Lake Koocanusa. Those Wigwam waters then flow into Montana's Kootenai River, crossing into Idaho and winding through a series of lakes and reservoirs in British Columbia before joining the Columbia River in Washington on their ultimate race to the Pacific Ocean.

The valleys and streams of the CCE have little regard for the forty-ninth parallel that forms the border between Canada and the United States, nor do the fish and wildlife that instinctively follow these lifelines. Complex laws, practices, and institutions govern the management of water, fish, and wildlife within the boundaries of each province or state. This jurisdictional complexity is compounded once waters cross the international border between Montana and British Columbia or Alberta. Successful arrangements have been painstakingly negotiated in instances where resource development and extraction are at stake. Such agreements, however, tend to be fleeting. Common-ground, long-term solutions have been more elusive for advocates of transboundary environmental conservation. This explains the challenges that communities and natural resource managers face in resolving land-use conflicts in the transboundary Flathead. It also explains the formation of coalitions designed to alleviate potential threats to natural resources from surface mining of coal and coal-bed methane (CBM) production in the headwaters of the North Fork of the Flathead River.

Background

Governmental agencies and other public-sector institutions have been ill-suited to address cross-border environmental challenges. Bureaucratic caution and narrow agency mandates have hampered ecosystem management across the international border. International agreements are imprecise and their force in law is subject to political whims on both sides of the border. Even the vocabulary of scientists and wildlife managers has been a limiting factor. Montana biologists, for example, refer to their side of the CCE as the Northern Continental Divide Ecosystem (NCDE; see Chapter 17). Canadians, meanwhile, refer to their side of the CCE as the Southern Rocky Mountains. Erase the border, however, and the bears, trout, and mountains flow seamlessly across an unbroken landscape.

Dating back three-quarters of a century, local community advocates joined hands across borders to conserve the world-class wildlands and waters of the transboundary Flathead. Grassroots, cross-border citizen advocacy has been the single most important factor in protecting these values. Over the years, as the national ethos shifted from unbridled resource exploitation to environmental conservation, the importance of these international alliances deepened. More recently, as federal governments in both nations welcome decisionmaking input and authority from local, state, and provincial authorities, the importance of listening to local conservation voices has grown.

Looking into the future, community-based advocates have great potential to champion a transboundary conservation agenda. This opportunity is available through a combination of new communication technologies, shifting economies, and growing appreciation of CCE wildlands in a world of dwindling wilderness. Achieving conservation will be a race against the challenge of accommodating the throngs of new residents attracted to the region's great outdoors and the world's seemingly unquenchable thirst for the fossil-fuel energy found in abundance in the transboundary Flathead.

Confronted with a rash of proposals to extract energy in southeastern British Columbia, a unique transboundary Flathead alliance was forged in 2004 between community leaders and conservationists in British Columbia's Elk Valley and Montana's Flathead County. Their success in analyzing and responding to various energy-development schemes exemplifies the potential influence of motivated and organized community advocates. It also signals a departure from previous means of cross-border conflict resolution. The impermanence of this conservation alliance underscores the magnitude of the long-term challenge facing conservation advocates.

The 1909 Boundary Waters Treaty

High above the windswept plains of northern Montana, mountain glaciers and snowfields send cascades of fresh, cold water into the St. Mary River and Many Glacier valleys of Glacier National Park (GNP). The waters coalesce in the St. Mary River and move due north to Canada, the Saskatchewan River, and ultimately, Hudson Bay.

Nature's export of Montana's fresh mountain water to Canada became a source of frustration to a growing number of small farmers that began settling north–central Montana along the Milk River in the latter half of the nineteenth century. Farmers began building small irrigation projects and diversion dams in the 1880s to water their crops. These projects had a limited financial return. Unlike the St. Mary River with its steady, year-long flows, the Milk River was fed by ephemeral foothill streams that were swollen with spring runoff, but mostly dry during the hottest days of August. By 1900, Montana farmers set their gaze on the St. Mary River whose clear, cold waters flowed toward a different ocean less than 48 km across a low divide from the North Fork of the Milk River.

When the U.S. Bureau of Reclamation was created in 1902, one of its first projects was to construct a canal that would divert the St. Mary River into the Milk River. Canadians had a different view of the project (Simonds 1999). Whiskey was for drinking, but water was for fighting, and Canada had Montana over a barrel. In a twist of geopolitical fate, the Milk River flows north into Alberta before swinging back south into Montana where some of the flow is used to irrigate crops. Canada warned that if Montana robbed the St. Mary River, it would permit Alberta irrigators to divert the Milk River back into the St. Mary River. The Canadian diversion would be downstream of the U.S. diversion, clearly nullifying any advantage for Montana.

The only solution to this dispute was compromise. Water issues were a big deal across the entire northern frontier of the United States, but it was the St. Mary and Milk rivers dispute that instigated treaty discussions. After four years of negotiations, a deal was struck. In 1909, the United States and Great Britain, still the sovereign overlord of Canada, signed and ratified the Boundary Waters Treaty. Article VI of the treaty declared that the St. Mary and Milk rivers are to be treated as one river for purposes of irrigation and power, and divided the rights to that water equally between the two nations. The St. Mary Canal was constructed soon thereafter, diverting glacial runoff from the St. Mary River to the Milk River. To this day, the two nations spar over the proper interpretation and implementation of the treaty.

The situation is similar on the west side of the GNP, where another, more modern water dispute remains hotly contested and imperfectly resolved by the 1909 Boundary Waters Treaty.

The Battle over Cabin Creek

Ten kilometers north of the Montana–British Columbia border within view of the GNP, international battle lines formed in 1974 when one of Great Britain's largest mining companies, Rio Algom Mines Ltd., announced plans to construct an industrial complex in the Cabin Creek Valley, one of North America's most remote and wild valleys. The company proposed strip mining high-grade coking (metallurgical) coal from two mountains and shipping it to Japan. The plan called for a new railroad, a paved highway, a coal-fired power plant and, in a 1,620-km^2 Canadian valley with no residents, construction of a company town that would house up to 5,000 people.

Rio Algom's announcement initiated a 14-year process of diplomatic negotiations, scientific studies, and legal wrangling. Throughout the period, there was a steady drumbeat of citizen advocacy whose potency was fueled by a cross-border grassroots alliance. Within months of Rio Algom's announcement, a few Montana hunters and conservationists began contacting diverse community groups, ranging from chambers of commerce to the Lions Club to the local chapter of the American Association of University Women. They also connected with a group of Canadian sportsmen who were alarmed by the proposed industrialization of a wild valley. Together, they formed the Flathead Coalition, which adopted a policy of zero pollution on the river and in the Flathead Lake downstream. The coalition launched an effective advocacy campaign to educate the public and sway decisionmakers.

The transboundary citizens' alliance was not alone in its conservation campaign. Some government agencies weighed in with force. The Montana Department of Fish and Game in 1976 published an article in *Montana Outdoors*, a broadly circulated magazine, warning that the Cabin Creek Mine—combined with other developments—threatened to destroy one of America's purest and most beautiful river systems. The article stated, "The root cause of the impending destruction of the North Fork is an increasing demand for energy which is, for the most part, created by extravagant life styles" (Schneider 1976, 2).

The government of British Columbia regarded the downstream reaction in Montana as overwrought. As Rio Algom geared up for a massive undertaking, it was making progress on a four-stage environmental assessment required by British Columbia (Wilson 1984). Spurred by the Flathead Coalition, Montana state officials and the Montana congressional delegation got involved. Canadian regulations seemed tilted toward mine approval despite the lack of information about the impacts of the proposed mine on water quality, fish, and wildlife. Congress commissioned the Environmental Protection Agency to conduct an assessment of resource development on the U.S. side and the lower portions of the Canadian side of the watershed. Montana river advocates began making an annual pilgrimage to Rio Algom's stockholder meeting to register their opposition and keep the issue in the public eye (Frederick 2004). The Fernie Rod and Gun Club erected a large wooden sign along a sparsely traveled logging road urging hunters to contact the provincial government (a sign that is still standing more than 20 years later). Many of those Canadian hunter-conservationists were coal miners in the Elk Valley who wanted the wildlife and backcountry recreation in the adjacent North Fork valley to be protected.

Meanwhile, Congressman Max Baucus and others appealed to the U.S. State Department to invoke the 1909 Boundary Waters Treaty. It took eight years with the two mines on the threshold of construction before the two federal governments finally agreed that the risk posed by the two mines merited a closer look by the International Joint Commission (IJC). Historians of North American water resource issues have remarked on the foresight that led to the insertion of Article IV into the 1909 treaty (Wilson 1984). Most of the treaty focused on water-use rights between the two nations, but Article IV addressed the issue of in-stream water quality. At the time, water quality was either taken for granted or readily sacrificed on the altar of progress. Yet the treaty went so far as to prohibit any transboundary pollution that would injure the health or property of the neighboring country. Montanans insisted that this little-used clause of the treaty be applied to the Cabin Creek project before rather than after the fact. Finally, prodded to action in 1984, the Canadian and U.S. governments formally referred the Cabin Creek matter to the IJC.

The IJC's final report on the Cabin Creek project was significant in several ways. First, the six U.S. and Canadian commissioners concurred that the final mine permit should be denied because of likely impacts on the far-ranging bull trout. Assuming no failings in mine operations, the report found negligible impacts to actual water quality at the boundary. Migratory fish were the hook that caught Rio Algom. In addition, the IJC explicitly invoked the precautionary principle:

> The Commission believes that, to ensure that the provisions of the Boundary Waters Treaty are honoured, when any proposed development project has been shown to create an identified risk of a transboundary impact in contravention of Article IV, existence of that risk should be sufficient to prevent the development from proceeding. This principle should apply, even though the degree of the risk cannot be measured with certainty, unless and until it is agreed that such an impact—or the risk of it occurring—is acceptable to both parties. (International Joint Commission 1988, *11*)

The IJC also invoked the principle of reciprocity, which establishes an expectation that the downstream nation "take every possible measure to maintain that standard in its own territory." In other words, Montana should not permit new residential subdivisions or gas drilling to mar the North Fork Valley's wild character, fisheries, and water. Perhaps most far-reaching was the IJC's final recommendation (made in the 1988 report) that the two nations should create a process "for defining and implementing compatible, equitable and sustainable development activities and management strategies" in the upper Flathead River Basin based on the principles of binational studies, fact finding, planning, and mutually acceptable use of resources.

The government of British Columbia discounted the IJC report, arguing that the IJC did not have authority to enforce the treaty but only to advise the two sovereign nations. IJC jurisdiction is further clouded by the predominant power of provincial governments to control natural resource development, as compared to the United States where most public land in the West is owned by the federal government. The federal treaty did not rest well with Victoria and the government of Premier Bill Vander Zalm never accepted the IJC report. Conveniently, Rio Algom voluntarily decided to abandon the project ostensibly on the grounds that metallurgical coal prices were low in Asia. Later, Rio Algom sold its interests to a small mining company in Toronto. With this turn of events, the Cabin Creek controversy became a distant memory, swept away by a period of growth and rapid community changes in Fernie and Montana's Flathead Valley. The Flathead Coalition went into hibernation in 1989 and did not rouse again for another 15 years.

Transboundary Neighbors Reengage

During the 1990s, the neighboring communities of northwestern Montana and southeastern British Columbia largely disengaged from each other. Canadian skiers who once flocked to the ski town of Whitefish faded away as the American dollar increased in value relative to the Canadian dollar. The Fernie ski resort blossomed. There was not much in the way of controversy, and communities in both nations were more focused on their own domestic changes and challenges.

Nevertheless, a succession of Montana governors pressed British Columbia to accept the IJC's challenge to prevent future disputes by cooperatively managing the North Fork watershed. Flathead County imposed new land-use zoning regulations for residential and commercial uses in the North Fork watershed. The GNP designated the east side of the valley as a primitive backcountry area, and the Flathead National Forest abandoned the practice of industrial-style clear-cut logging. Congress designated the North Fork as a Wild and Scenic River. British Columbia, however, had little interest in transboundary collaboration and kindly urged Montana to mind its own business. All the while, the battle of the 1980s receded into a distant memory.

By the turn of the century, Fernie began to undergo distinct demographic changes. Long a working-class railroad and mining town, new faces and diverse businesses came to town. Ski and golf developments brought in more visitors, including a growing number of second-home owners from the booming metropolis of Calgary. A new breed of entrepreneurs was establishing itself that was more mobile, independent, and tied to information-age businesses that filled new niches in an increasingly global economy. Employment in small businesses was growing whereas employment in mining, timber, and railroad businesses stagnated or decreased with modern mechanization (BC Stats 2005).

Perhaps the strongest and most active social integration between adjacent communities in the region during the 1990s occurred within conservation circles. The most enduring of these was the annual Peace Park Assembly, organized by the Rotary Clubs in Alberta and Montana. Rotarians were the driving force behind the 1932 designation of the Waterton Glacier International Peace Park by the Canadian Parliament and U.S. Congress. It is the world's first international peace park. Every year since 1932, hundreds of Rotarians meet annually at the border in a solemn recommitment of peace between nations and a joyous celebration of the ability of private citizens to bind their governments to transboundary wilderness protection.

The spirit of the International Peace Park was rekindled on a larger scale in 1993 when conservationists introduced the Yellowstone to Yukon (Y2Y) Conservation Initiative to the Canadian and U.S. governments; see Chapter 15 and Chadwick (2000) for more details. The vision for the Y2Y initiative is to maintain a system of protected areas connected by compatible land uses that allow fish, wildlife, and birds to move freely across large landscapes in the heart of the Rocky Mountains. It is a vision of vibrant human communities that are proud stewards of the region's land and waters.

In 2002, a group of conservation leaders in Montana, British Columbia, and Alberta came together to discuss a proposal to include the southeast corner of British Columbia in the Peace Park, thus aligning the western border of Waterton Lakes and Glacier national parks at the North Fork of the Flathead River. The Canadian Parks and Wilderness Society had just negotiated an agreement with Tembec Inc., a timber company, to support park expansion. Tembec would give up its logging tenure on the provincially owned land, clearing the way for its transfer to Parks Canada. In exchange, Tembec would receive funding to upgrade a sawmill that would produce the same amount of lumber from fewer logs. In October 2002, Canadian Prime Minister Jean Chrétien announced his intention to expand Waterton Lakes National Park (WLNP) pending completion of a feasibility study and consultations with the province and the Ktunaxa-Kinbasket First Nations.

The pace of cross-border discussions increased when a group of Montana mayors, Rotarians, conservationists, and business leaders journeyed to Fernie in January 2003 to discuss changes and challenges in their respective communities. They also shared information and opinions about the proposed park expansion in Canada's lower Flathead Valley. The 30 participants left the meeting with a newfound understanding of the commonalities between their respective communities.

A few months later, Fernie Mayor Randall Macnair, two city councilors, and several business owners visited Kalispell as guests of Kalispell Mayor Pam Kennedy. During the visit, they heard a new regional economic report at a Chamber of Commerce luncheon. The report (Swanson et al. 2003) was presented by economist Larry Swanson of the University of Montana's Center for the Rocky Mountain West. The report found that the Flathead economy was diverse and vibrant because of the presence of the GNP and the general attractiveness and recreational opportunities of the Flathead landscape. The natural landscape is the Flathead's most important economic asset, Swanson concluded, and the business community would be wise to work with public land managers to conserve the region's clean waters, wildlife, and wildland recreation. For the Fernie contingent, the report offered a glimpse of the opportunities and challenges that were quickly coming their way.

The proposed expansions of the WLNP and the Peace Park were applauded by community and business leaders in Montana, who echoed the local newspaper's belief that the international peace park is "our region's most important economic engine" (Jamison 2002). The Fernie City Council and mayors and rural district leaders in the East Kootenay region requested a formal feasibility study for the proposed park expansion.

The Province of British Columbia, however, saw no need for such information. Fernie was told that there would be no feasibility study because politicians in Victoria did not favor park expansion. Victoria had a different scenario in mind for the Canadian portion of the Flathead Valley.

Industrial Expansion or Wilderness Protection?

In 1902, British Columbia Mines Inspector Archibald Dick reported "everything in good order" in the tunnels and galleries of the collieries, deep beneath Coal Creek near the town of Fernie. King Coal's competent engineers and diligent regulators would ensure a safe and prosperous future for Fernie. The government man signed his reassuring report on May 20, 1902. Two days later, Coal Creek's Number 2 pit exploded, killing 128 miners. The culprit was the coalminers' constant dread—"fire damp"—the invisible, odorless gas that could kill them stealthily by suffocation, or explosively when ignited by a faulty lamp, or even a strike of a summer lightning storm that arced down the steel rails used by the mine's draft horses to haul heavy carts from the coal face. It was the miner's own drills, picks, and dynamite that released the volatile killer properly called methane, but better known today as "natural gas."

Mining ceased at Coal Creek near the end of the 1950s, as Canada retired its coal-burning steam locomotives. In the late twentieth century, new markets emerged in Asia for the low-sulfur coal from the Canadian Rockies. This metallurgical coal is first coked to purify its carbon and then cooked with iron to make steel. Modern metallurgical coal is mined with massive mechanical shovels that remove the porous, black mineral from mountains. These coal deposits originated as peat eons ago, before the Rocky Mountains rose from what was a flat, sweltering swamp of accumulating vegetation.

Coal Creek was not strip mined—that takes place further up the Elk Valley along British Columbia's boundary with Alberta. Grizzly bears, elk, lynx, and cougar reclaimed the Elk Valley nearly half a century after the last mining families moved 10 km downstream to the town of Fernie where Coal Creek empties into the Elk River. Coal Creek runs clear and pure again, offering spawning sites for runs of cutthroat trout and a spring getaway for harlequin ducks that fly in from the Pacific coast to pair off and lay their eggs amid streamside grasses and cottonwoods.

Although nature has reclaimed much of Coal Creek, miners have hardly tapped its coal potential. A few kilometers south of downtown Fernie is the Crowsnest Coalfield, a sprawling subterranean labyrinth of fractured, folding, and inverted coal and sedimentary rock layers. Exceeding 101,000 ha in size, the Crowsnest Coalfield underlies the headwaters of two transboundary watersheds, the Flathead and Elk valleys. It dwarfs the 12,000-ha Flathead Coalfield, which lies much closer to the international border and was the site of the Cabin Creek battle.

The Crowsnest Coalfield straddles a narrow section of the Rocky Mountain cordillera, a slender biological lifeline that is as vulnerable as it is rich. The transboundary Flathead is unique for its unimpeded floodplain ecosystem and dynamic predator–prey relations, which contribute to its renown among biologists as perhaps the single, most-important basin for carnivores in North America (Weaver 2001). Similarly, a scientific analysis by The Nature Conservancy (TNC) in the United States and Canada found that the adjacent Elk Valley in British Columbia is a biologically diverse landscape "that emerges as a keystone to the overall health and viability of the ecoregion, as well as the entire Rocky Mountain ecosystem" (NCC 2001). The Elk and Flathead River valleys are connected by the Crowsnest Coalfield.

Negotiations were expanded to include other strategies for maintaining the conservation values of the two mountain valleys during discussions between Tembec Inc. and conservationists about expanding the Peace Park. TNC's analysis led to an historic 40,000-ha agreement in which Tembec sold key riparian lands to TNC and maintained ownership of other low-elevation lands, but sold the development rights to those lands to TNC to preclude future subdivisions or residential development in key wildlife corridors. In addition, Tembec voluntarily placed a 10-year moratorium on development of other timberlands, and agreed to pursue independent environmental certification of its forestry practices.

The promise of sustainable forest management based on principles of environmental stewardship was warmly greeted in nearby Fernie, which saw its economic future tied to the long-term health of the environment. Fernie Mayor Randal Macnair stated:

> This initiative establishes a productive new model for conservation and economic development in and around our community. It affirms our community's heritage of maintaining a balanced connection to our landscape, wildlife and natural resources. Hunting, fishing, and forestry are ensured as honoured economic and recreational pursuits, while the new conservation status of the surrounding lands will provide a fresh economic boost to tourism and residential development within our urban concentrations. (Burwell 2003)

In late 2003, the 15-year logjam between Montana and British Columbia appeared to have broken. Suddenly, British Columbia Premier Gordon Campbell became responsive to Montana's long-standing request for a cross-border environmental agreement by proposing an environmental cooperative agreement to Montana Governor Judy Martz. The two leaders signed the agreement in September 2003, pledging "to identify, coordinate and promote mutual efforts to ensure the protection, conservation and enhancement of our shared environment for the benefit of current and future generations." The low-key announcement was greeted with enthusiasm by residents on both sides of the border.

Optimism quickly soured, however, when Fernie city officials learned in February 2004 that British Columbia planned to auction the drilling rights for CBM within two months. The auction covered 20,000 ha in the Crowsnest Coalfield, including the Coal Creek area near Fernie. The Fernie City Council called a town meeting with provincial officials to learn about and inform town residents about the auction. The provincial government agreed to postpone the auction for a short time to allow for public discussion of the proposal, but its pro-drilling assurances did not rest well with the 300 people attending the community meeting. The next day the City Council passed the following cautionary resolution:

> Be it resolved that the government of British Columbia defer any auction of coal-bed methane tenures for the Crowsnest Coalfield until completion of a comprehensive assessment of the potential impacts of large-scale gas exploration and production on the environment, economy and human community of the Elk Valley.
> —Fernie City Council Minutes, March 17, 2004

A New Transboundary Alliance Emerges

Nearly half of the area proposed for auction was within the Flathead watershed. News of the proposed auction quickly reached Montana. Conservationists in Fernie and Montana who had been working to support the Peace Park expansion refocused their attention on the auction. They contacted a recent University of Montana graduate student, Erin Sexton, whose master's thesis examined the potential water and wildlife impacts of CBM development based on a review of the industry in Wyoming, Colorado, and Alabama. Sexton's research concluded that the infrastructure of wells, compressors, pipelines, water treatment plants, and roads needed for CBM development in the Crowsnest Coalfield would effectively displace numerous wildlife species, particularly far-ranging large carnivores such as grizzly bears and wolverines (Sexton 2002). North–south connectivity for key wildlife populations could be severed in the heart of the transboundary Flathead. Sexton also found that water quality, native trout, and other aquatic life would likely be degraded by the disposal of huge volumes of wastewater pumped out of coal seams to release the flow of methane gas. Although British Columbia withheld water chemistry data for several test wells drilled on provincial land, citing the proprietary rights of corporations such as Chevron Texaco, coalfield wastewater associated with the CBM development in other areas was invariably contaminated with heavy metals, ammonia, salts, or all three.

A graceful and well-spoken presenter, Sexton was soon pressed into service. A loose alliance of Fernie residents organized Citizens Concerned about Coalbed Methane (CCCBM), and invited Sexton to share her findings with the alliance. She informed the group that CBM production in the East Kootenay region would be an "unparalleled experiment" on the landscape. A CBM industry had never been launched in such remote, wild, and mountainous territory. CBM development is expected to have a much greater impact than conventional natural gas production because of the far greater density of roads and drilling pads required. A higher concentration of wells must be drilled because coal seams must be dewatered for 6–24 months before a commercial gas flow can be achieved. The low-pressure flow of methane requires a higher density of wells and compressor stations than conventional gas fields. Although the end product is similar in coal-bed and conventional gas production, geological differences make for significant operational differences. Most natural gas comes from deep, naturally pressurized deposits that cause gas to flow up well holes under its own pressure, sometimes mixed with manageable amounts of water. On the other hand, CBM is in more shallow deposits that are bound in place by the pressure of water saturating the coal seams. Pressure over many millions of years causes the methane molecules to become physically bonded to the coal through adsorption. To release the gas molecules, the pressure must be released, which requires removing the water.

Sexton told CCCBM that the cumulative impacts of coalfield development, if permitted, could only be assessed against a comprehensive baseline assessment of pre-drilling water, fish, and wildlife conditions. British Columbia was proceeding with commercial production without a baseline assessment, cumulative impact study, or any semblance of a regulatory regime to limit well densities, wastewater disposal, and the zone of disturbance.

The methane auction received considerable attention in Montana, but was overshadowed by the concurrent announcement that Toronto-based Cline Mining Corporation was asking for permission to develop an open-pit coal mine at Cabin Creek in the Canadian Flathead, the same mine rejected by the IJC 16 years earlier. The state of Montana was given less than a week's notice to submit comments on the initial mining permit. Concerned by British Columbia's cavalier approach to the resurrected Cabin Creek Mine, Montana residents were alarmed at the prospect of coal mining and CBM development in the Canadian Flathead. Canadian officials discounted the need for substantive consultation with Montana and referred to the request for an exploratory permit as a "minor" consideration not worthy of international attention (Mann 2004). A provincial official told a Missoula newspaper reporter that although the province was willing to delay the CBM auction for a couple months to consult with local stakeholders, Montana was considered too far removed to qualify as a stakeholder (Jamison 2004a).

Caught unawares was the Flathead Basin Commission (FBC), a quasigovernmental agency composed of county, state, and federal agencies with land and water management responsibility in the Flathead Basin, a few citizen members appointed by Montana's governor, and a representative of the British Columbia government. The commission was created by the Montana legislature in the early 1980s in response to the original Cabin Creek Mine proposal. The FBC spent years pursuing the IJC's third recommendation, made in a 1988 report, to establish a comanagement agreement in the transboundary Flathead watershed. The FBC was politely rebuffed in this pursuit. With no legal authority to intervene in either the Cabin Creek Mine or CBM proposals, the FBC wrote a letter reminding British Columbia of the 1988 IJC recommendation against the same proposed coal mine.

Meanwhile, remnants of the old Flathead Coalition reformed to address the resurrected mine proposal. They decided to rejuvenate the coalition and ally with the Fernie-based CCCBM. The coalition asked for a meeting with Senator Max Baucus, who had led Montana's opposition to the Cabin Creek Mine decades earlier as a young congressman. Coalition members urged Senator Baucus to ask Secretary of State Colin Powell about the status of the 1988 IJC ruling, particularly whether there was any reason that the 1988 ruling should not apply to the resurrected mine proposal. Baucus immediately sent a letter to Powell (Baucus 2004) and received a response that the U.S. government considered the 1988 ruling to be valid and applicable (Kelly 2004). Canada had never formally accepted the earlier IJC ruling, and the U.S. State Department letter did not suggest that the IJC should be given authority to review Cline Mining Corporation's latest proposal. The Canadian government did not want to trigger the mechanisms of a diplomatic treaty. Andre Lemay, spokesperson for Foreign Affairs in Ottawa, told the *Missoulian* newspaper "Any kind of diplomatic request such as this takes time" (Jamison 2004b). Indeed, the two federal governments did not refer to the earlier Cabin Creek proposal until the final regulatory stage, ten years after the project was initiated. Even though diplomatic procedures remained uncertain and slow, there had been at least two significant changes in the diplomatic arena since the 1980s. First, British Columbia had become openly hostile to IJC intervention in its territory. Second, a weak Canadian federal government, battered by a decade of growing hostility from western provinces, was less inclined than ever to insert itself into provincial resource-management affairs.

Meanwhile, the Cline Mining Company was emboldened to pursue a much more aggressive schedule of development because of British Columbia's streamlined permitting process and pro-mining policies. Although provincial officials downplayed the international significance of Cline's proposal, company CEO Ken Bates publicly announced his intentions to achieve full capacity of the mine within two years in an effort to meet soaring Chinese demand for metallurgical coal (Barschel 2004). Bates brushed off the concerns expressed by Americans and local Canadian residents, assuring them that the mining project received exhaustive scrutiny for environmental protection in the 1980s (Cline Mining Company 2003).

Concerned that various government agencies and elected officials did not have sufficient clout or political will to address community concerns, the Montana-based Flathead Coalition and the Fernie-based CCCBM combined efforts in urging the Fernie City Council to remain steadfast in supporting a comprehensive baseline assessment before energy development. In addition, they urged Montana officials, including Governor Judy Martz and the bipartisan congressional delegation, to adopt Fernie's position. Nevertheless, the IJC remained aloof and hamstrung because the matter had not been formally referred by Ottawa and Washington for IJC action. With no judicial court willing to intervene, the transboundary citizens' alliance appealed to public opinion and linked with like-minded business groups, tribal governments, and hunting clubs. The alliance contacted a legion of newspaper and electronic journalists with an unusual story that combined international controversy and transboundary community harmony. For the first time ever, the obscure Flathead Valley received public attention throughout British Columbia because of a steady stream of reports in Vancouver and Victoria newspapers, the national press, and the Canadian Broadcasting Corporation's radio broadcasts. Numerous editorials appeared in the newspapers of both nations in support of taking a precautionary approach that included completing a comprehensive assessment before mining and drilling.

In late May 2004, the provincial government decided there was too much pressure to proceed with both an open-pit coal mine and methane drilling in the transboundary Flathead. During a visit to Fernie to promote the methane-drilling plan, B.C. Minister of Energy and Mines Richard Neufeld announced that Cline's mining permit had been denied. This came as a surprise to community groups because they had never gone so far as to oppose mining outright. Neufeld said the Cabin Creek area was off limits to coal mining although the government could just as quickly lift the restriction. The province would proceed aggressively to develop the gas resource in the Crowsnest Coalfield. If Neufeld thought his plan would mollify critics, he was mistaken. In the absence of a government press release or other documentation, the transboundary community alliance turned Neufeld's low-key Fernie announcement into a major media story. Their message was clear: Thank you, and now the same precaution must be taken with CBM.

Advocates Target Campaign to Private Industry

Emboldened by the success of their unusual cross-border campaign, the community alliance became more proactive. In July 2004, British Columbia announced it would auction methane-drilling rights, and provincial officials traveled to Helena to meet with Montana officials to inform them that no pre-drilling studies were needed. The citizen groups learned about the meeting a couple of days in advance. The provincial government delegation was surprised and angered to be greeted in Helena by leaders of the transboundary alliance and newspaper reporters. Unaccustomed to Montana's open-meeting law, they were unhappy to learn that the citizens and media would attend the meeting. Rather than placate Montana, the meeting stiffened Montana's resolve. B.C. Oil and Gas Commissioner Derek Doyle assured state officials that the province had sufficient baseline data to interpret monitoring results for future drilling. This was contrary to what was reported in a leaked copy of an internal government report obtained by the Fernie group. Four months earlier, a government contractor told the province there was a "critical information gap" and suggested the need for at least three years of intensive water quality monitoring. The government was also told that little was known about the fish or water conditions in the streams that would receive CBM wastewater (Summit Environmental Consultants Ltd. 2004).

Unsatisfied with the contradictory information and the polite hostility of B.C. energy officials, Montana Governor Judy Martz pressed Ottawa to invoke federal jurisdiction in the transboundary Flathead under both Canadian federal law and the 1909 treaty with the United States. Noting that the sale of drilling rights carried an obligation to actually begin drilling within a few years, she asked for a second cancellation of the methane auction until a transboundary assessment could be conducted. She was rebuffed by Ottawa, which said that the project was not ready for review. British Columbia refused to stop the auction.

With no other recourse, the transboundary citizens' alliance decided to target potential bidders directly. They assembled a delegation from both sides of the border to meet with oil and gas executives in Calgary, the heart of Canada's energy industry. Energy-industry newsletters and magazines were intrigued with what was happening and continued to cover the story. Soon, the popular mass media started covering the story and pressed industry executives for a reaction. In the course of five meetings with approximately 25 company executives from various companies, the delegation asked companies not to bid on the auction. Although the methane resources were rich, the cost of developing them in an environmentally responsible manner was unknown. British Columbia's refusal to conduct a pre-drilling assessment would impose this risk on the winning bidder. Were potential bidders prepared and willing to step into the middle of an international dispute? Although the companies were familiar with risk and controversy, they were dissatisfied with the way British Columbia managed the auction and associated publicity. Local approval, or social license, was an important consideration for potential bidders.

In an interview with Canada's *National Post* newspaper, Michael Gatens, chairman of the Canadian Society for Unconventional Gas, said "The concept of fast-tracking a project like this makes no sense. These kinds of projects in such environmentally sensitive areas are going to take a lot of time to even make the decision to develop" (Greenwood 2004). Gatens' statement outraged B.C. officials, who urged him to retract it. He declined, but did submit a "clarification" letter to the newspaper lauding B.C.'s "rigorous and sophisticated" regulatory system. Three weeks later, on August 25, 2004, potential bidders issued a united vote of no confidence in the development when the auction did not receive any bids.

Temporary Victories, Uncertain Future

Transboundary conservation efforts are rarely as successful as the 1932 initiative by Rotary clubs to establish Waterton Glacier International Peace Park. As the history of events in the transboundary Flathead indicate, conservation initiatives in a shared international watershed are more likely to be overshadowed by controversy and dispute than success. Growing global demand for fossil fuels will increase pressure to extract coal and CBM and stimulate controversy about industrial energy development, especially in environmentally sensitive areas like the upper Flathead Basin. Under British Columbia's regulatory system, energy development could take place in a series of staggered, unconnected decisions to mine coal and drill for methane gas. Citizen's groups on both sides of the border are concerned enough about the cumulative environmental impacts of energy development to demand that comprehensive baseline assessment of environmental and socioeconomic data be collected before a decision is made to develop energy resources. The expense and complexity of comprehensive assessment may be daunting for rural communities and volunteer-driven nongovernmental groups. Nevertheless, in the absence of a diplomatic agreement between the two nations, citizens' groups are taking steps to initiate the study themselves.

Community groups, local elected officials, conservationists, tribal leaders, and far-sighted business leaders have demonstrated the need and effectiveness of building transboundary alliances to alleviate environmental deficiencies in fast-track energy development in the transboundary Flathead. Successes in this endeavor are by nature temporary. Although community groups have been successful in slowing environmentally damaging development, securing a lasting conservation framework for the region has been more elusive. Having developed strong cross-border relationships and communication, community groups continue to wrestle with the institutional challenges posed by environmental management across an international border. They continue to strive to establish an effective institutional framework for dispute prevention and resolution that promotes conservation of the transboundary Flathead's remarkable natural resources and environment.

In 1988, the IJC urged both nations to adopt a creative, binational approach to protect environmental resources in the transboundary Flathead. Ultimately, only government resolve will bring about a lasting resolution to the transboundary region's perennial debates. Creative resolution of potential conflicts between the two nations is likely to emerge only as a result of sustained community and citizen efforts on behalf of landscape conservation in one of North America's wildest landscapes. In the muddle of competing governmental jurisdictions across an international boundary, the greatest power may lie with a resolute alliance of citizens who have learned to use the tools of persuasion and countervailing power inherent in a democratic society.

References

Barschel, M. 2004. Transcript of live radio interview by Marion Barschel, Canadian Broadcasting Corporation, with Ken Bates, CEO of Cline Mining Company, April 13.

Baucus, M. 2004. Letter to U.S. Secretary of State Colin Powell by U.S. Senator Max Baucus, April 30.

BC Stats. 2005. Community Facts, Fernie. Ministry of Management Services.

http://www.bcstats.gov.bc.ca/data/dd/facsheet/CF133.pdf (accessed December 5, 2006).

Burwell, J. 2003. Tembec, NCC Preserving Wildlife Corridors. *The Fernie Free Press*, December 17.

Chadwick, D.H. 2000. *Yellowstone to Yukon*. Washington, DC: National Geographic Society.

Cline Mining Company. 2003. Annual Information Form, filed with the BC Securities Commission, November 20.

Frederick, J. 2004. Personal communication with authors.

Greenwood, M. 2004. Go Slow on Methane, B.C. Told. *National Post*, August 3.

IJC (International Joint Commission). 1988. *Impacts of a Proposed Coal Mine in the Flathead River Basin*. Washington, DC: IJC.

Jamison, M. 2002. State Business Groups Call for Park Expansion. *Missoulian*, September 22.

———. 2004a. Canadian Mining Plans Pose Threat to Flathead Watersheds. *Missoulian*, March 17.

———. 2004b. Mine Plan Dead, U.S. Contends. *Missoulian*, May 21.

Kelly, P. 2004. Letter to Senator Max Baucus from Paul Kelly, assistant secretary of Legislative Affairs, United States State Department, May 19.

NCC (Nature Conservancy of Canada). 2001. Conserving Habitat in the Rocky Mountains. http://www.natureconservancy.ca/templates/template015.asp?lang=e_&region=8&sec=bc_welcome&screen=222&id=&order=&action (accessed December 5, 2006).

Mann, J. 2004. Coal Exec Optimistic about Mine Near Glacier. *Kalispell Daily Inter Lake*, April 22.

Schneider, B. 1976. Good-bye North Fork? *Montana Outdoors* 7: 2–7, January/February.

Sexton, E. 2002. Land Use in the Crown of the Continent Ecosystem: The Potential Disturbance Resulting from Coal-Bed Methane Production in the East Kootenay Coalfields of Southeast British Columbia. MS thesis, Department of Environmental Studies, University of Montana-Missoula.

Simonds, W.J. 1999. The Milk River. Bureau of Reclamation History Program, Denver. http://www.usbr.gov/dataweb/html/milkrive.html (accessed December 5, 2006).

Summit Environmental Consultants Ltd. 2004. *Bibliography of Environmental Resource Information for the Crowsnest and Bowron River Coalfields, B.C.* Prepared for the Ministry of Energy and Mines, Resource Development and Geosciences Branch, March.

Swanson, L.D., N. Nickerson, J. Lathrop, M.L. Archie, and H.D. Terry. 2003. *Gateway to Glacier: The Emerging Economy of Flathead County*. Washington, DC: National Parks Conservation Association.

Weaver, J. 2001. *The Transboundary Flathead: A Critical Landscape for Carnivores in the Rocky Mountains*. Working Paper No. 18. New York: Wildlife Conservation Society.

Wilson, D.K.W. 1984. Cabin Creek and International Law—An Overview. *The Public Land Law Review* 5: Spring.

19
Achieving Ecosystem Sustainability

Tony Prato and Dan Fagre

Although the CCE is internationally recognized for its ecological values, those very values that we so prize are being threatened by population and economic growth, energy development, changes in technology and climate, invasive species, and other factors. In particular, wildlife habitat is becoming more fragmented, biodiversity is decreasing, air and water pollution are increasing, and environmental amenities are being degraded in the region. Although some progress has been made in achieving sustainable landscape management, these threats will continue to expand, further degrading ecological functions and processes and impairing ecological values in the region. Other ecosystems in the Rocky Mountain West are also experiencing these negative effects (Best 2005).

This book is built on one major premise—that although human activities have altered its natural landscapes, the CCE is still intact and worth preserving because it provides valuable consumer and ecological goods and ecosystem services. An additional premise is that public-land-based recreational opportunities and environmental amenities that attract new people to the CCE and encourage long-term residents to stay pose significant threats to ecological integrity. In light of these premises, the goal of the book is to increase the capacity of stakeholders (planning and zoning commissions, natural resource managers, business leaders, scientists, environmental groups, and others) to manage the transformations occurring in the CCE in ways that maintain social, cultural, and economic values while minimizing further environmental degradation. Chapters 2 through 18 cover a wide range of topics, identifying key social, cultural, economic, and environmental determinants of past, current, and future land use, land cover, natural resource conditions, and ecological integrity of the CCE. In this last chapter, we summarize what has already been accomplished in terms of sustainable landscape management, and propose an adaptive ecosystem management approach for the CCE that draws on the concepts, knowledge, and insights contained in this book.

Sustainability Starts at Home

Ongoing efforts to improve ecosystem science, encourage cooperation across management agencies and other entities, and promote multiscale thinking about resource problems have already contributed to sustainable management of CCE landscapes. Efforts such as the Climate Landscape Interactions—The Mountain Ecosystem Transect (CLIMET) project move the scientific community toward conducting highly interdisciplinary regional ecosystem science (Fagre et al. 2003). One outcome of CLIMET is to place possible future climate change impacts to the CCE water supply in a regional context to better enable broad decisionmaking about natural resources. Joint research on snow and hydrology across the U.S.–Canada border is another example of better regional coordination. The CCE is also included in the Western Mountain Initiative, a research project that focuses on resource issues in mountain protected areas, and the Consortium for Integrated Climate Research in Western Mountains (CIRMOUNT), a collaborative, interdisciplinary consortium dedicated to understanding climates and ecosystems of western North American mountains. A better understanding of mountain processes is emerging simply by comparing and contrasting mountain systems. In a similar vein, management agencies are coordinating responses to resource issues through better information sharing and periodic joint meetings (e.g., the Crown Managers Partnership [CMP]).

Finally, the Climate Friendly Park Initiative is a U.S. National Park Service (NPS) initiative to inventory the impact that national parks have on the environment and to seek ways of mitigating those impacts. We hope that other entities—communities, agencies, and corporations—within the CCE will emulate Glacier National Park's (GNP) efforts to become a climate-friendly park and take similar steps to maximize sustainability.

A Proposal for Enhancing Ecosystem Sustainability in the CCE

Ecosystem management is not a new concept. It has been implemented in other ecosystems and regions, including British Columbia (Millennium Ecosystem Assessment 2005), the Florida Everglades (USACE 2005), the Great Lakes (Great Lakes Strategy 2002), and the Interior Columbia Basin (Interior Columbia Basin Ecosystem Management Project 2005). Ecosystem management, though, is not yet widely practiced in the CCE.

To enhance the sustainability of the CCE, we must (1) improve data collection and analysis; (2) expand landowner incentives to conserve natural, cultural, and environmental resources; (3) increase our scientific understanding and analysis of social, economic, and ecological processes; (4) develop better analytical tools; and (5) create new institutions that facilitate and improve natural resource and community planning and management. Adaptive ecosystem management (AEM) incorporates all these elements. The scientific concepts, ecosystem values, processes, and conditions described in this book form a solid foundation on which to develop and implement AEM in the CCE.

In this section, we propose an AEM approach that stakeholders can use to make more informed, balanced, and sound management decisions for alleviating threats to the sustainability of the CCE. The approach has five components:

- Developing a common vision of ecosystem sustainability
- Assessing ecosystem sustainability
- Identifying alternative management actions
- Selecting and implementing best management actions
- Monitoring and adjusting management actions as needed.

We describe each component in the sections that follow.

Developing a Common Vision of Ecosystem Sustainability

Stakeholder efforts to achieve and maintain the sustainability of the CCE should start by developing a common or collective vision that identifies the ecosystem values that are worth preserving, and describes the means by which those values would be sustained. Possible ecosystem values include a healthy environment, economic well-being, abundant wildlife, and ample recreational opportunities. Establishing a common vision does not require stakeholders to agree on the relative importance of ecosystem values.

Starting with a common vision is important for several reasons. First, it builds stakeholder consensus around the most important ecosystem values. Second, it creates the mutual respect among stakeholders that is essential for collaborative decisionmaking. Third, it allows us to develop indicators for each ecosystem value with which to assess ecosystem sustainability. Fourth, it is an important first step when there are significant threats to ecosystem sustainability. In the CCE, for example, consider the negative impacts of energy development on water quality and wildlife habitat in the North Fork of the Flathead River (as described in Chapter 18). When ecosystem threats are significant and widely recognized, stakeholders are often more motivated to develop a common vision.

CCE stakeholders have already worked together to preserve ecosystem values. Chapter 7 discusses how conservation easements were purchased on private land adjacent to Waterton Lakes National Park (WLNP) in Alberta in an effort to protect critical lower-elevation habitat for grizzly bears, trumpeter swans, and other species. Chapter 14 describes the CMP's efforts to cooperatively manage public lands in the transboundary CCE. Chapter 15 describes the

Yellowstone to Yukon (Y2Y) Conservation Initiative, an endeavor designed to maintain and restore biological diversity and habitat connectivity in a large ecoregion that stretches from the state of Wyoming to the north–central Yukon Territory.

We offer a word of caution about developing a common vision of ecosystem sustainability. If a separate vision is developed for each threat and management actions for alleviating threats are developed independently of one another, the cumulative effects of threats are not likely to receive adequate attention. This can decrease ecological integrity. Consequently, it is best for stakeholders to develop a unified vision of sustainability for the *entire* ecosystem.

Assessing Ecosystem Sustainability

In this area, stakeholders need to select one or more measurable indicators for each ecosystem value identified in the common vision and specify sustainability ranges for those indicators. If measured values of one or more indicators fall outside their respective sustainability ranges, the ecosystem is unsustainable and management actions designed to enhance ecosystem sustainability should be selected and implemented. Conversely, if measured values of all indicators fall within the respective sustainability ranges, the ecosystem is sustainable. This method is similar to the "limits of acceptable change" method for assessing recreational carrying capacities described by McCool and Cole (1997).

To illustrate the assessment of ecosystem sustainability, suppose stakeholders select two ecosystem values and two indicators for each value—economic well-being with unemployment rate and personal income as indicators, and preserving biodiversity with recovery success for threatened and endangered species and the presence of invasive species as indicators. Stakeholders would need to specify sustainability ranges for the four indicators. Indicators would be measured based on the best scientific and socioeconomic data and information. Selecting ecosystem indicators and their sustainability ranges does not require that stakeholders agree on the relative importance of indicators. If sustainability ranges for competing indicators include, for example, very high values for personal income and very low values for the presence of invasive species, it is more difficult to achieve sustainability. For this reason, sustainability ranges should represent the minimum acceptable amounts when more of the ecosystem value increases the welfare of stakeholders (e.g., personal income), and maximum amounts when less of the ecosystem value increases the welfare of stakeholders (e.g., the presence of invasive species).

The AEM approach can be formulated based on two different assumptions about inferring ecosystem sustainability from the indicators. The first assumption is that ecosystem sustainability can be accurately determined based on the measured values of indicators (i.e., ecosystem sustainability can be determined with certainty). The second assumption is that inferring ecosystem sustainability from measured values of the indicators is subject to uncertainty, which implies that decision errors can be made (i.e., deciding that the ecosystem is sustainable when it is not and vice versa). To minimize such decision errors, we can use a Bayesian approach similar to that described in Chapter 16.

Knowing whether external or internal threats have rendered an ecosystem unsustainable is important. Consider the nature of two external threats to the CCE: (1) persistent organic pollutants attached to precipitation enter alpine lakes and become biomagnified in fishes near the top of the food web, as explained in Chapter 9; and (2) climate warming causes glaciers in the GNP to melt, as described in Chapter 12. Both threats arise from events and conditions external to the CCE. We cannot alleviate external threats by taking actions within the ecosystem. Instead, national or global action is required. The Kyoto Treaty of 2005, which requires signatories to reduce greenhouse gas emissions to agreed-upon levels, is an example of a global action to mitigate climate warming.

Next, consider two internal threats to the CCE: (1) negative environmental impacts in the Montana portion of the North Fork Valley from energy development in the British Columbia portion of the valley, as explained in Chapter 18; and (2) decreased availability of corridors for linking secure core areas for grizzly bears in the Y2Y ecoregion as a result of urban sprawl, as described in Chapter 15. Although these internal threats are influenced by factors external to the ecosystem (e.g., increased demand and prices for metallurgical coal are stimulating energy development in the valley, higher personal incomes are increasing demand for environmental amenities, and pending retirements of baby boomers are increasing urban sprawl in the Rocky Mountain West), we can alleviate the threats by modifying land-use regulations within the ecosystem.

After determining which threats are external or internal to the ecosystem, we can rank the internal threats from most to least severe in terms of potential adverse impacts on ecosystem sustainability. Because top-ranked internal threats have a high likelihood of rendering the ecosystem unsustainable and we can take actions within the ecosystem to alleviate them, these threats should receive high priority.

Assessing ecosystem sustainability is facilitated by using simulation models such as the Flathead Landscape Ecosystem Model (FLEM) described in Chapter 4, the Regional Hydro-Ecological Simulation System (RHESSys) explained in Chapter 11, and A Landscape Cumulative Effects Simulator (ALCES) model elucidated in Chapter 14.

Identifying Alternative Management Actions

If one or more internal threats are causing the CCE to be unsustainable, stakeholders must identify feasible and efficient management actions for enhancing sustainability. Feasibility implies that the benefits exceed the costs of the actions, and feasible management actions are technically possible based on current technologies, politically acceptable based on current laws and regulations, and financially affordable in terms of the budgets available to management agencies. An efficient action provides combinations of indicators such that one indicator cannot be increased without decreasing another indicator. For example, the production possibility frontier in Figure 4-2 indicates that there are trade-offs between two indicators; wildlife benefits and economic benefits. We can use the procedure described in Chapter 16 to identify these actions.

Selecting and Implementing Best Management Actions

Stakeholders reaching consensus on a best management action is perhaps the most challenging component of AEM. Different feasible and efficient management actions accrue from different combinations of ecosystem indicators, and stakeholders generally have disparate preferences for indicators. This is certainly true in the CCE where stakeholders have different preferences for residential and commercial development; extraction of coal, oil, and gas resources; timber harvesting; recreation and tourism; wildlife habitat; air and water quality; and other environmental amenities. In particular, recreation and tourism, which generate income, employment, and tax revenues, can have negative impacts on the natural environment, as explicated in Chapter 5. Chapter 14 describes conflicts between residential and commercial development and the protection of wildlife habitat. Chapter 18 explains U.S. and Canadian tensions over the environmental impacts of energy development. As we can see in this example, stakeholders in the CCE are likely to prefer different best management actions for achieving ecosystem sustainability.

Stakeholders can use several approaches to reach consensus about the best management action for achieving ecosystem sustainability. Consider the following four methods. First, stakeholders can reach consensus on the relative importance of ecosystem indicators by developing a compromise set of weights for indicators. Feasible and efficient management actions are then scored using the compromise set of weights, and the best management action is the one with the highest score. Second, each stakeholder can rank management actions from most preferred to least preferred. A nominal group technique involving facilitated responses, voting, and discussion (Meffe et al. 2002) can then be used to develop a consensus ranking.

Third, stakeholders can vote on the weights (relative importance) assigned to ecosystem values and indicators or to management actions. If stakeholders vote on the weights assigned to ecosystem values and indicators, the set of weights receiving the most votes is used to score and rank management actions. Once again, the best management action is the one with the highest rank. Alternatively, stakeholders can cast votes for management actions, and the best management action is the one that receives a majority of the votes.

Voting requires that we determine which weights or management actions to include on the ballot. This requirement can be satisfied by using a nomination process for weights or management actions. With this method, we must also determine who is eligible to vote. If each stakeholder has one vote and there is a disproportionate number of stakeholders in one stakeholder group (i.e., a group of stakeholders having similar preferences for ecosystem indicators), ballot results would favor that stakeholder group. Voting could reduce the cost of resolving differences in stakeholders' preferences for ecosystem values and indicators, especially when there are a large number of stakeholders and management actions. The first three methods utilize multiple-attribute evaluation techniques (Prato 2004).

Fourth, stakeholders can collectively determine the best management action using the "analytical hierarchy process." This method attempts "to develop compromises between competing interests by pointing out areas of agreement, helping to isolate the areas of conflict, and illustrating the trade-offs between different options" (Kangas 1994, 75).

Monitoring and Adjusting Management Actions

Because a priori knowledge about the likely impacts of management actions on ecosystem indicators is imperfect, there is no guarantee that implementing best management actions determined by consensus will enhance ecosystem sustainability. For this reason, we must monitor ecosystem responses to implemented management actions to determine whether ecosystem sustainability is improving. If it is not, we may need to adjust or change management actions. The process of selecting, implementing, monitoring, assessing, and adjusting management actions is essentially AEM, as explained in Chapter 16.

If we apply passive adaptive management, we decide whether to adjust management actions solely on whether ecosystem indicators fall within the respective sustainability ranges. If applying active adaptive management is feasible, we base our decision about whether to adjust management actions on experimental testing of hypotheses about how the ecosystem responds to implemented management actions. Unlike passive adaptive management, active adaptive management yields statistically reliable information for making decisions about management actions. Active adaptive management is, however, more expensive and difficult to apply than passive adaptive management.

Feasibility of the AEM Approach

AEM involves a very different decisionmaking process than the one currently used in the CCE. Currently, stakeholders file lawsuits to "get their way," use the political system to outwit their "opponents," and impose their preferences for management actions on those opponents regardless of social and economic consequences. This process generally produces win–lose solutions (i.e., the welfare of one stakeholder group increases and the welfare of another decreases). In contrast, the AEM approach we propose requires that stakeholders develop mutual respect, work together to develop a common vision of ecosystem sustainability, and use consensus-based methods to resolve conflicts and determine best management actions. Generally, the AEM approach produces win–win solutions.

The current institutional framework (i.e., laws, rules, regulations, and norms) for making public land-management decisions in the CCE is strongly influenced by relevant statutes. For example, relevant statutes in the United States include the Clean Air Act (passed in 1970), the National Environmental Policy Act (NEPA; 1969), the Clean Water Act (1972), the Endangered Species Act (1973), and the National Forest Management Act (1976). These statutes and all their subsequent amendments are not particularly supportive of AEM. In fact, they could actually impede implementation of AEM (e.g., NEPA), and collaborative decision making (e.g., the Federal Advisory Committee Act of 1972 and subsequent amendments). A 1998 amendment to Canada's National Parks Act, on the other hand, is compatible with AEM. The amendment established the maintenance of ecological integrity through the protection of natural resources as the first priority of park management in Canadian national parks (Dearden and Rollins 2002). Consideration should be given to amending current laws and regulations as needed to make them more supportive of AEM and consensus-based decisionmaking.

Perhaps the biggest obstacle to implementing AEM in the CCE is getting stakeholders to change their adversarial behavior in which individuals and organizations use the political system to achieve self-serving management outcomes (i.e., the "rent-seeking behavior" described by Anderson [2005]). For example, private economic interests attempt to influence political decisions through corporate entities, legislation, and executive leaders. Environmental groups exert political influence by lobbying for actions and legislation that support their cause, by filing lawsuits that are resolved through the judicial system, and through media communications. Government agencies exert influence on resource management through their statutory authority. Scientific and professional organizations influence policy and legislation by issuing position statements. The political approach is not only contentious and divisive, but also costly and often counterproductive.

Facilitated and collaborative decisionmaking processes have been used successfully in areas that contain significant amounts of public land (Wondolleck and Yaffee 2000). The Keystone Center for Science and Public Policy (2005) has helped numerous stakeholders to gather the scientific, economic, and political information needed to reach consensus-driven decisions, plans, and agreements related to energy, environment, and health and social policy.

Consensus-building processes have been used successfully to make public land-management decisions in California, Montana, and North Dakota.

A New Paradigm for CCE Sustainability

Although implementing AEM in the CCE would be challenging, it affords a potentially viable and productive alternative to the political approach. The proposed approach draws stakeholders together rather than pulling them apart, reduces reliance on the political system, and possibly lowers the social cost of resolving land-management disagreements among stakeholders. Unfortunately, there is no guarantee that AEM would enhance the sustainability of the CCE. Because we can more accurately assess its potential for doing so through on-the-ground testing, we recommend the development of a pilot program that would test the feasibility of AEM in various management settings and relatively small areas. If the pilot program is determined to be feasible, we can use its results to fine-tune the approach before applying and testing it in progressively larger areas of the ecosystem.

References

Anderson, T.L. 2005. From the Old West to the New West and Back Again. 2005. In *State of the Rockies Report Card.* Colorado Springs, CO: Colorado College Economics and Business Department, 15–20.

Best, A. 2005. How Dense Can We Be? *High Country News* 37: 8–15.

Dearden, P., and R. Rollins. 2002. *Parks and Protected Areas in Canada: Planning and Management.* 2nd ed. Don Mills, ON: Oxford University Press Canada.

Fagre, D.B., D.L. Peterson, and A.E. Hessl. 2003. Taking the Pulse of Mountains: Ecosystem Responses to Climatic Variability. *Climatic Change* 59: 263–82.

Great Lakes Strategy. 2002. A Plan for the New Millennium: A Strategic Plan for the Great Lakes Ecosystem. http://gleams.altarum.org/glwatershed/strategy/gls05.html (accessed December 7, 2006).

Interior Columbia Basin Ecosystem Management Project. 2005. http://www.icbemp.gov (accessed December 7, 2006).

Kangas, J. 1994. An Approach to Public Participation in Strategic Forest Management Planning. *Forest Ecology and Management* 70: 75–88.

Keystone Center for Science and Public Policy. 2005. http://www.keystone.org/spp/index.html (accessed December 12, 2006).

McCool, S., and D. Cole (eds.). 1997. *Proceedings, Limits of Acceptable Change and Related Planning Processes: Progress and Future Directions.* General Technical Report INT-371. Washington, DC: U.S. Department of Agriculture Forest Service.

Meffe, G.K., L.A. Nielsen, R.L. Knight, and D.A. Schenborn. 2002. *Ecosystem Management: Adaptive, Community-Based Conservation.* Washington, DC: Island Press.

Millennium Ecosystem Assessment. 2005. British Columbia, Canada. http://www.millenniumassessment.org/en/subglobal.britishcolumbia.aspx (accessed December 7, 2006).

Prato, T. 2004. Multiple Attribute Evaluation for National Parks and Protected Areas. *George Wright Forum* 21: 62–75.

USACE (U.S. Army Corps of Engineers). 2005. Comprehensive Everglades Restoration Plan. http://www.evergladesplan.org/ (accessed December 7, 2006).

Wondolleck, J.M., and S.L. Yaffee. 2000. *Making Collaboration Work: Lessons from Innovation in Natural Resource Management.* Washington, DC: Island Press.

Index

Acid pollutants, 141
Active adaptive ecosystem management, 252, 253–57
Adaptive ecosystem management (AEM)
  applying, 253–57, 308
  assessing, 305–6
  benefits, 251, 303–4, 310
  common visions, 304–5
  defined, 11
  feasibility, 308–10
  passive *vs.* active, 252
  stakeholders' importance, 253, 307–8, 309
Aerial photography, 172–73, 279–80
Aesthetics as natural resource, 93
Agencies
  CEA involvement, 223–26
  changing role, 70
  cooperation among, 268
  grizzly conservation and, 107
  recreation and, 79–80
Agriculture and water resources, 142
Air pollution, 6–7
Akanahonek (Tobacco Plains), 40–41
Akiyinik (Jennings), 40–41
Alberta (Canada) land management, 30–31
ALCES model (A Landscape Cumulative Effects Simulator), 219–22
ALFA (alternative futures analysis), 58–59, 61–64
Algae, 129–30

Alpine ecosystems
  changes, 93–96, 194–95
  climate, 85–86
  future concerns, 97–98
  geomorphology, 86–87
  natural resources, 92–93
  plants, 24, 87
  treeline ecotones, 87, 90–92, 194–95
  wildlife, 87–88
Alternative futures analysis (ALFA), 58–59, 61–64
Altithermal climate, 45, 46–47
American Indian reservations, 32–33
  See also Native peoples
American Wildlife organization, 113
Amphibians, 27
Animals. *See* Wildlife
Anthropods, 27–28
Aquatic ecosystems
  health of, 117–20
  human impact on, 119–20, 128–30, 262–65
  nutrient dynamics, 126–33
  pathogens, 141–42
  species and habitats, 25–26, 241–45, 262, 264, 265
  strategies for restoring, 265–66
  See also Water resources
Ashley Creek, 124
Athabasca reconstruction, 161–63
Atmospheric deposition in snow, 177–78
Avalanches, 197–98

Backcountry use, 75, 76
Base-case modeling, 221–22
Bayesian statistics, 254–56
Bear management units, 107
　See also Grizzly bears
Bellevue Hill, 46, 47
Big Prairie ecosystem
　fire regimes, 205, 206, 208
　site description, 201–2
Biodiversity
　of birds, 26, 103–4, 111–12, 238–41
　defined, 102–3
　emerging issues, 112–14
　of fish, 110–11, 241–45
　overview, 114–15, 245–46
　See also Conservation; Restoration of ecological communities
BIOME-BGC, 183–84
Biophysical attributes and land use, 7–8
Biosphere Reserves, 31–32
Birds
　biodiversity, 26, 103–4, 111–12, 238–41
　migration, 26–27, 230, 238–41, 267
　northern goshawk, 111–12
Bison driving, 50–52
Black cottonwood forests, 241
Blackfeet people, 32, 39–40
Blackfeet Trust, 13–14
Blue Slate Canyon, 47, 48
Bob Marshall Wilderness Area. See Big Prairie ecosystem
Boundary Waters Treaty, 287, 289
British Columbia land management, 30
Built-up landscapes, 7
Bull trout conservation, 110–11, 242

Cabin Creek project, 288–90
Canada
　Indian reservations, 32–33
　land management, 30–31
　laws/strategies, 105–6, 108–9, 234
　tourism, 78–79
　See also Transboundary conflicts
Carbon-balance of treeline ecotones, 91
Carbon loading, 125–26, 127
Carnivore species, 268
CBM (coal-bed methane), 294–300
CCE (Crown of the Continent Ecosystem). See Crown of the Continent Ecosystem (CCE)
CEA (cumulative effects assessment). See Crown Managers Partnership (CMP)
Ceded Strip, 41–42
Charcoal analysis, 154–55, 159
Chinook winds, 137
Classification of landscapes, 7
Clean Water Act, 118
Climate
　alpine ecosystems, 85–86
　impact on culture, 50–52
　overview, 21
　in precontact times, 42–50
　Rocky Mountain interactions, 86, 136, 187–88
　temperatures, 161–63, 165–66, 188
Climate change
　complexity of, 167–68
　decadal shifts, 152–53
　droughts, 164–65, 198
　fire effects, 203
　global warming evidence, 93, 188
　history, 137–38, 152–53, 156–57, 167–68
　impacts, 93, 143–44
　indicators, 155, 156, 157–60, 165–66, 187–88
　monitoring, 94–96
　nonlinear phenomena, 189
　reconstruction, 161–65
　water resource threat, 264
Climate Landscape Interactions-The Mountain Ecosystem Transect (CLIMET), 303
Cline Mining Company, 297
Clovis Complex, 43–44, 47
CMP. See Crown Managers Partnership (CMP)
Coal mining, 292–300
Cody people, 44, 45, 47
Collaboration
　among AEM stakeholders, 253
　common vision development, 304–5
　in plant restoration, 276, 277–79
　transboundary alliance, 300
　in weed management, 273
　See also Crown Managers Partnership (CMP)

Index 315

Commercial parcels, 60–61
Community-based advocacy. *See* Transboundary conflicts
Connectivity
　of habitats, 245–46
　of landscapes, 233
　of species, 112–13
Conservation
　on continental scale, 230–31
　for grizzlies, 106–10, 109
　on private land, 29, 109–10, 111
　protected areas, 72, 104–5, 233
　USDA programs, 13–14
　See also Biodiversity; Restoration of ecological communities
Contact complex, 47
Continental-scale conservation, 230–31
Crandell Mountain Subphase, 47, 49
Critical habitat designation, 110–11
Crowfoot Advance, 157
Crown fires, 204
Crown Managers Partnership (CMP)
　assessing cumulative effects, 217–19
　lessons learned, 222–27
　modeling and framework development, 219–22
　monitoring changes, 269
　overview, 114, 215–17, 226
　weed management, 273
Crown of the Continent Ecosystem (CCE)
　assumption of intactness, 6–7, 268–69
　defined, 3
　natural resources of, 71
　origin of name, 19
　site description, 7, 17–18
　threats to, 269–70, 302
Crowsnest Coalfield, 293, 294, 295, 298
Cultural landscapes, 7
Cumulative effects assessment (CEA). *See* Crown Managers Partnership (CMP)
CWA (Clean Water Act), 118

Dalimata family, 12
Dams, 131, 143, 242
Decisionmaking, disjointed nature of, 216
Designated wilderness areas, 105
Development attractiveness, 60–61
Douglas fir, 202
Drilling rights, 294–300
Droughts, 164–65, 198
Dwarf trees, 91

Early precontact period, 43–46, 47
Earnings' shifts, 4–5
Easements for grizzlies, 109
Ecological modeling, 56–64, 179–84, 219–22, 224
Economic issues
　alternative futures analyses, 58–64
　earnings' shifts, 4–5
　impact of growth on CCE, 55–56
　in recreation/tourism, 69–70, 73, 74
　value of services, 6
Ecosystem management. *See* Adaptive ecosystem management (AEM); Resource management
Ecosystem modeling, 180–84
EDS (endocrine-disrupting substances), 139–40, 145
El Niño Southern Oscillation (ENSO), 153
Elevation deposition, 138–39
Elk, 267
Employment
　changes in, 4–5
　futures analysis, 59–60
　in recreation and tourism, 73
Endangered Species Act (ESA), 105, 107, 266
Endocrine-disrupting substances (EDS), 139–40, 145
ENSO (El Niño Southern Oscillation), 153
Enteric infections, 141–42
Erosion, 120–22
ESA (Endangered Species Act), 105, 107, 266
Eutrophication, 125, 140
Exotic invasive species. *See* Invasive species
Explorers, early, 18–19
Extractive resource development, 292–300

Federal Water Pollution Control Act of 1972, 118
Fell-fields, 88, 89
Fernie (British Columbia), 290–300
Financial assistance through USDA, 13–14
Fire BioGeo Chemical (FIREBGC), 180–81, 182–83
Fire regimes, 203–10
Fires
    benefits, 201–2, 208–10
    challenges, 21–22
    charcoal as indicator, 154–55
    climate change and, 164–65, 197, 203
    forest pathogens in, 280–81
    impacts, 206, 263
    management, 180–81, 207–10, 280–81
    modeling as forecasting tool, 182–84
    pollen records, 47, 159
    suppression, 206–7, 209
Fish, 103–4
    conservation, 241–45
    natives *vs.* exotic, 26, 262–66
    recovery of, 123, 131–32
Fish and Wildlife Service (FWS), 112–13
Flathead Landscape Ecosystem Model (FLEM), 57–64
Flathead region
    economic transformation, 4–5
    Indian Reservation, 32
    water resources, 23, 120–24
    See also Transboundary conflicts
FLEM (Flathead Landscape Ecosystem Model), 57–64
Forecasting
    climate, 181–82
    expected gains/losses, 254–56
Forests
    black cottonwood, 241
    impact on water resources, 142–43
    montane, 23
    national, 32
    pathogens from fire, 280–81
    responses to glaciers, 193–94
    ribbon, 92
    subalpine, 23–24
    See also Trees
Freshwater systems. *See* Aquatic ecosystems; Water resources
Front country areas, 75
Fuel loadings, 203–4
Funding, 225
FWS (Fish and Wildlife Services), 112–13

Gakawakamitukinik (Michel Prairie), 40
Geochemical changes, 93–94
Geographic Information Systems (GIS), 178–79, 279–82
Geological foundation, 19–20
Geomorphology, 86–87, 94
Glacier fluctuation
    forest responses to, 193–94
    as indicator of climate change, 156, 165–66
    mapping/monitoring, 178–79, 280
    Neoglacial period, 160–62
    recession, 189–93
    See also Ice sheets
Glacier National Park (GNP)
    ground-based monitoring, 174–75
    history of resource management, 260–61
    as primary attraction, 72
    remote sensing, 172–73
    site description, 31, 260
Global Observation Research Initiative in Alpine Environments (GLORIA), 94
Global positioning systems (GPS), 279
Global warming evidence, 93, 188
    See also Climate change
    GLORIA (Global Observation Research Initiative in Alpine Environments), 94
    GNP. *See* Glacier National Park (GNP)
Goods, ecosystem, 5–6
Grasslands as zone, 23
Gray wolf's journey, 232
Grinnell, George Bird, 19, 51
Grizzly bears
    in alpine ecosystems, 88
    conservation management, 106–10, 237–38
    habitat needs, 105, 107, 235, 237–38
    mortality, 237, 238

Index

population, 235–37
testing management options, 254–56
as umbrella species, 235
Ground-based monitoring/testing, 173, 174–75, 310

Habitats. *See* Wildlife habitats
Health of ecosystems, defined, 117
Herptiles, 27
Hill, James J., 19
Holocene period, 157
Human activities
   expressed through SHM, 130–31
   impact on ecosystems, 5–7, 55, 119–20
   impact on nutrient loadings, 93–94, 122–24
   impact on plant species, 274–75
   impact on water resources, 132–33, 138–44
   logging, 128–30
   See also Employment; Fires
Hydrological monitoring, 175

Ice sheets, 20, 156–57
   See also Glacier fluctuation
IJC (International Joint Commission), 289–90, 296–97
IMPLAN software, 59–60
Indicators, measurable, 305–6
Infections, enteric, 141–42
Insects, 27–28
Institutional arrangements for land use, 8, 74, 77–78, 216
Integrated resource management, 76–77
Integrated Weed Management (IWM), 272–74
Interactive mapping, 11
International conflicts. *See* Transboundary conflicts
International Joint Commission (IJC), 289–90, 296–97
Invasive species, 113–14
   fish, 263–66
   plants, 24, 196, 270–74
Issue statements, 225–26

Johns Lake, 157–60
Jurisdiction fragmentation, 216

   See also Transboundary conflicts

Kainaiwa (Many Chiefs), 41–42
Kerr Dam, 120–21
Krummholz, 90, 91
K'tunaxa, 39–41, 49–50

Lake Linnet culture, 44–45, 47
Lakes, 23, 153–55
   Johns, 157–60
   McDonald, 127, 175
   Saint Mary, 175
   Upper Waterton, 23
   Whitefish, 124–26
Land cover, 55, 57–58
Land use
   analysis of future alternatives, 58–60
   conversion predictions, 60–61
   defined, 55
   factors impacting, 7–10
   institutional arrangements, 8, 74, 77–78, 216
   sustainability and, 5–7
Land values, 8–10
Landsat Thematic Mapper, 281
A Landscape Cumulative Effects Simulator (ALCES), 219–22
Late precontact period, 44, 47, 48–50
Laws
   Endangered Species Act, 105, 107, 266
   Federal Water Pollution Control Act of 1972, 118
   impact on AEM, 309
   National Environment Policy Act of 1969, 112
   Species at Risk Act, 105–6, 234
   Wild and Scenic Rivers Act, 111
Linkage-zone analysis, 112–13
Little Ice Age (LIA), 51, 152, 160–62, 194
Livestock grazing, 142
Lodgepole pine trees, 202, 204–5, 208
Logging, 128–30
Long-range transport (LRT) of pollutants, 139

Mammal species, 24–25, 103–4, 267
   See also specific mammals
Management of CCE. *See* Adaptive ecosystem management (AEM);

Resource management
Many Glacier Subphase, 47–48
McDonald, Lake, 127, 175
McDonald Creek, 127
Meadows, 195–96
Medieval Warm Period, 152
Middle precontact period, 44, 46–48
Migratory birds, 26–27, 230, 238–41, 267
Miistakis Institute, 39–40, 217, 219–22, 224–25
Milk River, 287
Mining, 292–300
Models, 56–64, 179–84, 219–22, 224
Montana-British Columbia border. *See* Transboundary conflicts
Montane forests as zone, 23
Mountain bison eradication, 19
Mountain goats, 88
Mountains, 12–13, 19–21, 86, 136, 187–88

National Environment Policy Act (NEPA) of 1969, 112
National Park Service (NPS), 265
National wildlife refuge, 33
Native peoples
  American Indian reservations, 32–33
  Blackfeet, 13–14, 32, 39–40
  impact of climate change on, 50–52
  K'tunaxa, 39–41
  Nitsitapii, 39–40, 41–42
  precontact culture, 42–50
  role in fire regimes, 204, 206, 209
  tribal history overview, 18, 42–43
Native plant restoration, 274–76
Native Plants Network Protocol Database, 276
The Nature Conservatory (TNC), 13, 109–10, 293
Needle remains and climate, 159
Neoglacial period, 160–61
NEPA (National Environment Policy Act of 1969), 112
New West, economics of, 4
Ninastakis (The Chief Mountain), 42
Nitrogen loadings, 93–94, 122–27
Nitsitapii people, 39–42
  See also Blackfeet people
Non-native fish, 113–14, 263–66

Non-native plants, 24, 113–14, 196, 270–74
Nonlinear phenomena (climate), 189
Nonprofit organizations, 113
Northern goshawk, 111–12
NPS (National Park Service), 265
Nutrient loading, 122–27, 140
Nyack Floodplain, 12

Occupancy patterns, 42–50
Organic pesticide residues in snow, 178
Ownership of land. *See* Private land
Ozone, 27, 141

Pacific Decadal Oscillation (PDO), 193
Paleoenvironmental research
  Johns Lake history, 157–60
  Neoglacial period, 160–61
  overview, 151–53, 167–68
  reconstructions, 161–66
  sedimentary data types, 153–56
Parklands, 23, 29–30
  See also Glacier National Park (GNP); Waterton Glacier International Peace Park; Waterton Lakes National Park
Partners in Flight (PIF), 239–40
Partnerships. *See* Collaboration; Crown Managers Partnership (CMP)
Pass Creek Valley culture, 47, 49
Passive adaptive ecosystem management, 252, 253–54, 256–57
Pathogens, 141–42, 280–81
PDO (Pacific Decadal Oscillation), 193
Peace Park, 31–32, 39–40, 51, 104–5
Peace Park Assembly, 291
Persistent organic pollutants (POPs), 138–39, 145
Petroleum industry, 216
Phosphorus loadings, 93–94, 122–27
Piikáni tribe (Peigan), 41–42, 49–50
Pine Butte Swamp Preserve, 13, 109–10
Plants. *See* Forests; Vegetation
Pluie's journey, 232
Policies
  compared to law, 105
  for land conversions, 60
  for parks/forests, 9, 265
Politics, 70, 71

Pollen and climate, 154
Pollutants
    acid emissions, 141
    in air, 6–7, 138
    endocrine-disrupting substances, 139–40
    organic pesticide residues in snow, 178
    ozone concentrations, 141
    pathogens released in fires, 280–81
    persistent organic pollutants (POPs), 138–39, 145
    wastewater effluents, 140
Ponderosa pine ecosystem, 206
POPs (persistent organic pollutants), 138–39, 145
Population growth, 3–4, 6, 68–69
Postcontact phase, 50
Precautionary principle, 289
Precipitation reconstruction, 163–65
Precontact times, 42–50
Primary productivity, 122–23, 125
Priority species, 239–40
Private land
    for conservation, 29, 109–10, 111
    developing, 122
    fire management and, 207–8
    ownership, 8, 28–29
    tourism and, 78
    value, 29, 60–61
    varieties of, 28, 31
Proactive fire treatments, 210
Production possibility frontier, 62
Project coordination issues, 224–25
Protected areas, 72, 104–5, 233
Provincial parks, 30
Public land, 29–31, 78

Radiocarbon dating, 52, 155
Railroads and development, 19
Rainfall. *See* Droughts; Precipitation reconstruction
Ranching practices, 12–14
Recovery. *See* Restoration of ecological communities
Recreation and tourism
    biodiversity and, 93
    changes impacting, 68–71
    conflicts, 75–79

economics, 73–74
    overview, 67–68, 79–80
    popular activities, 72–73
    statistical trends, 71–73
    as sustainable, 78–79, 80
Red Rock Canyon culture, 45–46, 47
Regional HydroEcological Simulation System (RHESSys), 180–81
Regional Landscape Analysis Project (RLAP), 219–22
Regional resource management, 77–78, 167–68, 216, 218–19
    See also Crown Managers Partnership (CMP)
Remote sensing, 171–73
Reptiles, 27
Residential-use parcels, 60
Resorts, 75
Resource benefits, 7
    See also Aesthetics as natural resource; Alpine ecosystems; Mountains; Water resources
Resource management
    complexity of, 76–77
    regional approaches, 77–78, 167–68, 216, 218–19
    traditional approaches, 249–51
    See also Adaptive ecosystem management (AEM)
Restoration of ecological communities, 113
    difficulty in evaluating, 131–32
    fire regimes, 207–10
    fish, 123, 132
    native plants/trees, 274–79
    See also Biodiversity; Conservation
Revegetation work, 275–76
Ribbon forest, 92
Rio Algom Mines, 288–90
Rivers
    Milk, 287
    Saint Mary, 287
    Stillwater, 124
    Swan, 123
    systems, 22–23, 135–36
Roads
    density of, 113
    grizzlies and, 237–38, 254–56
Rocky Mountain Front, 12–13

Rotary Clubs, 291
Rural landscapes, 7

Saint Mary Lake, 175
Saint Mary River, 287
Salmon, 123, 230
Satellite-based multispectral sensors, 172–73
Scale issues for study areas, 56–57
Scenario modeling, 222
Scenario planning, 58–59, 61–64
Schultz, James Willard, 51
SDST (spatial decision support tool), 11
Sedimentary data, 153–60, 157–60
Services, ecosystem, 6
Settlers, early, 19
Sexton, Erin, 295
Shepard Glacier, 189–91
Shifting habitat mosaic (SHM), 130–31
SHM (shifting habitat mosaic), 130–31
Shoreline erosion, 120–22
Shrimp, 123
Siksiká (Blackfoot), 41–42
Smallpox epidemics, 51, 52
SNOTEL (SNOwpack TELemetry), 176–78
Snow accumulations
    in alpine ecosystems, 92–93
    avalanches, 197–98
    climate change and, 144
    glaciers' effects, 191–92
    impact on water resources, 136–37
    monitoring, 175–79
Snow water equivalent (SWE), 127
SNOwpack TELemetry (SNOTEL), 176–78
Snowpack trends, 193
Social forces and recreation, 69
Soils, 21, 86–87, 96
Solifluction terraces, 87, 88–89
Southwest Alberta Grizzly Strategy, 108–9
Spatial data, 279–82
Spatial decision support tool (SDST), 11
Species at Risk Act (Canada), 105–6, 234
Stakeholders, 253, 307–9
State land, 29–31, 78
Stillwater River, 124
Strategic environmental assessment, 218–19
Subalpine forests, 23–24
Sublimation of snowpack, 137
Summer temperatures, 165–66, 188
Sustainability
    assessing, 10–11, 305–6
    identifying alternatives, 306–7
    local practices, 12–14
    proposals for, 303–4
    of tourism, 78–79, 80
    See also Adaptive ecosystem management (AEM)
Swan River, 123
Swift Creek, 125–26

Talus, 88
Technical assistance, 13–14
Technology
    GIS, 178–79, 279–82
    GPS, 279
    impact on recreation, 70–71
    remote sensing, 171–73
Tembec Inc., 291
Temperatures, 161–63, 165–66, 188
Tephra deposits, 155
Timberlands, 28, 216
TNC (The Nature Conservancy), 13, 109–10, 293
Tobacco Plains Phase, 47, 49–50
Total nitrogen (NP) concentrations, 127–29
Total phosphorus (TP) concentrations, 127–29
Tourism. *See* Recreation and tourism
Trade-offs in development, 76
Transboundary conflicts
    background, 285–86
    Boundary Waters Treaty, 287
    Cabin Creek project, 288–90
    with Fernie (British Columbia), 290–30
Transportation advances, 70–71
Treaties, international, 287
Tree-ring records, 155, 165–66
Treeline ecotones, 87, 90–92, 194–95
Trees
    in ecotones, 90–92
    fire and, 22, 201–7
    growth trends, 196–97

Index

by Johns Lake, 158–60
soil impact, 96
See also specific trees
Tribal history. *See* Native peoples
Tundra, 88–89, 196

Ultraviolet (UV) radiation, 141, 174–75
Umbrella species, 235
Ungulates, 103
Upper Waterton Lake, 23
Urban runoff, 140
U.S. Department of Agriculture (USDA), 13–14, 176–78
Use value of land, 9
UV (ultraviolet) radiation, 141, 174–75

Values of land, 8–10
Vegetation
classification of tundra, 88–89, 196
fire adaptations, 22
monitoring changes, 94–96
native plant restoration, 274–76
non-native species, 24, 113–14, 196, 270–74
Peace Park management, 42
pollen data evidence, 154, 157–60
weeds, 24, 273
in zones, 23–24
Volcanic ash, 155

Wastewater treatment, 145–46
Water resources, 22–23
in alpine ecosystems, 92–93
critical nature of, 144–45
dams, 131, 143, 242
hydrological monitoring, 175
impact of snow, 136–37
importance of mountains to, 187–88
quality, 122–26
See also Aquatic ecosystems; Rivers; *specific water resources*
Waterton Glacier International Peace Park, 31–32, 39–40, 51, 104–5

Waterton Lakes National Park, 31–32
Waterton River Subphase, 47, 48
Websites for spatial data, 281–82
Weeds, 24, 273
Western larch trees, 205–6
White pine blister rust, 277
Whitebark pine trees, 204, 205, 208, 276–79
Whitefish Lake, 124–26
Whitepine Bark Ecosystem Foundation, 113
Wilderness Act of 1964, 209
Wildfires. *See* Fires
Wildlife
adaptations to fire, 22
in alpine, 87–88
biodiversity, 6, 103–4
correlated with economic benefits, 61–63
as earthmovers, 88
insects/anthropods, 27–28
integrity of communities, 267–70
mammals, 24–25, 103–4, 267
management by Montana, 29–30
organizations, 113
priority species, 239–40
recovery of, 25
species at risk, 234
See also specific wildlife species
Wildlife habitats
biodiversity, 239–41
impact of land conversions on, 61–64
national refuges, 33
need for connectivity, 233
See also Alpine ecosystems; Aquatic ecosystems; Big Prairie ecosystem; Ponderosa pine ecosystem
World Heritage Sites, 18, 32

Yellowstone to Yukon (Y2Y) Conservation Initiative, 113, 229–32, 245–46, 291